Motion Picture Restoration

A.C. Kokaram

Motion Picture Restoration

Digital Algorithms for Artefact Suppression in Degraded Motion Picture Film and Video

Springer

Dr Anil Kokaram
University of Cambridge
Signal Processing and Communications Laboratory
Department of Engineering
Trumpington Street
Cambridge CB2 1PZ

ISBN 978-1-4471-3487-9 ISBN 978-1-4471-3485-5 (eBook)
DOI 10.1007/978-1-4471-3485-5

British Library Cataloguing in Publication Data
Kokaram, Anil. C.
Motion picture restoration : digital algorithms for artefact suppression in degraded motion picture film and video
1.Motion pictures - Conservation and restoration 2.Image processing - Digital techniques 3.Image reconstruction
I.Title
621.3'67

Library of Congress Cataloging-in-Publication Data
Kokaram, A. C., 1967-
Motion picture restoration : digital algorithms for artefact suppression in degraded motion picture film and video / A.C. Kokaram.
p. cm.
Includes bibliographical references.

1. Motion picture film--Conservation and restoration. 2. Image processing--Digital techniques. 3. Signal processing--Digital techniques.
I. Title.
TR886.3.K65 1998
778.5'3'0288--dc21 98-4855

Typesetting: Camera ready by author

34/3830-543210 Printed on acid-free paper

To my wife, Stefanie

Preface

The manipulation of pictures and video in digital form has been an established research activity for more than twenty years. It is only recently, however, that digital image and video processing equipment has been accessible to the general public. This is due in part to the rapidly growing economy of the home computer. A major contributing factor has been the marked rise in the presence of the non-academic user on the internet, particularly the World Wide Web (WWW). Manipulating digital imagery has become synonymous with the WWW. It is the drive to present audio and visual media to the home user in an interactive form and to increase the available range of choices, which has encouraged agreements to begin digital video television broadcasting before the turn of the century. With the increased demand for video material, there is a perceived increase in demand for material from archive sources and this has fuelled commercial interest in automatic digital restoration processes. Furthermore there is a continuing effort to design techniques for correcting errors in received compressed video bit streams for the purposes of live communications links over noisy channels e.g. mobile telephones and the internet.

This book introduces the reader to a range of digital restoration activities beyond the well traversed areas of noise reduction and deblurring. It describes a number of problems associated with archived film and video. Particular attention is given to missing data since that applies equally well to correcting packet loss in coded video bit streams. The book should appeal both to those in industry and to academic researchers.

In many ways the book is a culmination of eight years work in the area of video reconstruction. This began in 1989 when I was a Ph.D student in the Signal Processing Laboratory at the Cambridge University Engineering Department and continued as a post-doctoral fellow from 1993–1998. I am grateful for the financial support of Cambridge University, the British Library and Cable and Wireless during the early years and the European Union restoration project AURORA[1] during the last three years.

Many exceptional people influenced my thinking in this time and this book would be incomplete without suitable acknowledgement. I wish to thank Dr. Simon Godsill for many educational discussions about the Bayesian paradigm which led to the design of the JOMBADI algorithm. Much of the research performed would not have been possible were it not for the Ph.D students who gave of their time to help maintain the laboratory network: Pete Wilson, Dr. Adam Tibbalds, Ray Auchterlounie, Dr. Robert Young, Ian Calderbank, Dr. Robin Morris and Dr. M. Lai. I am grateful for many enlightening conversations with Dr. Joan Lasenby, Dr. Bill Fitzgerald, David Elias, Ben Bradshaw, Dr. Julian Magarey, Dr. Adam Tibbalds, Dr. Robin Morris, Dr. Nick Kingsbury and of course my Ph.D supervisor Dr. Peter Rayner who also arranged funding in the early years. The work of many willing proofreaders was essential in creating a coherent manuscript: Steve Armstrong, Adam Tibbalds, Ben Bradshaw, Rebecca Moore (of Springer-Verlag) and my wife, Stefanie. I am especially grateful for helpful comments from Prof. Petar M. Djurić during his short stay at Cambridge. I wish to thank every one of my colleagues in the laboratory for creating a very pleasant atmosphere for interaction.

Thanks are due to several members of the AURORA project for valuable discussions about restoration during my last few years as a post-doctoral fellow: Louis Laborelli and Jean-Hugues Chenot of I.N.A.; John Drewery, Jim Easterbrook and Theodore Vlachos at the BBC; Martin Weston and Stuart Sommerville of Snell and Wilcox and Peter van Roosmalen at Delft University. I am also grateful for the help of David Throup at Quantel (U.K.), Wolfgang Lempp and Amit Gupta at the Computer Film Company, London, for providing source material. Some of the pictures used in Chapters 5 and 8 are provided by INA, Paris, through the help of Jean-Hugues Chenot and I must also thank João Sequeira of RTP (Raotelevisão Portuguesa) for readily providing some of the material used in Chapters 8, 9 and 10.

Finally, I must thank my family in Trinidad for their constant support over the years. I am indebted to my wife for her encouragement, punctuation and patience while my world-view narrowed to these three hundred pages in the last few months.

I am currently in-between worlds as I take up a Lectureship in the Electrical Engineering Dept. of Trinity College, Dublin, Ireland. There I shall continue the work started in this book.

Anil Kokaram
Cambridge
January 1998

[1] **AU**tomatic **R**estoration of **OR**iginal **A**rchives, European Union ACTS Project AC072.

Glossary

2D	Two Dimensional
3D	Three Dimensional
AR	Autoregressive
BBC	British Broadcasting Corporation
ARMA	Autoregressive Moving Average
2DAR	Two Dimensional AR Model
3DAR	Three Dimensional AR Model
WBME	Wiener Based Motion Estimator
MWBME	Multiresolution Wiener Based Motion Estimator
AWBME	Adaptive Wiener Based Motion Estimator
BM	Block Matching
BBM	Boyce Block Matching
MRF	Markov Random Field
JOMBADI	Joint Model Based Detection and Interpolation
MCMC	Markov Chain Monte Carlo
MPEG	Motion Picture Experts Group
pdf	Probability Distribution Function
cdf	Cumulative Distribution Function
ML	Maximum Likelihood
MAP	Maximum a Posteriori
MBD	Model Based Detection
MBI	Model Based Interpolation
LS	Least Squares
SNR	Signal to Noise Ratio
MSE	Mean Square Error
MMSE	Minimum Mean Square Error
MAE	Mean Absolute Error
DFD	Displaced Frame Difference
DPD	Displaced Pixel Difference
ICM	Iterated Conditional Modes
SA	Simulated Annealing
VZO	Temporal zero-order hold interpolation for motion
Pel	Pixel
RMS	Root Mean Squared
SDI	Spike Detection Index
ROD	Rank Order Detector
ROC	Receiver Operating Characteristic
PMSE	Percentage Mean Squared Error
PIMSE	Percentage Improvement in Mean Squared Error
MMF	Multilevel Median Filter
DFT	Discrete Fourier Transform
FFT	Fast Fourier Transform
IDFT	Inverse Discrete Fourier Transform

Contents

1
Introduction

The concept of consumer digital television is no longer a theoretical abstraction [152]. Many of the major television and satellite broadcasters will begin broadcasting digital video before the turn of the century. SKY and the BBC have already begun digital transmissions although there are very few receivers available. Digital Video Disk (or Digital Versatile Disk) (DVD) [57] seems set to provide that missing link between television and the home computer which has evaded the consumer electronics business for some years. It is a perceived demand for consumer devices and services which manipulate digital pictures that has fuelled the rapidly growing interest in digital image sequence processing during the last fifteen years.

The agreement to use MPEG–2 (Motion Picture Experts Group) as the compression standard for digital television broadcasting [58, 158, 72, 16] has focused industry and university research on the various modules that must be optimized for the design of usable MPEG–2 systems. The MPEG–2 standard does not define the encoding strategy, it only defines the transmitted bit stream. This allows incorporation of new technology into consumer devices as they become available. There are two areas which drastically affect the performance of any MPEG–2 encoder: transmission errors and motion estimation. Research into error resilient coding and post processing of the received bitstream for error concealment is a direct response to the observation that MPEG–2 does not perform well in noisy channels, in particular across mobile communications networks. Motion estimation research has long been an active area both for video compression and computer vision applications, however the increased interest in consumer MPEG decoders has encouraged the development of real time hardware for motion estimation.

With these developments implying more available channels, there is now a demand from television broadcasters for increased television programming. There is also an increasing perception that the consumer will come to demand a higher quality picture from the home television set. Holders of large film and video archives such as the BBC (British Broadcasting Company), INA (Institut National Audiovisuel, Paris, France) and RTP (Raotelevisão Portuguesa, Lisbon, Portugal) are concerned that they may not be able to satisfy this demand for higher quality images from their aging archives[1]. Coupled with this is the observation that removing unwanted artefacts from video (which is then MPEG compressed) improves the compression ratio possible. In other words, it has been observed that broadcasting 'dirty' film material consumes more bandwidth than broadcasting 'clean' film material. These factors have combined to increase the requirement for restoring or 'cleaning up' archived material and heighten awareness in the digital film and video community about the ongoing research in the area of Digital Image Sequence Restoration. In a recent development (1995), several institutions have combined their resources to produce equipment for real time digital video restoration under the auspicies of the AURORA[2] European Union ACTS project. There are 8 institutions involved: INA (the coordinator), BBC, RTP, SGT (Société Générale de Téléinformatique, France), Snell and Wilcox (U.K.), The Digital Media Institute of Tampere (Finland), The Information Engineering group at Delft University of Technology (The Netherlands), and The Signal Processing Laboratory at Cambridge University Engineering Department (U.K.).

This book is concerned with the general area of Digital Video Restoration or Digital Image Sequence Restoration. It discusses several algorithms for the *automatic* treatment of various film and video defects. The number of published works in this field has traditionally been small, both because of the lack of fast and cheap processing architectures suitable for the task and because of the relatively specialist application areas, e.g. satellite remote sensing, medical imaging. However, processing speed is no longer an issue and thanks to the increasing use of digital techniques in film and television the area has attracted more attention. Digital video restoration has clear implications for the digital film industry where film restoration is still being performed principally by hand, even if with the aid of digital techniques for storage and retouching. There is a need for automatic restoration devices that can clean the massive amounts of data stored for possible transmission.

Typical film problems are 'Dirt and Sparkle' (called blotches in this book), due to missing information, and noise due both to the photographic process and the slowly decaying material itself. An interim solution to the problem would be to transfer the image onto video tape. However, the telecine process itself can introduce distortion. Scratches can be introduced due to dirt in the apparatus and noise can be introduced in the recording process. Modern film is also subject to degradation. The same effects of Dirt and Sparkle as well as

---

[1] The archives mentioned are all members of the AURORA consortium

[2] **AU**tomatic **R**estoration of **OR**iginal **A**rchives, European Union ACTS Project AC072.

noise, are found. Repeated projection will also damage the film since it must be physically transported through the projection apparatus each time. Similarly, video tape is also subject to degradation. Repeated playback will damage the quality of the tape. A loss of synchronization information will occur at some point, causing frames or lines to be displaced vertically or horizontally on the screen or lost altogether.

Of course the applications of the algorithms presented in this book are not limited to motion pictures. Any degraded image sequence presents itself for similar treatment. The noisy video sequences from electron and scanning tunelling microscopes are a good example [88] as well as the images produced in remote sensing applications where the environment is quite harsh with respect to image quality. An interesting niche application area (although not video) is the restoration of ancient manuscripts [82, 95].

Much of the work in image processing has concentrated on stills, with the notable exception of video compression research. Despite the fact that image sequence analysis for video compression is now a well established pursuit, work in image sequence restoration is still emerging and tends to concentrate on noise reduction and deblurring. Before 1991 the references to work in this area have been limited to a few attempts at motion compensated temporal filtering [66, 31, 36, 116, 115]. More recently, researchers such as Sezan, Lagendijk, Biemond, Katsagellos [42, 17, 37] have shown that restoration processes in particular have a great deal to gain from a treatment of the image sequence as a spatio–temporal signal. In this mode there is the chance to overcome the high spatial nonstationarity of images by taking advantage of the high temporal correlation in the image sequence.

It is hoped that this book will serve to broaden the horizons of digital image sequence restoration by considering some of the less often addressed problems in degraded video e.g. line scratch removal and line synchronization. Particular attention is paid to missing data reconstruction for image sequences, a much neglected area especially with regard to large missing regions. The book does not consider the problems of deblurring and noise reduction in image sequences; these topics have been considered in depth in several other books [102, 79, 157]. In fact it appears that blur is not a common problem in broadcasted material. The work stresses that in dealing with image sequences, the spatio-temporal evolution of the data must be taken into account. Thus similar areas are treated together, and it becomes necessary to track the motion of regions in a sequence and consider the same region in separate frames, paying attention to its position in each frame.

Traditionally, to achieve a restoration, the approach has been to find the motion of regions in each frame and then restore the information in a particular region by filtering along motion trajectories. Unfortunately, motion estimation is not a solved problem. There are several approaches to motion estimation and these can be divided into three groups

- Region Matching
- Gradient Based Methods

- Transform Methods

Because of the lack of a tractable model for the image sequence, no algorithm can claim perfect results. The book therefore spends some effort on assessing the advantages and disadvantages of a few of the more popular motion estimation techniques. However, it is not the role of this book to evaluate motion estimation techniques. These are of interest purely through their effect on the final restoration. The fact that one motion estimator may give more inaccurate motion vectors than another is secondary to the performance of the restoration system.

The underlying theme of the design of the algorithms discussed in this text is 'model based'. While it is acknowledged that there exists no complete model of the image sequence, it is still educational to follow through model statements from the observed signal model to the corresponding inverse operation. A better understanding is then had of the limitations of the final method. For missing data in particular, the work adopts a Bayesian approach and then illustrates how several classes of algorithms can result from various simplifying arguments. Some effort is also centred on illustrating that a truly joint approach to the solution of some of these problems is not as computationally complex as one might expect. In this way a new treatment of the 'motion estimation with discontinuities' problem is presented. Generally, a pragmatic view of algorithm design is adopted, computational requirements are discussed as each algorithm is presented and some suggestions for fast implementations are made. It is not expected that the algorithms discussed here can be easily made into hardware, on the contrary, the environment most suited to these algorithms (in the short term) is an interactive software toolbox available for the user of digital film manipulation workstations. The CD-ROM included with the book illustrates most of the work using HTML documents as well as video data files that can be played using standard hardware and software.

A book that discusses film and video restoration should outline something about the history of these media and this is discussed next.

1.1 A short history of film and video

Cinema has been with us for just over 100 years. The date usually cited as the birth of cinema is 28th December 1895 when Auguste and Louis Lumière showed their films to a paying audience in Paris [180]. The development of their camera and projector can be traced back via a long line of developments to the invention of Faraday's wheel in 1831 which illustrated the theory of the persistence of vision. This consisted of two toothed disks rotating in opposite directions on the same axis. At certain speeds the back disk seen through the front, appeared to remain stationary.

Photography was explored in the 1840's and photographs were being made in 1839 by William Henry Fox on silver halide paper. 1872 saw the first use of a series of images for recording movement when Eadweard Muybridge used a wet plate camera to win a bet about horses having all four feet off the ground at some point while cantering. When George Eastman produced paper roll film in 1885, Etienne-Jules Marey, inspired by Muybridge, built a camera that could take 100 pictures per second. Just 4 years later in 1889, Thomas Alva Edison developed a camera using celluloid film. He showed his films on a continuous loop in a peep-show type machine called the kinetoscope. It is this invention which impressed the Lumière brothers sufficiently to attempt a public showing of cinema in 1895. The first purpose-built cinema appeared in 1906 (the Omnia-Pathé) in Paris. In the U.S. by 1908 there were nearly 10,000 *nickelodeons* attracting 20 million people weekly.

The early films were silent, and sound made its first appearance in the Warner Brothers film 'The Jazz Singer' in 1927. The sound was recorded on a phonograph disk which was badly synchronized to the film. This was superseded in a few years by the Fox-Movietone system which recorded sound directly onto the film. Magnetic audio tape was used for sound in the 1950's and modern cinemas now employ digital sound.

By the mid-1950's cinema was being heavily challenged by television (TV), a rivalry which exists to this day. The cinema producers responded by concentrating on better image quality in the form of wide/giant screen formats e.g. CinemaScope and Cinerama as well as 3-D Cinema. Films like South Pacific (1956) were made with a stereo soundtrack and 70mm film to impress the audience with the 'presence' of cinema.

The first television broadcast was made on November 2nd 1936 by the BBC from Alexandra Palace. It was with the broadcast of the coronation of King George VI on 12th May of that year, that television showed potential for rapid information dissemination. After shutting down for World War II in September 1939, television broadcasting resumed on 7th June 1946. In 1953 the Coronation of the Queen was watched in 3,000,000 homes and that event proved TV to be a genuine mass media device.

Colour television came along in 1967 and by that time the PAL (Phase Alternate Line) broadcast standard was already adopted for TV broadcast. The colour signal format was chosen so that people with black and white TV sets could still receive the signal, hence the YUV colour components discussed in Chapter 11.

Video recording devices came along much later than the TV. The early devices were invented in the 1950's and RCA developed a machine with tape speeds of 6 metres/sec. They used longitudinal recording of the signal on the tape, so fast speeds were required to record the large bandwidth TV signal. The first practical video recorder was developed in 1953 by the Ampex Corporation. It used the helical scan arrangement which remains with us today and allows higher effective recording speeds without a frightening tape speed.

The first home video equipment was developed by Philips in 1972 which was quickly superseded by the VHS machines of Panasonic and the Betamax machines of Sony in 1978. The arrival of these formats marked the start of the home video era. The ability to record and edit programmes made by the home user caused the TV to be seen as a flexible medium. Sony lost the now famous marketing war by 1980 when VHS had become the main standard for home video users mainly due to its longer playing time. This was despite the superior quality of the Betamax format.

Now, in late 1997, there are an increasing number of digital video consumer products. JVC is marketing a digital camcorder and both Panasonic and Sony have released low cost digital video devices aimed at the semi-professional video production market. The development of DVD which allows more than 4Gb (up to 17Gb) of information to be packed onto a CD type device, has further fuelled the home digital video market. DVD players are available from Philips, Toshiba and Sony and it is expected that recorders will become more widely available soon. At the moment copyright issues are still being discussed by the various motion picture production houses but it is expected that an increased quantity of movies would be available on DVD in 1998. This is one hurdle which must be overcome before DVD has as wide an acceptance as VHS video tape.

1.2 Film and video artefacts

The various types of film and video artefacts arise because of the physical mechanisms for film and tape transport as well as the chemical process of image recording on film. This section gives the reader some feel for the range of degradations encountered in the visual media although not all the problems are treated in this book.

1.2.1 Missing data

Missing data appears in different forms on video tape and film. On film the problem manifests as flashes of dark and bright patches called Dirt and Sparkle. This is caused when dirt/dust adheres to the film as it passes through the transport mechanism; abrasion of the film emulsion results in 'Sparkle'. With time, the actual material of the film may become brittle and show crennelations or cracks along the surface. This severe degradation of the film sometimes manifests as huge missing regions. An example of such degradation is shown in figure 8.1 of Chapter 8. Line scratches, which persist from frame to frame are a consequence of material getting stuck in the film transport mechanism and either abrading a swathe of film frames or smearing itself over the frames as it melts due to the heat in the projector.

Telecine machines transfer film to video. This also causes the transfer of the film artefacts to video. Of course, the telecine itself introduces new degradations since it also involves a film transport mechanism. These are discussed separately. Information is recorded on video tape using a drum shaped rotating head assembly that moves the head diagonally across the tape. This produces a spiral pattern of video line data on the tape and is known as 'helical' scan [178, 74]. Digital video recorders also store information in this manner on the tape although the information recorded is numerical and not the analogue video signal itself. Simple error detection and concealment techniques are incorporated into all video recorder/playback devices [178] but after several generations of record/playback on the same tape these techniques fail.

Missing data that arises due to video transport problems can be seen when the tape is scratched. A single horizontal tape scratch causes several short horizontal lines of missing data to appear on the video output (following an approximately helical position arrangement) because the video line information is arranged diagonally across the tape. In digital video devices digital dropout generally occurs in an 8×8 block. In severe cases of loss of video signal, most video recorders simply repeat the last line which has been successfully displayed [178]. This form of line dropout is sometimes very severe and occurs most often in outside broadcasts which are spliced into news programmes. A good example of this is shown as figure 8.2 in Chapter 8.

Missing data also occurs due to lossy transmission of MPEG coded bit streams. The effects of errors in the bit stream can cause loss of entire horizontal stripes of image data as well as loss of entire frames. This problem has received much interest in recent years as the design of post processing algorithms for restoring degraded uncompressed video.

In this book missing data is generically referred to as 'blotches'.

1.2.2 Noise

Noise is a common problem in all recorded signals. Film grain noise is related to the image formation process on film. The grain is due to the individual silver halide image sites on the film image. The density of these sites is one factor determining the photographic properties of film. Grain noise is an important part of the film 'feel' and some care has to be taken in noise reduction of film since removal of all of this noise can be undesirable. The noise on film can be quite severe in the case of archived material since physical degradation of the film will increase the noise level. Figure 8.1 of Chapter 8 is a good example of several noise effects on a film frame which was transferred to video via the telecine process.

Film grain noise is normally a multiplicative effect. Also, at digital film resolution[3] the noise is not uncorrelated from pixel to pixel. Noise in video can be explained more generically as being caused by noise in the receiving and or recording apparatus. Chapter 10 introduces the use of motion compensated Wiener filters for treating the noise problem.

1.2.3 Image unsteadiness

The usual form of image unsteadiness encountered by the home user is the hand held video camera. It is difficult to keep the camera steady while filming and resulting video clips usually show some unwanted global motion. In film projection or telecine apparatus any slight distortion in the perforations at the side of the film (which are used to guide the film through the transport) cause each

[3]The Computer Film Company in London use digital frames in excess of 3120×2048.

frame to be slightly displaced from its optimal position when projecting. This causes film shake. Telecine machines often have mechanical means for keeping the film steady to reduce the observed shake, but sometimes post processing is still necessary to correct the distortion.

In general film shake is more random than hand held camera shake which makes it easier to separate the global shake component from the regular motion of objects. The problem has been treated by Uomori et al [181] (Panasonic) and Vlachos et al at the BBC [188, 186, 187]. These authors have proposed digital means for reducing camera shake of various forms. The success of standard schemes for estimating shake is evidenced by the fact that most hand held video cameras today incorporate some kind of shake suppression process, either mechanical or digital. Most post-production film houses have access to user interactive hardware for the compensation of film shake.

A very unusual manifestation of image unsteadiness is line jitter. This is discussed in Chapter 5. The problem occurs in video due to the loss of line synchronization pulses. The image display device is unable to properly locate lines relative to each other and lines in the image show random horizontal displacements. The observed effect is that lines that should be straight in the image appear jagged or 'wobbly'. Some good examples of the effect are shown in figures 5.6 and 5.10.

1.2.4 Image flicker

Image flicker is generally a film effect. Most viewers would be accustomed to seeing old black and white movie which show a change in image intensity from frame to frame. This can be possibly explained by physical degradation of the film as well as aberrations in the exposure time for each frame. Lesser known is the problem of Twin Lens flicker [188] which is a telecine effect discussed in the next section. This book gives no real treatment of flicker although there is some evidence that simple frame averaging can reduce the effect of flicker in *mild* cases. For a treatment of the problem, see Roosmalen et al [146, 147].

1.2.5 Telecine effects

The Telecine device has been in use since 1946 when Rank Precision Industries in the U.K. introduced the first flying spot film scanner for converting film to video. The basic idea of the early systems is to use a round CRT (Cathode Ray Tube) which moves an electron beam generated at its back so that it scans the phosphor coating on the inside of the tube. The beam of light emitted by the tube (due to the electron beam impinging on the phosphor) is projected onto the film plane and used to acquire the film information by virtue of the coloured areas of the emulsion on the film. The resulting colour image is then passed through a series of dichroic[4] mirrors that successively separate the red green and blue images to three photomultiplier tubes for conversion into video.

[4] Only reflective to one wavelength of light.

In early telecine devices the optical paths for the recording of each field of video may have been different. Also the film may have moved during the recording of each television field. This causes a displacement between each recorded field which manifests as a kind of 'flicker' effect, although it is not due to the change of light intensity between fields. This artefact, called Twin lens flicker is a subset of the image unsteadiness problem and a good algorithm for automatic removal was published by Vlachos et al [188].

Both film scanning (Telecine) and film recording devices (which record digital information on film) have been produced which use Charge Coupled Devices (CCD) for scanning the image plane. The light source is not a CRT and instead the beam from a lamp is focussed using an optical arrangement to scan the film one line at a time. Each line is recorded by a row of CCD devices. Both CCD telecines and flying spot scanners are in use today.

Before video recording devices were invented, delayed broadcast of TV programmes was performed by recording the output of a TV monitor on film. The film was then rebroadcast on TV. This early device which performed the video to film conversion was called a Kinescope. In archives today some of the only records of TV programmes are on film. The actual TV scan lines can be seen on the film. Therefore, modern digitization processes for transferring the recorded program onto digital video tape (via CCD Telecine devices, for instance) sometimes cause aliasing. This happens when the beam of light used to scan the image does not use a wide enough 'blur'. This is equivalent to sampling with too narrow a low pass function. The effect is that of disturbing dark rings called Kinescope Moire. A solution to this problem is currently being explored.

The reader is directed to the book by Kallenberger et al [74] for a further description of telecine machines.

1.3 A rough guide to the book

The chapters have been written to be fairly stand alone. Most of the tedious algebra has been placed in appendices to keep the arguments flowing without interruption. The following is a brief outline of the chapters to follow.

Chapter 2: Image Sequence Modelling and Motion Estimation

The book begins laying the foundation for the restoration algorithms by discussing models of the image sequence and thus considering motion estimation as a matter of course. The chapter reviews some of the approaches to motion estimation to date and shows how they can be linked through the use of a simple temporal model. More complex models are introduced, including the autoregressive (AR) model of the image sequence. The idea of estimating motion discontinuities is introduced and a brief review of the area is undertaken. The chapter sets the basis for the algorithms in the book since the performance of the restoration techniques depends on the behaviour of the motion estimator/image model. A basic introduction to Bayesian inference is included which employs the motion estimation problem as an example. It is shown that most motion estimation schemes can be derived using the Bayesian framework and an appropriate image sequence model.

Chapter 3: Three Dimensional Autoregressive Modelling

This work investigates and extends the technique developed in [37] to use the three dimensional autoregressive (3DAR) model as a motion estimator. The technique is shown to give a better prediction error than other estimators, but cannot produce reliable motion vector estimates (when Least Squares model parameter estimation is employed), particularly important for the interpolation of missing data.

Chapter 4: Image Sequence Restoration: A Review

A review of the few reported techniques are presented. Median filtering has been put forward as an efficient technique for removing impulsive noise. This is an example of a global filtering operation applied to a problem which involves only a local distortion, yet not much work has been done in the detection of such distortion. The lack of spatio–temporal algorithms is also highlighted. Attention is also given to viewing the motion estimation algorithm in perspective with the rest of the system.

Chapter 5: Line Registration

Before restoration can begin, the analogue sequence must be converted to a digital format. In order to capture a frame successfully, the digitizing equipment must recognize the synchronization information that is presented with the input video signal. Unfortunately, this information is sometimes lost due to noise, or playback from degraded video tape. The resulting inter-line jitter is disturbing to the viewer. A method using two dimensional autoregressive (2DAR) modelling is presented for removing this distortion.

Chapter 6: Heuristics for Missing Data Removal

Several low cost schemes are investigated for missing data detection and removal. These are designed by using the simplest translational model of the image sequence. A common technique for removing impulsive distortion in images is the median operation. Arce [6, 8] has presented multistage median structures for the image sequence that are robust to motion and maintain image detail. This chapter extends that approach by employing motion estimation as well as incorporating more spatio–temporal information. A detector for impulsive noise is also presented which allows the median operation to be controlled. This has the twofold benefit of improving image quality and decreasing the computation necessary.

Chapter 7: Model Based Reconstruction for Missing Data

This work reports on the use of a modelling technique that is able, ultimately, to both detect and remove impulsive noise in the same framework. The use of this technique was prompted by the success of the equivalent 1–D process reported by Veldhuis, Vaseghi, Godsill and Rayner. Veldhuis also showed good results for the removal of missing data in an image using a 2–D AR model. The method gives very robust behaviour. The necessity for interpolating both the image data *and* the motion data is illustrated. The Bayesian approach to the missing data problem in image sequences is extended to deal simultaneously with motion estimation as well as detection and removal of the corrupted data.

Chapter 8: Reconstructing Large Missing Areas
When large areas are missing there is insufficient uncorrupted spatial information to effect a restoration and the previous methods discussed become less useful. A good algorithm must therefore rely more heavily on temporal motion information to infer the correct organization of image data in the missing area. This chapter explores a scheme derived using a Bayesian approach which jointly estimates motion and occlusion information while reconstructing image data which could be as large as an entire frame.

Chapter 9: Line Scratch Detection and Removal
A very common degradation on film material is the persistent vertical scratch. An algorithm for detecting and interpolating the data in the scratched region is presented which relies on a Bayesian refinement strategy for rejecting false alarms. The problem is more complicated than the 'blotch' problem because lines exist as natural features in any image.

Chapter 10: Noise Suppression in Image Sequences
Some of the standard approaches to image sequence restoration are compared. The 3D IIR and FIR Wiener filter for image sequences are also presented. The chapter discusses the effects of the algorithms on real degraded motion pictures.

Chapter 11: Using Colour
It is possible to propose multichannel extensions to many of the algorithms proposed. The chapter proposes extensions to the work in the rest of the book for using colour.

Chapter 12: Parting Remarks
This chapter highlights the current achievements with respect to the solutions presented for problems in degraded film and video. It summarizes the joint estimation approach which proves very successful especially in line removal and large area reconstruction. The chapter considers that the higher fidelity allowed through the use of the spatio-temporal image data models (e.g. the 3DAR model) may be secondary to the importance of the robustness and reliability of the motion estimation algorithm.

2

Motion Estimation for Image Sequence Processing: An Overview

An image sequence, such as a motion picture, consists of a set of images of a scene recorded at regular intervals in time. For the case of television in PAL format this interval is $\frac{1}{25}$th of a second.

The restoration of degraded image sequences can be performed with the repeated execution of the same two dimensional process on the separate images. Viewed in this light, image sequence restoration can be achieved by drawing from the many solutions already proposed for image restoration [67, 106, 161, 47, 141, 71, 150]. However, the recording of each image in a motion picture, for instance, generally occurs more rapidly than the change of information in the scene[1]. Therefore consecutive images in the sequence contain similar information. Hence, an algorithm that can take advantage of the high temporal correlations that exist between frames, has the potential to reject more distortion than a 2D operation. It is the development of such 3D (space/time) algorithms which is of interest in this book.

Unfortunately the design of a spatio–temporal restoration algorithm is made difficult by the motion of objects in the recorded scene. This motion implies that there exists useful information only along the trajectory of motion of each object in the sequence of images. Therefore, a 3D algorithm needs to involve knowledge of motion in the image sequence.

An alternative view is to perform spatio–temporal processing only in those areas of the image that are not moving. This would remove the need for motion information. However, such algorithms are limited in performance especially with regard to blotch removal and more obviously in scenes with many moving objects.

---

[1] With the exception of scene cuts.

The subject of Image Sequence Processing, in this case Sequence Restoration, is therefore inextricably linked with Motion Estimation. The restoration algorithms are more effective than 2D processes only if the motion estimator performs satisfactorily. In addition, some robustness to erroneous motion estimation must be built into the restoration algorithm since the behaviour of the motion estimator cannot be guaranteed. There is also the consideration that the sequences being dealt with are degraded and so the motion estimator must be robust to noise.

The previous statements have implied that the motion estimator is a separate issue from the sequence restoration algorithm. Indeed, this has been the usual mode of image sequence processing. The motion in the sequence is first estimated by some algorithm and then processing is directed along the calculated motion trajectories. In many respects, the work presented in this book does not deviate from that approach although Chapter 7 does unify the extraction of motion with image reconstruction. This mode of thinking has been forced by the lack of effective image sequence models. If it were possible to develop a model for the image sequence that can predict the observed grey level intensities correctly, then the signal recovery problem would be better defined. Unfortunately the development of an image sequence model which begins at the source of the information, i.e. moving objects in a 3D world, is extremely complicated [130] and at the moment impractical. Chapter 7 shows that a combined model based restoration approach does have distinct advantages over separate motion estimation and restoration, even though the image sequence model employed is not complete.

In some sense much of the current thinking with respect to image sequences, particularly in commercial digital video applications, revolves around a translational model. The model used for image sequence processing is therefore

$$I_n(\mathbf{x}) = I_{n-1}(\mathbf{x} + \mathbf{d}_{n,n-1}) \tag{2.1}$$

where $I_n(\mathbf{x})$ is the grey level of the pixel at the location given by position vector $\mathbf{x}$ in the frame n, and $\mathbf{d}_{n,n-1}$ is a displacement mapping the region in the current frame n into the previous frame $n-1$. Under this model, the grey level at each pixel in the current frame can be predicted from a shifted version of levels at pixels in the previous frame. The only parameter that is needed for this model is the motion vector $\mathbf{d}$. Hence the problem of image sequence modelling, using this model, reduces to one of motion estimation.

This Chapter attempts to bridge the gap between the various ideas of image sequence modelling, motion estimation and image sequence processing. It points out that although no source model of the sequence has been employed to date, the current techniques can be related through the translational image model. The following sections outline some of the basic characteristics of the image sequence and also consider model based approaches that do not involve a translational approach. It is important to realize that motion estimation is a research topic in its own right. The various approaches to date are reviewed, the idea being to justify the approach taken to motion estimation in this book. An exhaustive review is not attempted since there have been many such articles

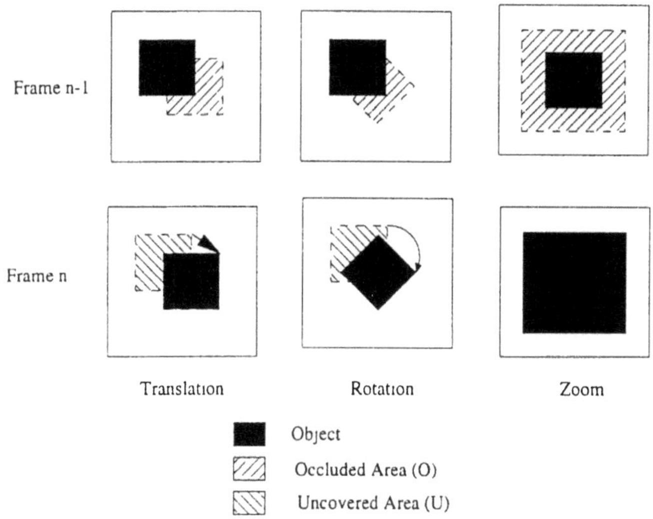

FIGURE 2.1. Examples of simple motion across two frames.

published [145, 126, 129, 127, 117, 144, 169]. The Chapter lays the foundation for some of the ideas developed in the rest of the book.

Before continuing the reader should note that the task addressed here is the extraction of motion from entirely unconstrained sequences. There is no prior knowledge about the composition of the scenes regarding the number, shape and elastic properties of objects and so feature based tracking of motion, typically employed in computer vision problems [153], is less effective here.

2.1 Image sequence modelling: The image centric view

There are two approaches to image sequence models. The source approach would begin with consideration of lighting on a 3D world and the subsequent projection of that scene onto a 2D plane. The image centric view disregards the source of the information and tries to place constraints on the behaviour of pixel intensities by observing the sequence itself. This latter approach has been the more tractable one and almost all the motion estimation techniques to date can be related via a model as below.

$$I_n(\mathbf{x}) = I_{n-1}(\mathbf{F}(\mathbf{x})) \tag{2.2}$$

The vector function $\mathbf{F}(\mathbf{x})$ represents a linear transformation of image coordinates to represent motion such as zooming, rotation and translation. One restriction of the model is that the intensity changes in the sequence are due only to motion and not to lighting effects such as shadow. The model is intuitively correct since it implies that one image can be predicted from the previous one by rearranging the positions of objects in the scene.

Figure 2.1 shows three different forms of motion typically encountered in an image sequence. When the motion in the 3D world represents completely unconstrained movement involving for instance motion perpendicular to the image plane (e.g. zooming), the vector transformation $\mathbf{F}(\mathbf{x})$ can become non-linear. However, if the motion is small enough between frames, zooming (and rotation) of small portions of an object can be approximated by translation. In this case the model is as stated in equation 2.1. The vector displacement, $\mathbf{d}_{n,n-1}$, is called the motion vector representing motion between frames n and $n-1$. The modelling problem is reduced to finding the motion vector $\mathbf{d}$ for all the pixels in the image.

Although the model represented by equation 2.1 is at the root of most of the current techniques for image sequence processing, it ignores an important fact. This is well illustrated in Figure 2.1. Motion vectors are only defined for areas which can be found in both frames n and $n-1$. Whenever an object moves, some area must be covered and some other area must be uncovered. For these areas there can be no motion vector since these areas represent some newly formed region. The area, labelled 'U' in Figure 2.1, has been uncovered in frame n, and the area labelled 'O' has been occluded in frame n. This emphasizes the discontinuous nature of the motion field of a scene. Not only can there be many different objects showing different motion, but unless the entire image is occupied by a moving object, there will at least be uncovered and occluded regions. In estimating motion fields, this problem has been ignored until recently in works such as [34, 176, 162].

The different approaches to motion estimation are identified by the way in which they solve equation 2.1 for $\mathbf{d}$. Direct search techniques, called Block Matching have been used [18, 49]. There are also transform approaches which use some variations on the shift theorem [196, 174, 175, 173, 30]. Recently there has been some work on the use of the Wavelet transform for motion estimation [110, 111, 112]. There is a huge body of work using a linearization of equation 2.1 to solve for the motion vector directly [145]. Finally there has recently arisen work [1, 162] using Gibbs distributions and a Bayesian methodology as introduced by Geman and Geman [47]. The rest of this section reviews some of these approaches.

2.1.1 Correspondence matching

The most popular and to some extent the most robust technique to date for motion estimation is Block Matching (BM) [49, 197, 23]. The basic assumption in the technique is that pixels in some small region undergo the same translational motion. The image in frame n, is divided into blocks usually of the same size , $N \times N$. Each block is considered in turn and a motion vector is assigned to each. The motion vector is chosen by matching the block in frame n with a set of blocks of the same size at locations defined by some search pattern in the previous frame.

The separation of the candidate blocks in the search space determines the smallest vector that can be estimated. For integer accurate motion estimation

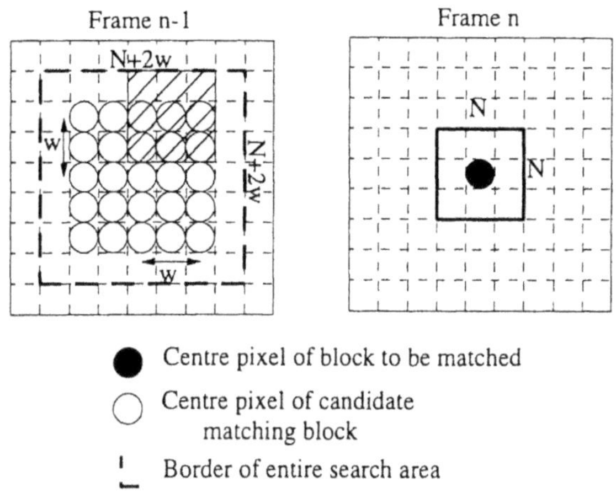

FIGURE 2.2. Motion estimation via Block Matching. The positions indicated by a ○ in frame $n-1$ are searched for a match with the $N \times N$ block in frame n. One block to be examined is located at displacement $[1 \ -2]$, and is shaded.

the position of each block coincides with the image grid. For fractional accuracy [50], blocks need to be extracted between locations on the image grid. This requires some interpolation. In most cases bilinear interpolation is sufficient.

Figure 2.2 shows the search space used in a full motion search technique. The current block is compared to every block of the same size in an area of size $(2w+N) \times (2w+N)$. The search[2] space is chosen by deciding on the maximum displacement allowed: in Figure 2.2 the maximum displacement estimated is $\pm w$ for both horizontal and vertical components. The method used for determining the best matching block varies, but usually either the Mean Squared Error (MSE) or Mean Absolute Error (MAE) of the pixel intensities is used. The best match is the one that minimizes the error measure used.

The technique arises from a direct solution of equation 2.1. Defining the Displaced Frame Difference as below,

$$\mathrm{DFD}(\mathbf{x}, \mathbf{v}) = I_n(\mathbf{x}) - I_{n-1}(\mathbf{x} + \mathbf{v}) \tag{2.3}$$

where $\mathbf{v}$ is some displacement vector, the BM solution can be seen to minimize the Mean Absolute DFD (or Mean Square DFD) with respect to $\mathbf{v}$, over the $N \times N$ block. The chosen displacement, $\mathbf{d}$ satisfies the model equation 2.1 in some 'average' sense.

The Full Motion Search is computationally demanding, requiring on the order of $N^2(2w+1)^2$ operations per block for an integer accurate motion estimate. Several reduced search techniques have been introduced which lessen this burden. They attempt to reduce the operations required either by reducing the locations searched [49] or by reducing the number of pixels sampled

[2]There are $(2w+1)^2$ searched locations.

in each block [197]. However, reduced searches may find local minima in the DFD function and yield spurious matches.

An important aspect of any motion estimator is the implementation of a motion detector. The motion detector attempts to limit searches to the regions where motion is suspected. If a search is undertaken at all locations spurious matches are likely. The motion detector employed usually takes the form of some threshold on the error measure recorded between two blocks at the same location in frames n and $n-1$. The usual mode of operation is to record the error measure at no displacement, and if this is greater than some set motion threshold, motion is flagged and a search is initiated.

The BM algorithm is noted for being a robust estimator of motion since noise effects tend to be averaged out over the block operations. In [23], Boyce introduces a technique to make the BM algorithm more robust to noise. The main error introduced by noise is that of spurious matches since the motion detector is likely to erroneously flag motion due to the presence of noise. To isolate this problem, Boyce compares the best match found (E_m), to the original 'no motion' match (E_0). If these matches are sufficiently different then the motion estimate is accepted otherwise no motion is assumed. The threshold acts on the ratio $r_b = \frac{E_0}{E_m}$. The error measure used is the MAE. If $r_b < t$, where t is some threshold chosen according to the noise level suspected, then no motion is assumed. This algorithm will be referred to as the Boyce Block Matching algorithm (BBM) in future references, and the threshold ratio called the *Boyce ratio*. This algorithm therefore verifies the validity of the motion estimate once motion is detected.

The main disadvantages of Block Matching are the heavy computation involved (although these are byte wise manipulations) and the motion averaging effect of the blocks. If the blocks chosen are too large then many differently moving objects may be enclosed by one block and the chosen motion vector is unlikely to match the motion of any of the objects. The advantages are that it is very simple to implement[3] and it is robust to noise due to the averaging over the blocks.

2.1.2 Gradient based approaches

The second school of thought regarding motion estimation evolved out of a closer look at the image model in equation 2.1. The techniques were initially developed independently by workers in Video Coding and those in Computer Vision. In 1976 Cafforio and Rocca [26], introduced a direct solution for motion using a least squares approach. They were motivated by the need to find a motion estimator that required less computation than BM and also to avoid the problem of the motion averaging of the blocks.

In 1978 Netravali and Robbins [132, 133, 145], improved upon that work, introducing the iterative, pel–recursive estimation scheme. Higgins et al [63, 62], in 1980, were motivated by the study of the human perception of motion. Later

[3] It has been implemented on Silicon for video coding applications.

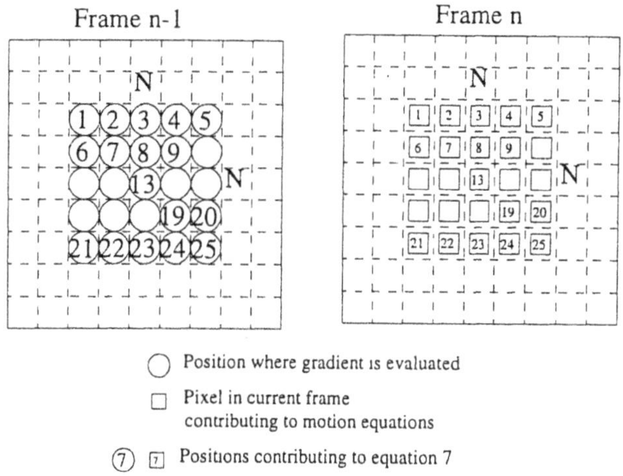

FIGURE 2.3. Motion estimation via spatial gradients. A block of pixels of size $N \times N$ in frame n is chosen to set up N^2 motion equations; $N = 5$. Each equation requires the evaluation of a gradient in the previous frame. The initial motion estimate is [0 0].

on, workers such as Walker and Rao [189], and Biemond et al [17] (1987), improved upon the technique introduced by Netravali et al. Efstratiadis and Katsagellos [37] have introduced a framework from which all the previous techniques can be derived.

The basic connection between all these techniques is that they rearrange the model equation 2.1 to have direct access to the motion parameter $\mathbf{d}$. The equation 2.1 may be linearized about $\mathbf{d}$ using a Taylor series expansion as follows:

$$I_n(\mathbf{x}) = I_{n-1}(\mathbf{x}) + \mathbf{d}^T \nabla I_{n-1}(\mathbf{x}) + e_{n-1}(\mathbf{x}) \tag{2.4}$$

Here $e_{n-1}(\mathbf{x})$ represents the higher order terms of the expansion, and the ∇ operator is the usual multidimensional gradient operator. The equation can be rearranged to yield an expression involving the Displaced Frame Difference with zero displacement.

$$\begin{aligned} \mathrm{DFD}(\mathbf{x}, 0) &= I_n(\mathbf{x}) - I_{n-1}(\mathbf{x}) \\ &= \mathbf{d}^T \nabla I_{n-1}(\mathbf{x}) + e_{n-1}(\mathbf{x}) \end{aligned} \tag{2.5}$$

This equation represents the relation between a number of observables and $\mathbf{d}$. However, there is one equation and 2 variables. To solve the equation therefore, at least one more equation is necessary. If the assumption is made that within a small region all the pixels behave the same with respect to motion, then several more equations can be set up. This assumption will be referred to as the Smooth Local Flow assumption. Therefore, neglecting the higher order terms allows an equation to be set up at each pixel in some defined region to yield a vector equation as below. The situation is illustrated in Figure 2.3.

$$\mathbf{z}_0 = \mathbf{G}\mathbf{d} \tag{2.6}$$

$$\text{where } \mathbf{z}_0 = \begin{bmatrix} DFD(\mathbf{x}_1, 0) \\ DFD(\mathbf{x}_2, 0) \\ \vdots \\ DFD(\mathbf{x}_{N^2}, 0) \end{bmatrix}$$

$$\text{and } \mathbf{G} = \begin{bmatrix} \frac{\partial}{\partial x} I_{n-1}(\mathbf{x}_1) & \frac{\partial}{\partial y} I_{n-1}(\mathbf{x}_1) \\ \frac{\partial}{\partial x} I_{n-1}(\mathbf{x}_2) & \frac{\partial}{\partial y} I_{n-1}(\mathbf{x}_2) \\ \vdots & \vdots \\ \frac{\partial}{\partial x} I_{n-1}(\mathbf{x}_{N^2}) & \frac{\partial}{\partial y} I_{n-1}(\mathbf{x}_{N^2}) \end{bmatrix}$$

In Figure 2.3, the set of N^2 equations is set up using the block indicated. The vectors are therefore generated by assembling the row–wise observations into columns. The gradient measurements are made using a simple difference technique, for example the horizontal gradient at pixel 2 in Figure 2.3 is $\frac{I(3)-I(1)}{2}$. This is admittedly a noisy process and a more robust class of gradient estimation techniques, using curve fitting, was presented in [117].

A solution for $\mathbf{d}$ can then be generated using a pseudoinverse approach.

$$\mathbf{d} = [\mathbf{G}^T\mathbf{G}]^{-1}\mathbf{G}^T\mathbf{z}_0 \tag{2.7}$$

This solution yields $\mathbf{d}$ in one step by using gradient measurements in some region as in Figure 2.3. It is one of the early solutions proposed by Cafforio et al, Netravali et al, Martinez [117] and others. The important drawback with this approach, recognized in [132], is that the Taylor series expansion is only valid over very small distances. To overcome this problem, a recursive solution was introduced called the pel–recursive approach. This new solution refines an initial estimate for $\mathbf{d}$ until no further reduction in DFD is observed.

Rather than expand the model equation about $\mathbf{x}$, the image function is linearized about a current guess for $\mathbf{d}$, say $\mathbf{d}_i$. The idea is to generate an update, $\mathbf{u}$ for the current estimate that will eventually force the estimation process to converge on the correct displacement. The update is intended to be a small one to maintain the validity of the Taylor series expansion. Therefore, equation 2.4 becomes

$$\begin{aligned} \mathbf{u}_i &= \mathbf{d} - \mathbf{d}_i \\ I_n(\mathbf{x}) &= I_{n-1}(\mathbf{x} + \mathbf{d}_i) + \mathbf{u}_i^T \nabla I_{n-1}(\mathbf{x} + \mathbf{d}_i) + e_{n-1}(\mathbf{x} + \mathbf{d}_i) \end{aligned} \tag{2.8}$$

The solution may then continue as before. However, Biemond [17] presented a Wiener solution for the displacement which is more robust to noise. It handles the higher order terms more effectively by considering their effect on the solution to be the same as Gaussian white noise. The new vector equation to be solved becomes

$$\mathbf{z}_i = \mathbf{G}\mathbf{u}_i + \mathbf{e} \tag{2.9}$$

$e(\mathbf{x}+\mathbf{d}_i)$ represents the higher order terms which are considered to be Gaussian white noise of variance σ_{ee}^2. Note that the matrices $\mathbf{G}$, $\mathbf{z}_i$, now consist of terms similar to those defined earlier in equation 2.6, but the arguments are offset by the current estimated displacement, $\mathbf{d}_i$. The Wiener solution chooses the update $\hat{\mathbf{u}}_i = \mathbf{L}\mathbf{z}_i$ which minimizes the expected value of the squared error between the true update $\mathbf{u}_i$ and the estimate $\hat{\mathbf{u}}_i$. The objective function is therefore,

$$E[(\mathbf{u}_i - \hat{\mathbf{u}}_i)^T(\mathbf{u}_i - \hat{\mathbf{u}}_i)] \tag{2.10}$$

Minimizing this function with respect to $\mathbf{L}$ yields the following solution for $\hat{\mathbf{u}}_i$. The full derivation may be found in [17].

$$\begin{aligned} \hat{\mathbf{u}}_i &= [\mathbf{G}^T\mathbf{G} + \mu\mathbf{I}]^{-1}\mathbf{G}^T\mathbf{z} \\ \mu &= \frac{\sigma_{ee}^2}{\sigma_{uu}^2} \end{aligned} \tag{2.11}$$

Here, σ_{uu}^2 is the variance of the estimate for $\hat{\mathbf{u}}_i$, provided that both components of the motion have the same variance. Also, note the use of μ as a regularizing term in the required matrix inverse. The success of the method lies in treating the observables $\mathbf{z}, \mathbf{G}, \mathbf{e}$ as random variables. The estimator will be referred to as the Wiener based motion estimator (WBME) in future references.

After the update is estimated, $\mathbf{d}_i$ is refined to yield the displacement that is used in the next iteration, $\mathbf{d}_{i+1} = \mathbf{d}_i + \mathbf{u}_i$, and the process continues again. During the iterative process, the observables may have to be evaluated at fractional displacements. Usually bilinear interpolation is employed to keep computation at a minimum. The iteration is terminated when the algorithm converges. This is usually detected by thresholding the size of the update vector and the size of the current Mean Square Error. When either of these two quantities are below their respective thresholds, convergence is assumed.

The main advantage of the Wiener based pel–recursive estimator is that it requires much less computation than the Block Matching algorithm[4]. It is also capable of resolving both fractional and integer displacements without changing the number of operations involved. Because the computational complexity is so low, many authors in the Video Coding field have used the technique to estimate a motion vector at each pixel. This is done so that there is less chance of the 'motion averaging' effect of blockwise manipulations. The approach is of limited use because some finite support region must always be employed to set up the necessary equations. Depending on the size of that region, errors are likely. The most important disadvantage of the estimator is that even as an iterative process, it cannot estimate large displacements because of the limited effectiveness of the first order Taylor series expansion. This is in contrast to BM which is only limited by the defined search space.

---

[4]When convergence is achieved it is typically within 10 iterations.

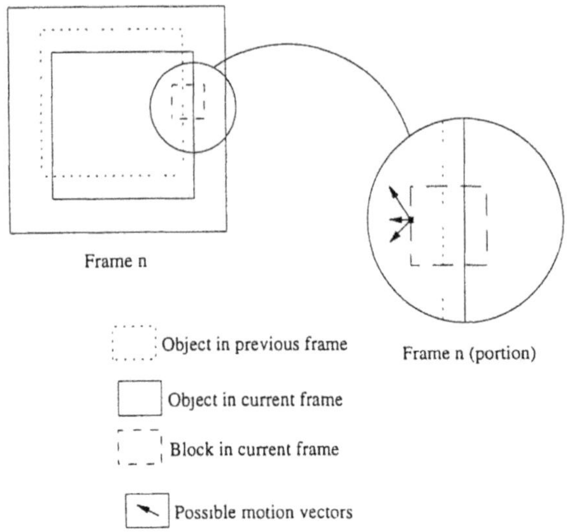

FIGURE 2.4. Motion ambiguity at an edge. *The motion estimate is most accurate along the direction of maximum image gradient.*

2.1.3 Ambiguity in motion estimation: Adaptive solutions and additional constraints

A fundamental fact about the translational model of motion is that it cannot be solved using one observation at one pixel alone. To overcome this difficulty the Smooth Local Flow assumption must be made to allow observations at several pixels to be used. In order to improve motion estimation, additional constraints must be found. A closer look at the gradient based approach to motion estimation led several authors to introduce further improvements to the basic estimator. Biemond et al [17, 20, 21, 35, 34, 33] altered the WBME by introducing adaptive schemes. A separate approach was employed by Nagel [131, 128, 124, 126, 125, 130] and others [44, 65, 63, 62, 154, 155] who further refined the motion estimation solution by employing constraints on the motion itself. The ambiguity involved in motion estimation is illustrated in Figure 2.4. The rectangular object of uniform intensity is moving diagonally and so has two non zero components of motion. If the block used for estimating motion in either a BM or WBME algorithm is located at an edge as shown in the Figure then several motion estimates can be found which will cause a zero DFD. Only one of these is the true motion vector. The situation illustrates two points,

- The motion estimate is most accurate in the direction of maximum image gradient (or contrast). At edges this direction is perpendicular to the edge, at corners there are two directions along which motion estimation is accurate. Hence, for corners both components of motion (horizontal and vertical), can be estimated accurately.
- The size of the block used for estimating motion affects the locality of the motion estimate. Because the block in Figure 2.4 is small compared

to the size of the object, a local motion estimate results which may not reflect the true motion of the object. It is this effect which causes the motion estimated in that block to be different from the global motion of the object.

If the block used for motion estimation is located in the centre of the moving object in Figure 2.4, then it is likely that neither estimator will estimate any motion at all. There will be zero error for zero displacement and so the motion detector will not flag motion. Even if motion estimation were to be engaged then the zero or near zero DFD coupled with the zero gradient in the previous frame will bias the result toward a zero motion vector. Martinez [117] and Kearney at al. [78] quantitatively investigated this problem with respect to the gradient based estimation process. Those findings are discussed next.

Improved pel–recursion

Errors in the WBME can be linked directly to the extent of ill–conditioning in the matrix $\mathbf{M}_g = [\mathbf{G}^T\mathbf{G}]$. In this respect the WBME can be viewed as a regularized least squares solution for motion. The μ acts as a damper on the system, preventing the inverse from being unstable. Large μ encourages small updates and small μ allows large updates. The WBME is a regularized version of solution 2.7.

It is clear that μ should respond to ill–conditioning in $\mathbf{M}_g$. This can be measured by the ratio of the eigenvalues of the matrix. Martinez was able to prove that this ratio was governed by the relative magnitudes of the two components of average spatial contrast (or gradient). His analysis is repeated in Appendix E. He showed that the solution for the displacement was ill–conditioned when the size of one component of spatial gradient was much larger than the other; i.e. at an edge. The corollary to this result is that the motion estimate is most accurate in the direction of maximum image gradient. This is self evident from Figure 2.4, but the investigation was then able to employ the SVD decomposition of $\mathbf{M}_g$ intelligently.

Martinez presented in [117] a motion estimator, which was not pel–recursive, that was sensitive to this problem. It is stated below.

$$\begin{aligned}
\mathbf{d} &= \begin{cases} \alpha_{\max}\mathbf{e}_{\max} & \text{if } \lambda_{\max} \gg \lambda_{\min} \\ \left[\mathbf{G}^T\mathbf{G}\right]^{-1}\mathbf{G}^T\mathbf{z} & \textit{otherwise} \end{cases} \qquad (2.12)\\
\alpha_{\max} &= \frac{\mathbf{e}_{\max}^T}{\lambda_{\max}}\mathbf{G}^T\mathbf{z}
\end{aligned}$$

Here, λ and $\mathbf{e}$ refer to the eigen values and eigen vectors of $\mathbf{G}^T\mathbf{G}$, and $\alpha_{\max}$ is a scalar variable used to simplify the final expression. The solution presents a specific response when the inverse operation is ill–conditioned. This response involves aligning the motion estimate along the direction of maximum contrast in order to make the most of an uncertain situation.

Researchers such as Efstratiadis et al [37], Driessen et al [35, 34] and Böröczky [20, 21] have concentrated on adapting μ in the WBME to the ill–conditioning of $\mathbf{M}_g$ during the pel–recursive process. The estimator proposed by Driessen is shown here.

$$\begin{aligned} \mathbf{d} &= [\mathbf{G}^T\mathbf{G} + \mu\mathbf{I}]^{-1}\mathbf{G}^T\mathbf{z} \\ \textit{where } \mu &= |\mathbf{z}|\frac{\lambda_{max}}{\lambda_{min}} \end{aligned} \tag{2.13}$$

Therefore, as the required inverse solution becomes ill–conditioned (measured as a ratio of eigenvalues), μ increases, thus increasing the damping in the system. The additional multiplying factor of $|\mathbf{z}|$ helps to further stabilize the situation by relaxing the damping when the error is small and the motion estimate is close to convergence. There is obviously some scope here for a joint Driessen/Martinez algorithm and this is investigated in the later chapters of this book.

Additional constraints

Rather than salvage the motion estimate when the solution is ill–conditioned, Nagel [131, 128, 124, 126, 125, 130] and others [44, 65, 63, 62, 154, 155, 41] used the Local Smooth Flow assumption itself as a constraint in the solution. These studies attempted to identify the true motion in a scene. They addressed the remaining problem facing standard gradient based motion estimators. When the textural information in a region is low, e.g. within the rectangular shape in Figure 2.4, the gradient information is insufficient to provide any motion estimate. The resulting vector field is then non–zero only across the edges of the object, where there is some textural information.

To get around this problem, the motion estimate was constrained to be some value which did not violate the smoothness of the motion field. Therefore after several iterations, motion vectors could be propagated from the boundaries of moving objects into the interior where there was less gradient information. The standard approach in these multiple constraint techniques is to consider a rearrangement of the expression 2.8 to yield an objective function $F(\mathbf{x})$ defined as below.

$$F(\mathbf{x}) = DFD(\mathbf{x}, \mathbf{d}_i) - u_i(x)\frac{\partial I(\mathbf{x} + \mathbf{d}_i)}{\partial x} - u_i(y)\frac{\partial I(\mathbf{x} + \mathbf{d}_i)}{\partial y} \tag{2.14}$$

The truncation error term is ignored. $F(\mathbf{x})$ is zero if the correct displacement is found and so to find this value the function $F(\mathbf{x})^2$ is minimized with respect to the update vector $\mathbf{u}_i = [u_i(x)\ u_i(y)]$ and a constraint as below, following Horn et al [65].

$$(\frac{\partial u_i(x)}{\partial x})^2 + (\frac{\partial u_i(x)}{\partial y})^2 + (\frac{\partial u_i(y)}{\partial x})^2 + (\frac{\partial u_i(y)}{\partial y})^2 = K \tag{2.15}$$

Using Lagrange multipliers, the solution was then able to produce a smooth vector field since the constraint restricted the rate of change of the motion estimate with distance.

Nagel [131, 128, 124] considered the phenomenon of directional accuracy in motion estimation, as discussed earlier. He argued that

1. Motion estimation is most accurate in the direction of maximum image gradient, i.e. when aligned with the principal direction of image gradient given by the eigenvectors of the gradient matrix considered previously.

2. Motion field smoothness is violated at motion boundaries[5]. These motion boundaries would in general correspond to boundaries of objects in the image and so motion flow smoothness over such boundaries should be discouraged.

His ideas were implemented as a directed smoothness constraint as opposed to the homogeneous smoothness constraint of Horn. The constraints introduced were sensitive to the direction of image gradient and to edges in images.

These gradient based estimators improve upon the basic idea of the previous section, but also introduce another level of complexity into the motion estimation process. Although these estimators have better potential for yielding a true motion field it is questionable to what extent this is necessary for the application in this book. The fact that BM or the WBME cannot give true motion estimates in some image regions is perhaps secondary to their role in providing a low prediction error. When the choice of several vectors gives the same low prediction error, does it matter that several of these vectors do not reflect the true motion? The answer to this question depends on the application. When the motion field is manipulated over several frames, for example to reconstruct missing motion fields, it is better to employ 'true' motion rather than use motion giving low prediction error. This is particularly important in Chapter 8. In Chapters 6 and 7 the use of 'true' motion is less important since only spatial motion reconstruction is performed. The reason for this is simply that temporal smoothness constraints can only be applied successfully to 'true' motion and not to motion giving low prediction error. Chapters 6, 7 and 8 use multiple constraint techniques for motion estimation/reconstruction.

2.2 Block Matching vs. WBME

The two more popular techniques to date are Block Matching and Gradient based motion estimation as presented by Biemond [17]. Block Matching is still the more popular for Video Coding techniques despite its greater computation requirement, for two reasons.

1. It is simple to implement and requires only integer operations.

2. It can handle any displacement size depending on the search space used.

[5]Local Smoothness versus Global Discontinuity.

BLOCK MATCHING	GRADIENT BASED
Heavy Computation	Low Computation
Large motion OK	Small motion only
Fixed resolution motion[6]	Variable resolution
	Ill–conditioned when no texture
Only the actual data is used	Gradient measurements are noisy

TABLE 2.1. Comparing Block Matching and Gradient Based motion estimation.

The WBME algorithm is computationally lighter, but it cannot handle displacements greater than 5 pixels in practice. This is because, for large displacements, the two regions being compared in the two frames may have no relation at all. Therefore the Taylor series expansion of the region in the previous image will have no bearing on the current region. This accounts for the lower popularity of the WBME. The basic tradeoffs between the two techniques are outlined in table 2.1.

To increase the importance of the first order term in the Taylor series expansion, Kearney et al [78] have proposed blurring the image using some low–pass filter prior to estimation. The method has merit and tends to stabilize the algorithm. There is some loss in precision but this is not catastrophic provided the size of the blur is controlled. The technique will be referred to again in later discussions.

2.3 Image sequence modelling: Alternatives

The previous sections have discussed the solutions proposed for the purely translational model for image sequences. The model represents an image centric view in that the constraints employed have not been derived from a consideration of the image formation process. This section attempts to briefly outline some of the work done in this area. It also reviews an Image Centric Model which breaks from the tradition of the translational model.

Schunck in [154, 155] shows that the image flow equation 2.14 is valid also across image discontinuities i.e. edges. In [154] he points out that it is a simple relation that does not consider lighting effects such as shadow, and depth motion effects such as zoom. In an attempt to incorporate such observations he defined a measure of 'feature' density in preference to grey level. He suggested this density to describe the occurrence of important structures e.g edges, corners etc. that were of use in motion estimation. His arguments were speculative, but gave rise to a modified flow equation. The equation expresses the conservation of feature density in a unit element of image. The equation statement is as follows:

---

[6] The error field from the BM algorithm can be interpolated to yield an arbitrary resolution vector field, but this is not usually done.

- *The flux of feature fragments through a given unit image element must be balanced by the accumulation of such fragments within the element.*

The modified flow equation is as follows:

$$N\nabla.\mathbf{v} + \mathbf{v}^T\nabla N + \frac{\partial N}{\partial t} = 0 \tag{2.16}$$

Here N is the feature density, ∇N is the spatial gradient $\nabla N = [\frac{\partial N}{\partial x}\ \frac{\partial N}{\partial y}]$, and $\nabla.\mathbf{v} = [\frac{\partial v_x}{\partial x} + \frac{\partial v_y}{\partial y}]$ is the term representing the divergence of the flow field. Schunck recognized that only the consideration of perspective projection of a unit area in a real scene with certain irradiance and reflectance properties, would yield a true constraint.

Nagel [130] considered just that problem. He derived a displacement rate constraint equation by considering the projection of a planar surface element having some reflectance properties, onto an image element in the image plane. The surface element was presumed to rotate and translate in the 3D scene and the element was considered as a Lambertian source of reflected light. The constraint equation was derived on the basis of equating the instantaneous temporal change of intensities at a point in the 3D scene with respect to the spatial change of intensities at the image element. Nagel's equation was as follows:

$$\mathbf{v}^T\nabla I(\mathbf{x},t) + \mathrm{DFD}(\mathbf{x},t) = I'(\mathbf{x},t)f(\mathbf{x},t) \tag{2.17}$$

Here $I'(\mathbf{x},t)$ is not the grey level at $\mathbf{x}$ in the image plane, but a ratio between the power received by an image element divided by the size of the element. $f(\mathbf{x},t)$ refers to a combination of effects dealing with irradiance factors and the velocity of the object element in scene space. He found that although a divergence term similar to the one suggested by Schunck did arise, it was cancelled by the effects due to changes in size of the element in projection onto the image plane. The essential statement, however, is that the conservation of image grey level described by the simple flow equation is not satisfied. The right hand side of 2.16 does not vanish. Nagel concluded his findings by saying that his derivation was based on simplistic assumptions regarding the nature of image formation. Therefore more effort was necessary to derive a more general constraint.

These two contributions by Schunck and Nagel serve to highlight how difficult it is to derive image sequence constraints based on scene modelling. Particularly troublesome is the depth ambiguity problem. Any motion in a direction perpendicular to the image plane causes a change in size of the object. This implies that constraint equations such as these invariably incorporate depth motion as an independent variable. The ambiguity arises in attributing image displacement to motion or a zooming effect. Both of these could yield similar image transformations over small regions.

The application of this book is, however, is sequence restoration. Although it is probably better in the long term to have an estimate of the true motion in the image, it is not of primary concern. If over small displacements in scene space, depth motion is indistinguishable from motion parallel to the image plane, then an image centric motion model will suffice. Of particular importance to a model used for sequence restoration is the ability of that model to predict a frame. The translational model is sufficiently accurate in this respect for its use in Video Coding[7], and the next section reviews an Image Centric model which improves upon this idea.

2.3.1 A stochastic image centric sequence model

One of the shortcomings of the basic translational model is that it is unable to explain the variation in image intensity from frame to frame in the same object. In 1990, Efstratiadis et al [37] introduced the use of the Three Dimensional Autoregressive (3DAR) Model for image sequence processing. This model can deal with intensity changes and can be stated in words as follows:

- *A prediction of the grey level at a pixel in frame n can be given by a linear weighted combination of pixels in some local region surrounding the pixel to be predicted.*

The model was actually presented earlier by Strobach [165], but that implementation used the model primarily for image segmentation rather than motion estimation.

The model tries to make the best prediction of a pel in the current frame based on a combination of intensities at pels in a predefined support region. This support region may occupy pels in the current frame as well as future and past frames. Alternatively the model may be thought of as a stochastic model for the image generation process. Such models have traditionally been used for audio processing to good effect.

Because of motion between frames in a sequence, the support region in each frame must be offset from the predicted pel by a distance related to that motion. Therefore, as opposed to the two dimensional AR model [71, 149, 70, 69, 192, 193, 171], a third parameter, considered to be the relative motion between frames, must be introduced.

In order to best explain the form of the final equations, the first consideration is given to describing a 3D model which does not involve motion. Following the notation of Kashyap [149], the location of each pixel in the support region is defined by an offset vector $\hat{\mathbf{q}}_k$. This vector consists of three components, the horizontal offset and the vertical offset within a frame and then the temporal offset between frames. There are N vectors in the support which may define a completely general region in a (space/time) volume around the predicted position. If the grey level at a particular location $\mathbf{x}$ in frame n is defined as

[7] Of course in Video Coding applications, the error signal is transmitted to further improve the prediction, but the application of the model for that purpose still indicates some usefulness, especially with respect to robust performance, which cannot be ignored.

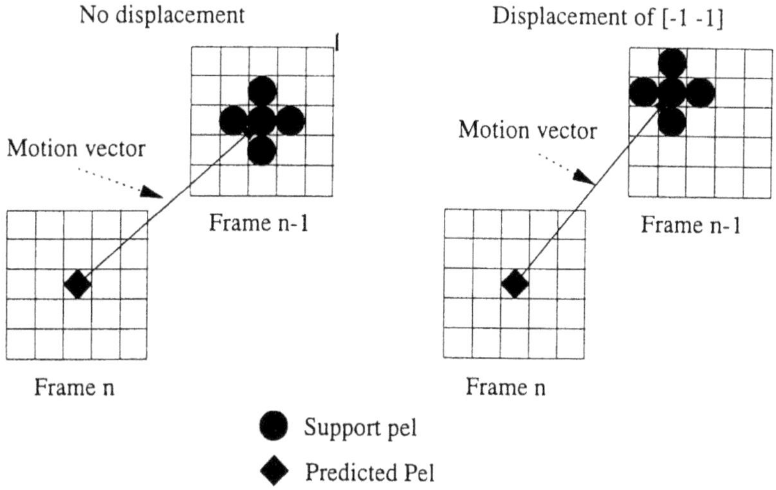

FIGURE 2.5. Handling motion with the 3DAR model.

$I(\mathbf{x})$, then the model is

$$I(\mathbf{x}) = \sum_{k=1}^{N} a(\mathbf{q_k})\mathbf{I}(\mathbf{x} + \mathbf{q_k}) + \epsilon(\mathbf{x}) \tag{2.18}$$

$\epsilon(x)$ is a Gaussian random variable whose correlation properties depend on the structure of the support for the AR model (See Appendix B). The $a(\mathbf{q_k})$ are the weights corresponding to each particular offset vector. The Figure 2.5 shows a typical configuration where the grey level at $\mathbf{x}$ is a linear combination of those levels in a + shape in the preceding frame.

In the case of motion however, the supports in various frames are necessarily displaced relative to each other. In order to describe the altered model, a 'motion' parameter, $\mathbf{d}_{k,l} = [sx_{k,l}, sy_{k,l}]$ is introduced[8]. The 2-D vector $\mathbf{d}$ is the relative spatial displacement between the same region in frames k and l. Because $\mathbf{d}_{k,l}$ is a function of time (the k,l), it becomes necessary to separate the various components of the arguments in equation 2.18. Therefore a position vector $\mathbf{x} = [i, j, n]$ is defined, where i,j,n are the horizontal, vertical and temporal components respectively. Similarly, the offset vector $\mathbf{q}_k$ consists of the three components, $[q_k(x), q_k(y), q_k(n)]$. In Chapter 7 $\mathbf{q}_k = [\mathbf{q}_k^s,\ q_k^t]$, where $\mathbf{q}_k^s = [q_k(x), q_k(y)]$ is the spatial component and q_k^t is the temporal component.

[8]It will become clear why the word motion is placed in quotes after a few results from experiments have been presented in Chapter 3.

The three dimensional autoregressive model incorporating motion now becomes

$$I(i,j,n) = \sum_{k=1}^{N} a_k I(i + q_k(x) + sx_{n,n+q_k(n)}, j + q_k(y) + sy_{n,n+q_k(n)}, n + q_k(n)) + \epsilon(i,j,n) \tag{2.19}$$

Examination of the above equation will show that the correct relation between points in the support of the AR model has been achieved. That is, the support pel in a particular frame p say, must be offset by the relative displacement between the current frame n and p i.e. by $\mathbf{d}_{n,p}$. Figure 2.5 illustrates this point. Note that in the equation above, a_k has been used as shorthand for $a(\mathbf{q_k})$.

The model is plainly a superset of the translational model 2.1. If $a_1 = -1.0$, and there was only one tap at $\mathbf{q}_1 = [0\ 0\ -1]$, then model 2.1 results. $\epsilon(i,j,n)$ would then be some measure of the deviation of the real world behavior from the translational ideal. For the case of the general formulation, instead of modelling image behavior as a pure translation of the same grey level value, it is modelled as a translation and a modification by some gain factor. This modification attempts to explain effects which cannot be handled by intensity constant translation. To a certain extent, the model is partially able to handle effects such as occlusion by the ability to alter the model coefficients to make the best prediction.

Parameter estimation

The parameters which are required are the weights, a_k and the displacement $\mathbf{d}_{k,l}$. Unfortunately, the $\mathbf{d}_{k,l}$ is an argument in the intensity function, $I(\cdot)$. If it were possible to write $I(\cdot)$ explicitly as a function of $\mathbf{d}$, then one would have ready access to this parameter. However, $I(\cdot)$ is only known by observation of its behaviour and in any typical informative scene, cannot be described easily by an analytic function. Unless some effective approximation to $I(\cdot)$ can be found which is expressed explicitly in terms of $\mathbf{d}_{k,l}$, this parameter can only be evaluated on the basis of a direct search which minimizes some cost function.

The coefficients $\mathbf{a}$[9] themselves can always be estimated from equation 2.19 on either a Maximum Likelihood basis or a least squared error basis if the displacement is known. The fact that both a and $\mathbf{d}$ are unknown then, necessitates some approximation that will make the solution tractable. The alternative is to minimize some cost function over the extended space $[a\ \mathbf{d}]$, a task which involves quite high dimensionality. In [37], the correlation structure of the image was assumed to be some known function. Given this function, the Normal Equations could then be solved to yield some fixed set of model coefficients, $\mathbf{a}$. Knowing the coefficients, therefore, the task of modelling reduces to estimation of the displacement parameter.

[9] Vector notation is used here to represent the vector of model coefficients.

In the derivation of a closed form solution for $\mathbf{d}$, the modelling equation is considered in its prediction mode. The 'noise' $\epsilon(i,j,n)$ is considered to be the error between the predicted grey level intensity, $\hat{I}$ and the actual intensity $I(i,j,n)$, where

$$\hat{I}(i,j,n) = \sum_{k=1}^{N} a_k I(i + q_k(x) + sx_{n,n+q_k(n)}, j + q_k(y) + sy_{n,n+q_k(n)}, n + q_k(n)) \tag{2.20}$$

The task then becomes that of choosing the parameters to minimize some function of the error, or residual,

$$\epsilon(i,j,n) = I(i,j,n) - \hat{I}(i,j,n) \tag{2.21}$$

The estimation of the displacement parameter (described in [37]) proceeded in a manner analogous to the approach by Biemond et al discussed previously. The derivation is outlined in Appendix C. A Wiener based estimate for an update vector was found, and the final solution is stated here.

$$\hat{\mathbf{u}}_w = [\mathbf{G}_w^T \mathbf{R}_{vv}^{-1} \mathbf{G}_w + R_{uu}^{-1}]^{-1} \mathbf{G}_w^T \mathbf{R}_{vv}^{-1} \mathbf{z}_w \tag{2.22}$$

The form of the final solution is similar to the Biemond solution with two important differences. The gradient matrix, $\mathbf{G}_w$ consists of terms that result from a weighted[10] combination of gradients defined by the support region of the model. The error terms involved are a summation of both the truncation error of the Taylor expansion and the actual prediction error $\epsilon(i,j,n)$. In fact, the prediction error $\epsilon(i,j,n)$ is not white for non–causal model support (see Appendix B). However, as long as the model support does not incorporate non–causal taps in the current frame[11] , the correlation structure of the error within the current frame (i.e. for the purposes of the solution 2.22), can be assumed to consist of one non–zero term at the zero lag position. In such a situation the update equation becomes,

$$\hat{\mathbf{u}}_w = [\mathbf{G}_w^T \mathbf{G}_w + \mu \mathbf{I}]^{-1} \mathbf{G}_w^T \mathbf{z}_w \tag{2.23}$$

where $\mu = (\sigma_{\epsilon\epsilon}^2 + \sigma_{ee}^2 \sum_{k=1}^{N} a_k^2)/\sigma_{uu}^2$. ($\sigma_{ee}^2$ = truncation error variance.) For a more complete treatment, see Appendix C.

The fact that the matrices used for this update have terms which are a weighted combination of observed values, may imply that the solution is more robust to noise. Efstratiadis et al have indeed reported an increased robustness with this algorithm. The algorithm is more stable than the standard WBME and a smoother motion field was reported. However, an analysis similar to that carried out by Martinez [117], in Appendix E, shows that the conditions for ill–conditioning are still the same. If the points used to set up equation 2.23 lie along an edge, then the solution is ill–conditioned.

---

[10] Hence $\mathbf{G}_w$ for weighted gradient.

[11] That is, taps to the right or below the predicted location.

The work presented in [37] is extended in Chapter 3. In that Chapter, consideration is given to adapting the coefficients to the image data by observing the image correlation structure at each iteration. One remaining point about the model is that the displacement parameter may not represent the true displacement of the scene. This is investigated further in Chapter 3 and the resulting observations have implications for the use of the model in a multiresolution scheme. This model is used later in Chapter 7 for blotch detection and removal as an alternative to the heuristic solution of Chapter 6. A Bayesian framework is also employed in order to unify the model parameter and motion estimation problems in the presence of missing data.

2.4 Estimating large displacements

Throughout the discussion so far it has been noted that gradient based techniques can only effectively estimate small displacements. A correspondence technique, such as BM, is only limited in this respect by the extent of the search space used. However, increasing the search space to deal with a large displacement, increases the computational requirement. Considering that motion pictures can involve displacements in excess of 10 pixels per frame, increasing the search space of BM to deal with this motion results in a huge increase in computation.

It has been widely accepted that the multiresolution schemes provide the most practical way of dealing with this problem effectively. The idea of a multiresolution representation was presented early on for use in Image Coding by Burt and Adelson [25]. The technique can reduce the computational requirement to estimate a particular displacement in the case of BM [18]. In the case of gradient based techniques, the scheme can make the problem better conditioned and increase convergence rates [41, 117].

The multiresolution scheme begins by building successively smaller versions of the images in terms of spatial extent, while still maintaining useful image information in each version. The original images are low pass filtered and subsampled until the displacements measured in the subsampled frames are small enough to allow reasonably accurate pel–recursive, or BM motion estimation with a small search space. The motion vectors found at the lower resolutions are then projected onto the higher resolutions where the motion estimation process is allowed to continue. Successive projections of the estimated motion field eventually leads to an initial motion field at the highest resolution and then it is presumed that only a few iterations will be needed for a pel–recursive algorithm to converge. In the case of BM the search space at each level can be drastically reduced, because a small displacement at the higher levels represents a large displacement at the original level. It is typical to subsample by a factor of 2 so that if the original level is of resolution $H \times V$ then level l of N levels has a size $\frac{H}{2^l} \times \frac{V}{2^l}$. Motion of magnitude k pixels at level 0 (the original resolution level), is then reduced to $k \times 2^{-(N-1)}$ at level $N-1$. Figure 2.6 illustrates the pyramidal representation.

Many schemes have been proposed that generate multiresolution pyramids, they all employ different low pass filters. The Wavelet transform [113] in particular has been demonstrated to provide an analytic basis for multiresolution analysis and fast algorithms have been developed for its implementation. It is not the purpose here to launch into a discussion of the merits of the various multiresolution schemes. It must be noted that the underlying motivation for a multiresolution approach here is to create a representation of the original image in which the motion is small enough to allow a motion estimation algorithm to terminate successfully. To this end the basic character of the image at the varying scales must be maintained. This is necessary to ensure that the motion fields estimated at the lower resolutions bear some resemblance to the actual motion field at the original resolution.

The multiresolution algorithm of Enkelmann and Nagel [41] is adopted for use in this book. It uses a Gaussian shaped low pass filter which is able to provide a basis for the pyramid and also slightly blur the image at each scale. This latter point is useful in that it artificially increases the proportion of the first order term in the Taylor series expansion of the image function, thus stabilizing the iterative process, (see Kearney et al [78]). Further, the filter is non directional and so does not favour motion estimation in any direction. The implementation used here is the standard refinement strategy common to multiresolution motion estimators. The low pass filter is defined as

$$\begin{aligned} f(x,y) &= w(x,y)\frac{1}{A}\exp^{-\left(\frac{r^2}{2\sigma^2}\right)} \\ r &= \sqrt{x^2+y^2} \\ w(x,y) &= \left\{ \begin{array}{ll} 1 & \text{for } x^2+y^2 \leq R^2 \\ 0 & \text{otherwise} \end{array} \right\} \\ A &= \sum_{|x|\leq R}\sum_{|y|\leq R} w(x,y)\exp^{-\frac{r^2}{2\sigma^2}} \end{aligned} \tag{2.24}$$

The multigrid algorithm is enumerated below

1. Generate L levels of the multiresolution pyramid, such that level $l = 0$ is the original resolution image, and level $l = L - 1$ is the image with the smallest resolution.

2. Set the initial level to be the level with the most coarse resolution, level $l = L - 1$. Set the initial field at this level to be a set of zero vectors.

3. Generate an estimate of the motion field at the resolution level l using the current starting field.

4. If $l = 0$ then goto 8.

5. Propagate the motion field from the coarse resolution level l down to the next higher resolution level $l - 1$.

6. Set the current resolution level to be $l = l - 1$.

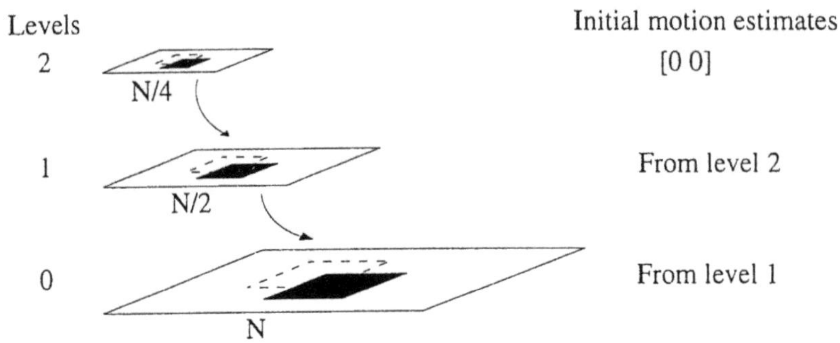

FIGURE 2.6. Multiresolution pyramid showing 3 level decomposition of an $N \times N$ frame. The dotted lines indicate the position of the object in the previous frame. Note this displacement is reduced at the same rate as the level dimensions.

7. Goto 3.

8. Stop.

Bilinear interpolation is used to propagate the motion field one level down, i.e. onto the higher resolution level. This will assign all pixel sites at the new resolution with a start vector. It is arguable what interpolating function would perform best, but it is found that the bilinear operation gives satisfactory results.

There is a drawback with the multiresolution motion estimation scheme. It brings similar regions, not related by motion, closer together. A typical effect is the assignment of displacement vectors to stationary regions just outside moving regions. For gradient based algorithms this could cause a lack of convergence at the lower levels. The effect is called the vector halo effect. Bierling [18] solves the problem with respect to BM by using a motion detector at the lowest level, regardless of the final vector estimate. Therefore, before assigning a motion vector to a region, Bierling verifies that it is in fact moving.

An alternative approach was used by Enkelmann[41] to improve the convergence of his pel–recursive algorithm. When it is detected that at a particular level, level l say, the algorithm does not converge quickly enough, the vectors are propagated up to the coarser level $l + 1$. Estimation then proceeds at that level. He found that it was better to propagate just the update vectors between levels rather than the entire motion displacement. This bi–directional control strategy, as opposed to top down only, was found to give good results for real scenes.

In the interests of computation, this book employs only the top down approach with motion detection as implemented by Bierling. This is illustrated in Figure 2.6.

2.5 Motion prediction

A few authors, dealing with the gradient based motion estimators, have used the continuity of motion flow to generate an initial estimate for the iterative schemes. The phenomenon is easily understood by considering a region, at position $\mathbf{x}$ in image n say, moving with a constant velocity $\mathbf{v}$ pixels/frame. Suppose that using frames n and $n-1$, the velocity of this region was correctly estimated in frame n. Given the constant velocity assumption, the region at position $\mathbf{x}+\mathbf{v}$ in frame $n+1$ should also be moving at $\mathbf{v}$ pixels/frame. Hence, an initial guess for the velocity of that location in frame $n+1$ would be $\mathbf{v}$.

The initial estimate can be predicted both by temporal means, as just outlined, and also through some spatial support [35, 37]. This phenomenon arises directly from the Local Smooth Flow assumption outlined earlier. In practice, of course, objects will have some acceleration and so the temporal prediction of motion may not be accurate. One of the advantages of incorporating this scheme is that it enables the motion estimator to converge quickly, and can reduce the search space required in BM for instance.

However, occlusion phenomena can adversely affect the motion prediction. Depending on the magnitude of the motion involved, this could mean abandoning the prediction over large areas of the image. Note also that at every scene change, there will be no available motion estimates and the algorithm will necessarily increase in computation at that point. This has interesting implications for real time applications. In such applications the motion estimator needs to generate a motion field for each frame in one frame interval. Therefore, even at a scene change, the algorithm would need to operate in one frame interval[12] starting with no previous estimate. If this is the case, then one could use the zero estimate at all frames. The advantage then, of using the motion prediction idea, would be one of more accurate motion estimation rather than reduction in computation required for an estimate.

Consider the multiresolution scheme discussed previously. Given that displacements are large, the multiresolution approach is appropriate. To use temporal motion prediction, initial estimates for the vectors at the smallest resolution image would need to be generated by propagating initial vectors up from level 0. However, at that level, all motion is quite small. Given that errors will arise in projecting vectors up the pyramid, and that the constant velocity assumption will be violated, it may well be that an initial motion estimate of $\mathbf{v} = 0$ is as useful as one projected up from level 0. Of course the alternative approach would be to use the multiresolution scheme at scene changes to reduce computation and help estimate large displacements, then depend upon temporal motion prediction to give useful estimates of motion thereafter, without using the multiresolution approach.

---

[12] Unless, of course enough frame buffer memory was available to hold the incoming frames. This can involve an impractical amount of memory considering a data rate of $720 \times 576 \times 25$ bytes/sec for PAL television.

Some experimental work is necessary to verify how much is gained by using some motion prediction scheme, but this is outside the scope of this book. Instead, the motion estimators in this work use multiresolution schemes that have no 'motion memory' and use zero initial motion estimates at the top level of the pyramid. This keeps the motion estimation aspect of the work as simple as possible.

2.6 A unifying theory

It is typical to introduce the idea of motion estimation using Bayesian inference as a separate class of motion estimation algorithms, for example in [169]. The Bayesian paradigm is more general than this might imply to the first time reader. It is certainly not true that a solution derived using a Bayesian methodology is automatically better than other approaches. It is often the case that the different concepts of Markov Chain Monte Carlo (MCMC) algorithms and Markov Random Fields (MRFs) are considered as having an umbilical link with Bayesian inference. The tremendous influence of the papers by Geman and Geman [47], Besag [15] and Konrad et. al [98, 157] can cause new workers in the area to expect that employing a Bayesian approach necessarily requires the use of Gibbs energy priors and stochastic MCMC searches like the Gibbs sampler for instance. Recently, some excellent texts by Winkler [191] and Li [104] help to clarify the situation with regard to the use of 'probabilistic' analysis as applied to image processing. The book by O'Ruanaidh [151] also gives a very good description of the overall methodology of Bayesian signal processing. In some sense, the exercise of applying a Bayesian methodology to any problem can be separated into an initial *conceptual* stage, followed by a more quantitative design stage involving a choice of probability densities and choosing a search strategy. In many cases, specific image processing algorithms, for example motion estimation algorithms, can be shown to arise as a consequence of specific choices for particular probability distributions that have been articulated under the same Bayesian framework.

This section introduces some concepts of probabilistic approaches to motion estimation and image sequence processing. In so doing, it lays the groundwork for the ideas explored in Chapters 7 and 8.

2.6.1 Conceptual motion estimation

This Chapter makes some overview of motion estimation algorithms available to the reader and it is natural to consider the problem as a specific example of the use of the Bayesian paradigm. The problem is to estimate the motion field $\mathbf{D}_{n,n-1}$ mapping frame n into frame $n-1$. The arguments to follow apply equally well to estimation of forward motion, $\mathbf{D}_{n,n+1}$. $\mathbf{D}_{n,n-1}$ represents the *entire* motion field at every pixel in the image. For the sake of simplicity, the problem is set up using just two frames, thus avoiding the general concepts introduced in [98]. Information about the motion field must be obtained by processing the observed image data in the two frames, I_n and I_{n+1}. The probabilistic approach proceeds to extract this information by somehow measuring and manipulating the probability distribution $p(\mathbf{D}_{n,n-1}|I_n, I_{n-1}, \mathbf{K})$ which is the probability of the motion field *given* the observed data and any other parameters required, collected in the vector $\mathbf{K}$. The choice of which motion field is best depends on the requirement of the algorithm designer. If the requirement is to pick the most likely motion field then the algorithm must maximize $p(\mathbf{D}_{n,n-1}|I_n, I_{n-1}, \mathbf{K})$ with respect to $\mathbf{D}_{n,n-1}$. If the requirement is to explore the *uncertainty* in the estimate for $\mathbf{D}_{n,n-1}$ then the algorithm must somehow explore the probability distribution $p(\mathbf{D}_{n,n-1}|I_n, I_{n-1}, \mathbf{K})$ itself and assess, for instance, the spread of possible estimates for $\mathbf{D}_{n,n-1}$.

Choosing the motion field which maximizes $p(\mathbf{D}_{n,n-1}|I_n, I_{n-1}, \mathbf{K})$ can involve the use of deterministic, well known optimization strategies such as Newton-Rhapson or Conjugate Gradient [142]. In general $p(\mathbf{D}_{n,n-1}|\cdot)$ is multimodal and so these strategies are likely to perform well only when a good starting point can be provided for an iterative update. Otherwise these methods tend to yield estimates at local minima which may not be useful solutions [142]. The alternative class of strategies attempt to generate random samples of $\mathbf{D}_{n,n-1}$ from $p(\mathbf{D}_{n,n-1}|I_n, I_{n-1}, \mathbf{K})$ and then process these samples numerically to yield the required solution. For instance a hyper-histogram[13] of $\mathbf{D}_{n,n-1}$ could be measured and then the modal value is the most likely estimate as required. These random search strategies fall into the class of MCMC algorithms. Generating random samples also has the additional effect of allowing the user to assess the uncertainty in the solution since access to the distribution is already available by collecting the samples generated.

Generating random samples of $\mathbf{D}_{n,n-1}$ is a problem. It is a vector valued matrix of estimates ultimately assigned to every pixel in the image. One could directly evaluate $p(\mathbf{D}_{n,n-1}|I_n, I_{n-1}, \mathbf{K})$ at every possible value of $\mathbf{D}_{n,n-1}$ and then directly evaluate the Cumulative Density. The transformation required to map a uniformly distributed random variable onto the distribution for $\mathbf{D}_{n,n-1}$ could then be numerically evaluated [142]. But there are too many sites in a particular image and too many possible values for $\mathbf{D}_{n,n-1}$ for this to be feasible. However, drawing samples from a complicated distribution is precisely the kind of problem for which the Gibbs sampler [47] was designed. The Gibbs sampler decomposes a draw from a joint multidimensional distribution into a draw from a number of conditional distributions of smaller dimension. In this case the multidimensional draw can be decomposed into draws from local conditional distributions at every pixel site. Of course, there is no need for the conditional draws to be defined on as fine a grid as a pixel basis, in fact Chapter 7 uses a block based decomposition instead. The 'resolution' of the conditionals depends on the application.

For a pixel accurate motion estimation field, the Gibbs sampler proceeds iteratively, starting from an initial estimate for $\mathbf{D}_{n,n-1}$, defined as $\mathbf{D}^0_{n,n-1}$

[13] Because $\mathbf{D}_{n,n-1}$ is a vector valued function at many pixel sites.

which may be a zero estimate. Samples are drawn as follows:

$$\begin{aligned}
\mathbf{d}^1_{n,n-1}(\mathbf{x}_1) &\sim p(\mathbf{d}_{n,n-1}(\mathbf{x}_1)|I_n, I_{n-1}, \mathbf{D}^0_{n,n-1}(-\mathbf{x}_1)) \\
\mathbf{d}^1_{n,n-1}(\mathbf{x}_2) &\sim p(\mathbf{d}_{n,n-1}(\mathbf{x}_2)|I_n, I_{n-1}, \mathbf{D}^0_{n,n-1}(-\mathbf{x}_2), \mathbf{d}^1_{n,n-1}(\mathbf{x}_1)) \\
&\vdots \\
\mathbf{d}^2_{n,n-1}(\mathbf{x}_1) &\sim p(\mathbf{d}_{n,n-1}(\mathbf{x}_1)|I_n, I_{n-1}, \mathbf{D}^1_{n,n-1}(-\mathbf{x}_1)) \\
&\vdots
\end{aligned}$$

Where $\mathbf{d}^i_{n,n-1}(\mathbf{x}_j)$ denotes the ith sample of $\mathbf{d}_{n,n-1}$ at site $\mathbf{x}_j$, $\mathbf{D}^0_{n,n-1}(-\mathbf{x}_1)$ denotes all the motion vectors in the motion field which are *not* at site $\mathbf{x}_1$ and $\mathbf{K}$ has been dropped for brevity. The mechanics of the iterations are simple. At each pixel (or block site) in the image, a sample for $\mathbf{d}_{n,n-1}$ is drawn *given* the current state of samples for motion in the image. The sample drawn at $\mathbf{x}_1$ (for instance) replaces the previous one at $\mathbf{x}_1$ and the process continues at the next site, $\mathbf{x}_2$, where a motion sample is drawn based on the current state of the motion field which now includes the sample just drawn at the previous site $\mathbf{x}_1$. The process is repeated iteratively until convergence when the samples being generated are samples from the multidimensional distribution $p(\mathbf{D}_{n,n-1}|\cdot)$ as required.

The problem now is to manipulate the conditionals $p(\mathbf{d}_{n,n-1}(\mathbf{x}_1)|\cdot)$ in order to draw samples. Bayes theorem allows the conditional to be written using terms which can be derived from data observation models such as equation 2.1 and constraints such as local motion smoothness. The conditional can be expressed as follows:

$$\begin{aligned}
&p(\mathbf{d}_{n,n-1}(\mathbf{x}_1)|I_n, I_{n-1}, \mathbf{D}_{n,n-1}(-\mathbf{x}_1), \mathbf{K}) \propto \\
&\quad p(I_n|I_{n-1}, \mathbf{d}_{n,n-1}(\mathbf{x}_1), \mathbf{D}_{n,n-1}(-\mathbf{x}_1), \mathbf{K})p(\mathbf{d}_{n,n-1}(\mathbf{x}_1)|\mathbf{D}_{n,n-1}(-\mathbf{x}_1), \mathbf{K})
\end{aligned} \tag{2.25}$$

The first term on the right hand side is called the likelihood of the data given the unknown motion parameter and the last term is the prior probability of motion. The entire product is called the posterior distribution for the unknowns. The likelihood expression arises directly out of the data observation model and the prior allows the injection of constraints on the solution. At a basic level, this is the end of the conceptual design stage. It now remains to decide on actual distributions and then manipulate them to gain some advantage over deterministic design.

2.6.2 Quantitative analysis

Consider that the image sequence obeys the simple translational model as in equation 2.1. It is possible to allow for some deviation from that model, i.e. incorporate uncertainty, by stating the model as

$$I_n(\mathbf{x}) = I_{n-1}(\mathbf{x} + \mathbf{d}_{n,n-1}(\mathbf{x})) + e_n(\mathbf{x}) \tag{2.26}$$

where $e_n(\mathbf{x}) \sim \mathcal{N}(0, \sigma_e^2)$ which is shorthand for defining $e(\cdot)$ as a Normally distributed random variable with mean 0 and variance σ_e^2. Therefore it follows that the likelihood expression required in the assembly of the posterior is as follows:

$$p(I_n | I_{n-1}, \mathbf{d}_{n,n-1}(\mathbf{x}_1), \mathbf{D}_{n,n-1}(-\mathbf{x}_1), \sigma_e^2) = \frac{1}{\sqrt{2\pi\sigma_e^2}} \exp - \left(\frac{(I_n(\mathbf{x}) - I_{n-1}(\mathbf{x} + \mathbf{d}_{n,n-1}(\mathbf{x})))^2}{2\sigma_e^2} \right) \quad (2.27)$$

Where σ_e^2 is the only entry in the parameter vector $\mathbf{K}$ introduced previously. The likelihood is influenced directly by the choice of data model, thus affine motion can easily be incorporated using the same steps shown here.

The experience of the designer plays a part in the choice of a good prior probability distribution. It is here that information about the organization of the motion field can be inserted into the solution. There are no rules which govern the choice of prior. To a large extent it is the ability to design priors which concisely define desired features of motion fields that separate one motion estimation algorithm from another. A concept which is typically invoked is that of the Markov Random Field. The concept is quite a broad one and simply states that the probability of the state of a variable at a particular site in time and space depends only on the states of a local neighborhood of variables. This can be expressed more precisely in the context of motion estimation as follows:

$$p(\mathbf{d}_{n,n-1}(\mathbf{x}_1) | \mathbf{D}_{n,n-1}(-\mathbf{x}_1)) = p(\mathbf{d}_{n,n-1}(\mathbf{x}_1) | \mathbf{S}_{n,n-1}(\mathbf{x}_1)) \quad (2.28)$$

Where $\mathbf{S}_{n,n-1}(\mathbf{x}_1)$ denotes a local neighbourhood of vectors around the current location $\mathbf{x}_1$. The Hammersley-Clifford theorem [15, 169] allows this Markov property to be defined in terms of Gibbsian energy/probability distributions for the local conditionals. There is a one to one correspondence between an MRF on the whole multidimensional vector field and the use of Gibbs energy distributions for $p(\mathbf{d}_{n,n-1}(\mathbf{x}_1)|\mathbf{S}_{n,n-1}(\mathbf{x}_1))$. This is a very convenient state of affairs because it allows the use of priors which are defined using only the state of local variables without any loss in generality in defining the global MRF. To encode the local smoothness of the motion field as recognized by Horn [65], Nagel [129], Enkelmann [41] and others, a typical prior distribution for motion is as follows:

$$p_d(\mathbf{d}_{n,n-1}(\mathbf{x}) | \mathbf{S}_{n,n-1}(\mathbf{x})) \propto \exp - \left(\sum_{\mathbf{v} \in \mathbf{S}_{n,n-1}(\mathbf{x})} \lambda(\mathbf{v}) [\mathbf{d}_{n,n-1}(\mathbf{x}) - \mathbf{v}]^2 \right) \quad (2.29)$$

where $\mathbf{v}$ is each vector in the neighborhood represented by $\mathbf{S}_{n,n-1}(\mathbf{x})$, and $\lambda(\mathbf{v})$ is the weight associated with each clique or pair of neighbourhood interactions. The situation is illustrated in Figure 2.7 for an eight connected neighbourhood of vectors (labelled 1 to 8). This is the usual form of the motion prior as used in Chapters 7 and 8 in this book. The reader will recognize that the energy function that is the argument of the exponential acts as a loss or error function which penalizes vectors that deviate substantially from the surrounding neighborhood.

This is a very simple formulation of the motion prior. Chapters 7 and 8 explore extensions that allow the incorporation of Occlusion as well as spatial motion discontinuities similar to the work by Konrad et. al [98].

2.6.3 Deriving motion estimators

The final problem is now the actual implementation. Exactly how does one go about drawing samples for motion at each site? Before this is considered it is educational at this stage to recognize that it is already possible to show how correspondence matching motion estimators for example BM, WBME, arise from this approach. First of all note that none of these older motion estimation approaches made any allowance for smooth motion constraints. In essence no prior knowledge about the motion field was incorporated into the solution. Therefore the prior distribution required in specifying the posterior for the motion vector at each site is uniform, or constant. Thus the posterior distribution is only influenced by the likelihood as follows:

$$p(\mathbf{d}_{n,n-1}(\mathbf{x}_1)|I_n, I_{n-1}, \mathbf{S}_{n,n-1}(\mathbf{x}_1), \sigma_e^2) \propto \frac{1}{\sqrt{2\pi\sigma_e^2}} \exp - \left(\frac{(I_n(\mathbf{x}) - I_{n-1}(\mathbf{x} + \mathbf{d}_{n,n-1}(\mathbf{x})))^2}{2\sigma_e^2} \right) \tag{2.30}$$

Furthermore, this implies that each iteration in the Gibbs sampler uses no variables generated by the previous iteration, since motion vectors at each site are not interconnected because of the lack of a prior. Thus the global maximization or sampling problem is completely specified by manipulations at each local site independent of the other sites. In correspondence matching the task is to find the best motion vector at each site. This implies maximizing equation 2.30 with respect to $\mathbf{d}_{n,n-1}$. This requires either generating samples and finding the mode or directly maximizing the distribution. In general this is hard because the distribution with respect to the motion is defined only implicitly through the interaction of image data which itself is unlikely to be modelled easily by some simple linear function. A direct search technique gives rise to the BM algorithm. In that algorithm every possible vector is evaluated according to equation 2.30 and the best one chosen. Of course there is no need to evaluate the normalized probability since maximizing that expression is the same as minimizing the log probability. Furthermore, no knowledge of σ_e^2 is required since it is a constant.

A more subtle approach can be taken. The argument of the exponential can be linearized around a current estimate for the displacement. This provides a Gaussian approximation to the generally complicated likelihood function with respect to the motion unknowns. This was presented in [98, 13]. The manipulation requires completing the square to express the linearized argument of the exponential in terms of the square of two terms which explicitly involve motion. The first steps are as follows, where it is assumed that the higher order terms in the Taylor series expansion are negligible.

$$\begin{aligned}
&p(\mathbf{d}_{n,n-1}(\mathbf{x})|\mathbf{S}_{n,n-1}(\mathbf{x}), I_n, I_{n-1}) \\
&= \frac{1}{\sqrt{2\pi\sigma_e^2}} \exp - \left(\frac{(I_n(\mathbf{x}) - I_{n-1}(\mathbf{x} + \mathbf{d}_{n,n-1}(\mathbf{x})))^2}{2\sigma_e^2}\right) \\
&\propto \exp - \left(\frac{(I_n(\mathbf{x}) - I_{n-1}(\mathbf{x} + \mathbf{d}^0_{n,n-1}(\mathbf{x})) - \mathbf{u}^T \nabla I_{n-1}(\mathbf{x} + \mathbf{d}^0_{n,n-1}(\mathbf{x}))^2}{2\sigma_e^2}\right) \\
&\propto \exp - \left(\frac{(z_n(\mathbf{x}) - \mathbf{u}^T \nabla I_{n-1}(\mathbf{x} + \mathbf{d}^0_{n,n-1}(\mathbf{x}))^2}{2\sigma_e^2}\right)
\end{aligned}$$

where $\mathbf{u}$ is the same update vector as introduced previously for the WBME, and $\mathbf{d}^0_{n,n-1}$ is the current estimate. Assuming that the motion is constant in a block, the observations of displaced pixel difference $z(\mathbf{x})$ and gradient $\nabla I_{n-1}(\cdot)$ can be collected in the block and manipulated to form a Gaussian distribution for the update vector as follows:

$$\begin{aligned}
p(\mathbf{u}|\mathbf{S}_{n,n-1}(\mathbf{x}), I_n, I_{n-1}) &= \frac{1}{\sqrt{2\pi\sigma_e^2}} \exp - \left(\frac{(\mathbf{z} - \mathbf{G}\mathbf{u})^T(\mathbf{z} - \mathbf{G}\mathbf{u})}{2\sigma_e^2}\right) \\
&\propto \exp - \left(\frac{\mathbf{z}^T\mathbf{z} - 2\mathbf{z}^T\mathbf{G}\mathbf{u} + \mathbf{u}^T\mathbf{G}^T\mathbf{G}\mathbf{u}}{2\sigma_e^2}\right)
\end{aligned}$$

Completing the square gives

$$\begin{aligned}
&\propto \exp - \left(\frac{(\mathbf{u} - [\mathbf{G}^T\mathbf{G}]^{-1}\mathbf{G}^T\mathbf{z})^T[\mathbf{G}^T\mathbf{G}](\mathbf{u} - [\mathbf{G}^T\mathbf{G}]^{-1}\mathbf{G}^T\mathbf{z})}{2\sigma_e^2}\right) \\
&\quad \times \exp - \left(\frac{\mathbf{z}^T\mathbf{z} - \mathbf{z}^T\mathbf{G}(\mathbf{G}^T\mathbf{G})^{-1}\mathbf{G}^T\mathbf{z}}{2\sigma_e^2}\right) \qquad (2.31)
\end{aligned}$$

Since only $\mathbf{u}$ is of concern this is the same as

$$\propto \exp - \left(\frac{(\mathbf{u} - [\mathbf{G}^T\mathbf{G}]^{-1}\mathbf{G}^T\mathbf{z})^T[\mathbf{G}^T\mathbf{G}](\mathbf{u} - [\mathbf{G}^T\mathbf{G}]^{-1}\mathbf{G}^T\mathbf{z})}{2\sigma_e^2}\right)$$

hence

$$\mathbf{u} \sim \mathcal{N}((\mathbf{G}^T\mathbf{G})^{-1}\mathbf{G}^T\mathbf{z}, \sigma_e^2(\mathbf{G}^T\mathbf{G})^{-1}))$$

As can be seen, this Gaussian approximation to the likelihood changes the problem from that of sampling for $\mathbf{d}_{n,n-1}$ to that of sampling for $\mathbf{u}$, the update to $\mathbf{d}^0$. Thus the conditional distribution for $\mathbf{u}$ is a Normal distribution with mean $\bar{\mathbf{u}} = (\mathbf{G}^T\mathbf{G})^{-1}\mathbf{G}^T\mathbf{z}$ and correlation $\mathbf{R}_{uu} = \sigma_e^2(\mathbf{G}^T\mathbf{G})^{-1}$ as shown above.

If an estimate of $\mathbf{u}$, $\hat{\mathbf{u}}$ is required to maximize the local conditional then it follows that $\hat{\mathbf{u}} = \bar{\mathbf{u}}$. This solution for $\mathbf{u}$ does not immediately appear to be related to the WBME. It is certainly the same as the pseudoinverse solution presented by Martinez [117]. Firmer relationships can be established when it is recognized that equation 2.31 also expresses the uncertainty in $\hat{\mathbf{u}}$ through the correlation matrix $\mathbf{R}_{uu}$. This implies that the distribution for $\hat{\mathbf{u}}$ is arranged so that its major and minor axes are aligned with the eigenvectors of $\mathbf{G}^T\mathbf{G}$. It is precisely this information which Martinez in effect employed to overcome problems at edges. The WBME employs μ to regularize the solution and account for uncertainty.

Having established how this *Maximum Likelihood* solution is related to BM and gradient based motion estimators, it remains to be shown how links can be found with optical flow estimation.

Incorporating priors

Using the simple smoothness prior introduced in equation 2.29, the conditional at each site is as follows:

$$p(\mathbf{d}_{n,n-1}(\mathbf{x}_1)|I_n, I_{n-1}, \mathbf{S}_{n,n-1}(\mathbf{x}_1), \sigma_e^2) \propto$$
$$\frac{1}{\sqrt{2\pi\sigma_e^2}} \exp - \left(\frac{(I_n(\mathbf{x}) - I_{n-1}(\mathbf{x} + \mathbf{d}_{n,n-1}(\mathbf{x})))^2}{2\sigma_e^2}\right)$$
$$\times \exp - \left(\sum_{\mathbf{v}\in\mathbf{S}_{n,n-1}(\mathbf{x})} \lambda(\mathbf{v})[\mathbf{d}_{n,n-1}(\mathbf{x}) - \mathbf{v}]^2\right) \quad (2.32)$$

There is now interdependence between motion vectors at different sites because of the prior. Thus it is no longer possible that the final algorithm can proceed in the same manner as BM for instance. The complexity of the expression depends on the choice for the prior. Because of this, drawing samples for $\mathbf{d}_{n,n-1}$ at each site is not necessarily a trivial matter. The expression can still be approximated by a Gaussian but this is not discussed further here. In general there are two available options for sampling strategies at this point. Samples can be generated using either a direct method or using MCMC algorithms for example, the Metropolis algorithm [168].

The direct method involves numerically evaluating the probability expression at each possible value of $\mathbf{d}_{n,n-1}$. Then a sample is drawn by transforming a uniformly distributed random variable using the numerically evaluated cumulative density function [142]. This is computationally intensive. A more practical approach is to evaluate $p(\mathbf{d}_{n,n-1}|\cdot)$ only at highly probable values of $\mathbf{d}$ and use this as an approximation to the true density. A suitable region of values may be easily selected by adopting the maximum allowed motion region as used for BM. Alternatively the region may be defined by perturbing vectors predicted from previous frames. A version of this *Griddy Sampler* [168] is explored in Chapters 7 and 8.

The alternative sample generation strategy is to use the Metropolis algorithm. This is an example of the use of an MCMC subchain to provide samples from a local conditional distribution which is difficult to manipulate. The algorithm accepts or rejects random proposals from a particular 'proposal' distribution. For the problem of sampling from the expression 2.32, the Metropolis algorithm proceeds as follows:

1. Propose a random change to the current sample for motion, $\mathbf{d}^0_{n,n-1}$. Define the proposal to be $\mathbf{d}^*$.

2. If $p(\mathbf{d}_{n,n-1} = \mathbf{d}^*|\cdot) > p(\mathbf{d}_{n,n-1} = \mathbf{d}^0_{n,n-1}|\cdot)$ (using equation 2.32), then accept the change.

3. Otherwise accept the change with probability p_a where

$$p_a = \frac{p(\mathbf{d}_{n,n-1} = \mathbf{d}^*|\cdot)}{p(\mathbf{d}_{n,n-1} = \mathbf{d}^0_{n,n-1}|\cdot)} \tag{2.33}$$

After some iterations the resulting sample is that required. Of course the computational load of the technique depends on the ratio of accepted to rejected proposals which is influenced by the choice of proposal distribution. In general the proposal distribution is chosen to be easier to sample from than the required distribution for example Normal or Uniform. Tanner [168] gives a good discussion about the choice of proposal densities. If few evaluations are required before generating a new sample, then the algorithm is likely to be computationally cheaper than the direct method discussed previously.

This iterative procedure embedded within the overall Gibbs sampling strategy will allow (eventually) generation of samples from $\mathbf{D}_{n,n-1}$ as required. Typical optical flow algorithms try to estimate the motion field $\mathbf{D}_{n,n-1}$, which maximizes the posterior distribution. This Maximum-a-Posteriori (MAP) estimate can be made by processing the samples generated from the Gibbs sampler to select the mode of the joint distribution. As discussed previously, this evaluation of the mode is likely to be computationally intractable since there are many possible configurations of $\mathbf{D}_{n,n-1}$ which must be collected before a decision can be taken. Instead another MCMC algorithm can be used, for example Simulated Annealing (SA). No details will be given here except to say that it is a method for altering the local conditionals at each iteration of the Gibbs sampler. The conditionals are slowly altered to converge on a delta function centred at the MAP estimate. This alteration of the distribution is achieved by the introduction of a temperature variable [168, 191, 142, 151]. This book does not employ SA to generate MAP estimates for unknowns and further discussion is not required.

Besag [15] introduced the Iterated Conditional Modes (ICM) algorithm in an attempt to design a computationally more attractive search procedure for the MAP estimate. ICM proceeds iteratively by choosing at each site the local MAP estimate for the unknowns conditional on the current state of the rest of the image. This amounts to choosing the motion estimate (in this problem) at each site which maximizes the posterior defined in equation 2.32. This maximization can be done directly (as is the case in Chapters 7 and 8) or it can be done using

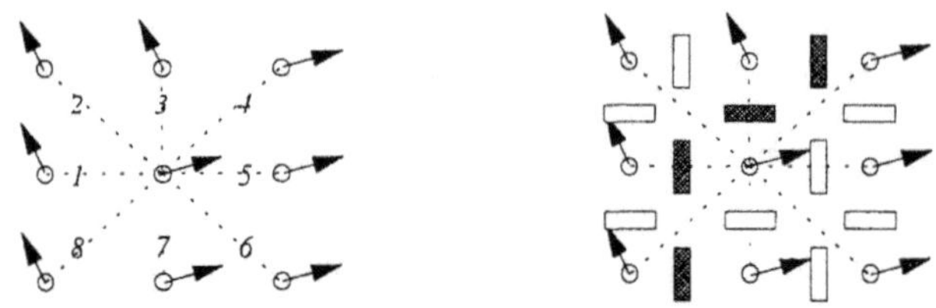

FIGURE 2.7. Left: Neighbourhood and cliques used for $p(\mathbf{d}_{n,n-1}(\mathbf{x}))$. Pixel sites marked with grey o. Right: Same cliques with the introduction of edge sites (marked with rectangles). An example edge configuration is shown. Edge sites set to 1 are shaded, otherwise they are 0.

a local gradient directed search for example a Newton-Rhapson technique [142]. The reader will appreciate that because the local conditionals are maximized at each iteration, the overall algorithm is more likely to get caught in local maxima of the posterior than SA. Whether this is detrimental or not depends on how good a start estimate is available.

It is now more easily seen that the resulting algorithm is related to the approaches of Enkelmann [41], Horn [65] etc. Those methods incorporated constraints using Lagrange optimization. In a sense, the Gibbs sampler iterations have replaced the unidimensional gradient based updates which are employed in [41, 65] for instance and the Lagrange multiplier has been replaced by $\lambda(\mathbf{v})$, the strength of the motion vector interactions. The resulting process, using ICM to find the MAP estimate of the motion field, yields virtually the same algorithm as discussed in section 2.1.3.

In general the different approaches to motion estimation can all be related by the incorporation of different priors or data observation models into the Bayesian framework discussed here. In Chapter 8 occlusion is introduced. Layer models [169, 190, 39] (not discussed here) for motion can also be dealt with under a Bayesian framework by including motion segmentation as part of the estimation framework.

As an example of incorporating an additional prior on motion information, consider the problem of introducing spatial discontinuities into the motion field. A useful form for the Gibbs energy prior for motion incorporating discontinuities is as follows:

$$p_d(\mathbf{d}_{n,n-1}(\mathbf{x})|\mathbf{S}_{n,n-1}(\mathbf{x})) \propto$$
$$\exp -\left(\sum_{\mathbf{v}\in\mathbf{S}_{n,n-1}(\mathbf{x})} \lambda(\mathbf{v})(1 - u(\mathbf{x},\mathbf{v})[\mathbf{d}_{n,n-1}(\mathbf{x}) - \mathbf{v}]^2\right) \quad (2.34)$$

A new set of variables, $u(\mathbf{v},\mathbf{x})$ has been introduced which represents an *edge* field [47] that exists in the interstices between image pixel sites. This is a binary field which is turned on when there is an edge between the current site $\mathbf{x}$ and the site from which the neighbourhood vector $\mathbf{v}$ is extracted. The situation is illustrated in Figure 2.7. When an edge site shown in Figure 2.7 is turned ON it removes the effect of the corresponding motion vector clique in the spatial smoothness energy. This will allow motion discontinuities at moving edges, for instance. Motion edges are more likely to occur at object edges [131] (which tend to correspond with image edges) and so an additional prior may be incorporated which encodes this knowledge. One form is as follows:

$$p(u(\mathbf{x}, \mathbf{v}) \propto \exp\left(-\alpha(1 - u(\mathbf{x}, \mathbf{v})\mathcal{E}(\mathbf{x}, \mathbf{v}))\right) \qquad (2.35)$$

where $\mathcal{E}(\mathbf{x}, \mathbf{v})$ is an image edge field that is observed using a typical edge detector for example Marr [114] or Canny [27], and α is a large positive constant. The prior therefore encourages motion edges at image edges. Other priors can be introduced which help to organize the edge field $u(\cdot)$ by discouraging isolated edge elements etc. This kind of strategy was investigated in [60, 61]. A simpler strategy for allowing motion discontinuities is pursued in Chapter 8 where the energy argument of the prior for motion smoothness is clipped.

Introduction of this prior for motion edges now requires extending the estimation algorithm to these new variables as well. This of course increases computational complexity but the advantage is now that the motion smoothness constraint is allowed to be violated at motion edges as required.

This completes the introduction to Bayesian Inference with respect to the motion estimation problem. Chapters 7 and 8 introduce new techniques for the manipulation of local conditional distributions to further simplify the sampling problem and improve convergence.

2.7 Final comments

It has been stated that image sequence restoration is invariably related to motion estimation. Several motion estimation approaches have been reviewed. It has been pointed out that BM, although robust, is computationally expensive and gradient based approaches provide an attractive alternative. Several gradient based approaches have been outlined and it was noted that there is scope for an adaptive gradient based scheme that incorporates the work of Biemond [17], Martinez [117], Driessen [35] and Böröczky [20, 21]. With the incorporation of multiresolution estimation it is found that the WBME is suitable for many of the algorithms in this book and for most of the book it is adopted for use. The next chapter explores adaptive versions of this motion estimator.

Image sequence processing in general is hampered by the lack of a general model. It is possible to treat the simple flow equation 2.1 as a sequence model, but the effect of lighting and 3D motion in general are then ignored. It has been stated that the development of 3D/perspective projection based models is fundamentally difficult. Instead it has been proposed to use the 3DAR model, previously presented in [165, 37], as a first approximation to a model of the image sequence. Further experimental work on the model is presented in Chapter 3 where a multiresolution adaptive 3DAR scheme is developed. Several interesting points arise about the nature of the displacement estimated from the model.

Difficulties with large displacements have been highlighted. The approach by Bierling [18] appears to be the simpler and more computationally effective. For work on real television sequences, multiresolution schemes are essential.

The last part of this chapter attempts to illustrate how a Bayesian framework for organizing the motion estimation problem can unify many of the proposals for motion estimation algorithms. The reader should appreciate that the advantages of a Bayesian methodology are in the ease of incorporating constraints through the use of different priors. It also allows access to MCMC techniques for assessing uncertainty. It is up to the algorithm designer to decide whether a deterministic scheme or MCMC exploration of probability surfaces is more useful in a particular situation. In many ways the improvements claimed using a Bayesian methodology are due only to the introduction of new priors which encode more complicated information into the final solution, for example motion edges and occlusion in the motion estimation problem. Chapters 7 and 8 present further ideas regarding the manipulation of probability distributions in a Bayesian framework. Chapter 7 unifies all the known algorithms for missing data manipulation under the Bayesian paradigm. It is hoped that the reader is now equipped to appreciate the arguments presented in later chapters.

As a final note, it is recognized that no discussion of the Layer model [190, 39, 169] has been provided here. This model is in increasing use particularly for simultaneous motion estimation and segmentation of image sequences. It is an extremely useful concept since it considers the image sequence as a volume of interacting components. It potentially represents a distinct advantage over the traditional use of two or three frames at a time for extracting motion information. The algorithms presented in this book do not employ this idea for manipulating image sequences; it is a topic of current research.

3

Three Dimensional Autoregressive Modelling for Image Sequences

In Chapter 2 various systems for motion estimation were reviewed. In particular, a model of the image sequence based on the 3DAR model was considered [37]. In that discussion the model coefficients were fixed for the entire sequence. Further, the implementation presented in [37] did not involve a multiresolution scheme, and as such it is difficult to use that scheme for motion estimation in a fast moving sequence.

Vaseghi and Rayner [183] have already introduced the use of the AR model for detection of impulsive noise in audio signals. They and Veldhuis [184], also show that the model can then be used for interpolation of the missing data. Veldhuis goes on to show that the 2DAR model is also effective in interpolating missing data in images.

In the light of these investigations, it is worthwhile to consider the usefulness of the 3DAR model as an image sequence model. In this chapter the model is compared with the WBME and integer accurate BM. Several improvements to the proposed scheme in [37] are made. The AR coefficients are made to change with the varying image information, and an adaptive scheme for estimating the motion parameter is presented which combines the efforts of Martinez [117] and Driessen et al [35, 34, 33]. Some important points are highlighted with respect to the accuracy of motion estimation using the adaptive scheme and this eventually leads to an adaptive multiresolution 3DAR algorithm for image sequence modelling.

This chapter takes a rather traditional approach to signal estimation and does not employ a probabilistic approach. Chapter 7 uses the model in a Bayesian framework to detect and remove missing data in motion pictures. It proposes a more coherent simultaneous motion and parameter estimation process than is considered here. The work discussed here should give the reader a feel for the potential advantages and disadvantages of the model.

3.1 The model

It is important to acknowledge from the onset that the AR model is not necessarily *the* model of the image sequence. In fact, the choice of this structure is motivated more by computational reasons and its spatio-temporal nature than by any physical basis.

Simply put, the model tries to make the best prediction of a pel in the current frame based on a weighted linear combination of intensities at pels in a predefined support region. This support region may occupy pels in the current frame as well as previous and past frames. The model was already described in Chapter 2, and the equation is repeated below.

$$I(i,j,n) = \sum_{k=1}^{N} a_k I(i+q_k(x)+sx_{n,n+q_k(n)}, j+q_k(y)+sy_{n,n+q_k(n)}, n+q_k(n)) + \epsilon(i,j,n) \tag{3.1}$$

For the purposes of parameter estimation, the model is considered in its prediction mode. In this mode the model is reconfigured as a predictor of images in the sequence. The predicted intensity is as follows:

$$\hat{I}(i,j,n) = \sum_{k=1}^{N} a_k I(i+q_k(x)+sx_{n,n+q_k(n)}, j+q_k(y)+sy_{n,n+q_k(n)}, n+q_k(n)) \tag{3.2}$$

The task then becomes to choose the parameters in order to minimize some function of the prediction error, or residual,

$$\epsilon(i,j,n) = I(i,j,n) - \hat{I}(i,j,n) \tag{3.3}$$

The equation 3.3 is just a rearrangement of the model equation 3.1 with the emphasis placed on the prediction error, $\epsilon(i,j,n)$.

3.2 Parameter estimation

As stated previously, the parameters which are required are the weights, a_k and the displacement $\mathbf{d}_{k,l}$. Because it is not possible to find a closed form solution for both the coefficients and the displacement at the same time, an iterative solution is proposed. The coefficients are estimated based on an initial guess for the displacement, this set of coefficients is used to generate an update to the initial guess for the displacement. The new displacement is then used to estimate a new set of coefficients and so on until the error is as low as required.

To this end therefore, the following sections treat the coefficient estimation and displacement estimation separately. The model is then investigated with respect to artificial sequences and finally a few different algorithms are compared.

3.2.1 Estimation of the AR coefficients

It was decided, in the interest of computational load, to use a least squared estimate for the model coefficients in order to adapt the coefficients to the image function prior to motion estimation. The coefficients are chosen therefore, to minimize the square of the error $\epsilon()$, above. This leads to the Normal equations. The derivation is the same as the one dimensional case and the solution can be arrived at by invoking the principle of orthogonality. Solving the Normal equations yields the model coefficients, a_k. Appendix A gives a more complete treatment, but the final set of equations to be solved is stated below.

$$\mathbf{C}\mathbf{a} = -\mathbf{k} \tag{3.4}$$

Here, $\mathbf{C}$ and $\mathbf{k}$ represent terms from the correlation function of the image sequence. $\mathbf{a}$ is the vector of model coefficients.

3.2.2 Estimating the displacement

In order to gain an explicit relation for $\epsilon()$, in terms of $\mathbf{d}$, the approach used by Biemond [17] and Efstratiadis [37], was to expand the image function, $I()$, in the previous frames, as a Taylor series about the current displacement guess. This effectively linearizes the equation for $\epsilon()$ and allows a closed form estimate for $\mathbf{d}$ through a recursive update process. It is this iterative solution that is used for estimating $\mathbf{d}$ in this work. The reader is referred to Appendix C for a derivation and definition of the terms.

Recall that the displacement for the next iteration, $\mathbf{d}_{i+1}$ is found by updating the current displacement, $\mathbf{d}_i$, using $\mathbf{d}_{i+1} = \mathbf{d}_i + \hat{\mathbf{u}}_w$. The final solution for the update is

$$\hat{\mathbf{u}}_w = [\mathbf{G}_w^T\mathbf{G}_w + \mu\mathbf{I}]^{-1}\mathbf{G}_w^T\mathbf{z}_w \tag{3.5}$$

The case treated in this work is a causal model, so that the support region for the model does not have any taps in the next frame. Therefore, only one displacement parameter mapping the current frame into the previous one is considered. The terms in the solution equation are defined as below[1]. The equations constituting the matrix framework are assembled from observations of pixels in a block of size $N_1 \times N_2$ in the current frame. (See Figure 2.3).

- $\mathbf{z}$, $(N_1 \times N_2 \times 1)$ is a column vector of prediction errors given the current displacement, as defined in 3.3, at all the points in the region used for estimation. Note that $\epsilon_o(\mathbf{x}_k, n)$ denotes the kth current observed prediction

[1] See Appendix C.

error at position $(\mathbf{x}_k, n)$ and not the actual model excitation $\epsilon(\mathbf{x}_k, n)$.

$$\mathbf{z}_w = \begin{bmatrix} \epsilon_0(\mathbf{x}_1, n) \\ \epsilon_0(\mathbf{x}_2, n) \\ \vdots \\ \epsilon_0(\mathbf{x}_{(N_1 N_2)}, n) \end{bmatrix} \tag{3.6}$$

- $\mathbf{G}_w$, $(N_1 \times N_2 \times 2)$ is a matrix of gradients at the support positions in the previous frame. In the following expression, $I_n()$ refers to the nth pixel in the block in the current frame used to set up the motion equations and $D()$ is a function representing the necessary offset terms corresponding to support vector $\mathbf{q}_k$. The model is assumed to be purely temporally causal and has no taps in the next frame.

$$\mathbf{G}_w = \begin{bmatrix} \sum\sum a_k \frac{\partial I_1(\mathbf{D}(f,k,n))}{\partial x}) & \sum\sum a_k \frac{\partial I_1(\mathbf{D}(f,k,n))}{\partial y} \\ \sum\sum a_k \frac{\partial I_2(\mathbf{D}(f,k,n))}{\partial x}) & \sum\sum a_k \frac{\partial I_2(\mathbf{D}(f,k,n))}{\partial y} \\ \vdots & \vdots \\ \sum\sum a_k \frac{\partial I_{(N_1 N_2)}(\mathbf{D}(f,k,n))}{\partial x}) & \sum\sum a_k \frac{\partial I_{(N_1 N_2)}(\mathbf{D}(f,k,n))}{\partial y} \end{bmatrix}$$

- $\mathbf{u}$ is the (2×1) vector of updates defined as

$$\mathbf{u} = \begin{bmatrix} u_{n,n-1}(x) \\ u_{n,n-1}(y) \end{bmatrix} \tag{3.7}$$

- μ is the regularizing parameter.

$$\mu = \frac{\sigma_{\epsilon\epsilon}^2 + \sigma_{\nu\nu}^2 \sum_{k=1}^{N} a_k^2}{\sigma_{uu}^2} \tag{3.8}$$

  Where $\sigma_{\epsilon\epsilon}^2$ is the variance of the actual model error, $\sigma_{\nu\nu}^2$ is the variance of the truncation noise, and σ_{uu}^2 is the variance of the estimated update components.

In practice, neither $\sigma_{\epsilon\epsilon}^2$ nor $\sigma_{\nu\nu}^2$ are available. The standard approach is to assume some constant value for μ or to use some adaptive scheme. Such an adaptive scheme is presented later in this chapter.

The algorithm for finding the displacement is recursive and some criterion must be used for halting the process when convergence is sufficient. Several criteria are normally used in conjunction and they are as follows:

- Threshold the estimated update magnitude. *if* $|\mathbf{u}| < t_u$ *then halt.*
- Threshold the MSE. *if* $|\mathbf{z}| < t_z$ *then halt.*
- Limit the number of iterations to ensure the process does terminate even if convergence satisfying the above two criteria are not satisfied.

3.3 Experiments on artificial autoregressive sequences

The previous arguments have introduced a number of assumptions. These are

- The image sequence obeys an autoregressive model.
- The model can be linearized effectively using the Taylor expansion.
- The higher order terms in the expansion can be considered to have the same effect as adding white noise to the linearized equations.
- That the components of the vector estimate are uncorrelated.

It is educational to consider the effects of these assumptions in a controlled environment where the model is known. To this end it is first necessary to generate a sequence which obeys a three dimensional autoregressive model.

3.3.1 Generating three dimensional autoregressive sequences

To create these sequences, the prediction equations must be viewed from a different perspective. The observed intensity can be considered to be the result of an IIR system driven by an 'innovations' sequence, $\epsilon(\mathbf{x}, n)$. The z-transform of the denominator of the IIR system transfer function is the z-transform of the prediction equation 3.3. Therefore the standard method of creating autoregressive signals is to drive some IIR filter with a Gaussian noise sequence.

In the multidimensional case it is more natural to consider non-causal models than causal ones. In this case of image sequence modelling, there does not exist any notion of causality in the spatial domain other than the direction of processing and the z-transform of the system. However, the generation of non-causal 3DAR sequences is complicated by the fact that the driving function required is not white noise as is the case with causal systems[2] (See Appendix B).

There is the further complication that stability criteria for multidimensional IIR filters are extremely convoluted and difficult to use for general filter design [107]. However, if the IIR filter is separable then the stability criteria involve only consideration of the n one dimensional filters involved. To simplify matters, a specific case is considered and this is used in the ensuing experiments. The model used is a purely causal one defined by equation 3.9 and illustrated in Figure 3.1.

$$\begin{aligned}
&I(i,j,n) = \\
&a_1 I(i-1,j,n) + a_2 I(i-1,j-1,n) + a_3 I(i,j-1,n) + a_4 I(i+\mathrm{sx}, j+\mathrm{sy}, n-1) \\
&\quad + a_5 I(i-1+\mathrm{sx}, j+\mathrm{sy}, n-1) + a_6 I(i-1+\mathrm{sx}, j-1+\mathrm{sy}, n-1) \\
&\quad + a_7 I(i+\mathrm{sx}, j-1+\mathrm{sy}, n-1) + \epsilon(i,j,n) \qquad (3.9)
\end{aligned}$$

[2] This is not unique for multidimensional systems, it is the same for the one dimensional case.

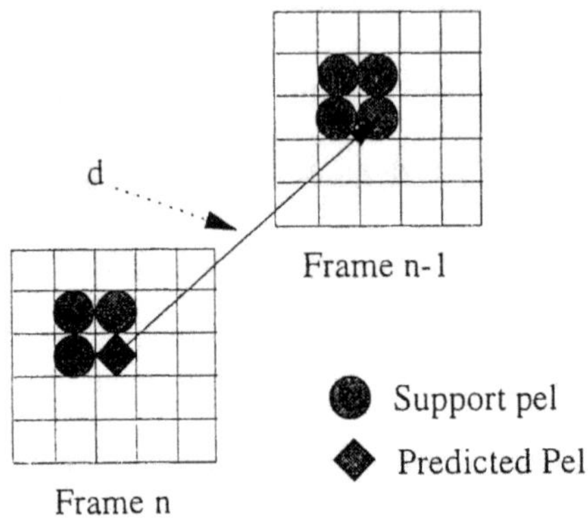

FIGURE 3.1. The structure of the AR system used for the synthetic sequences.

Taking the z–transform, the model equation becomes

$$\begin{aligned} I(z_1, z_2, z_3) &= \\ I(z_1, z_2, z_3)[a_1 z_1^{-1} &+ a_2 z_1^{-1} z_2^{-1} + a_3 z_2^{-1} + a_4 z_1^{sx} z_2^{sy} z_3^{-1} + a_5 z_1^{sx-1} z_2^{sy} z_3^{-1} \\ &+ a_6 z_1^{sx-1} z_2^{sy-1} z_3^{-1} + a_7 z_1^{sx} z_2^{sy-1} z_3^{-1}] + E(z_1, z_2, z_3) \end{aligned} \quad (3.10)$$

Defining

$$H(z_1, z_2, z_3) = \frac{I(z_1, z_2, z_3)}{E(z_1, z_2, z_3)} \quad (3.11)$$

makes it easier to realize that the original predictor equation 3.2, can be rearranged so that the image sequence $I(i, j, n)$ is the output of a system $H(z_1, z_2, z_3)$ driven by the function $E(z_1, z_2, z_3)$. The model transfer function is seen to be an all pole model. For this causal case the output is a Markov autoregressive sequence if the driving function is Gaussian white noise.

In order to generate an autoregressive sequence one must choose the coefficients **a** so that $H(z_1, z_2, z_3)$ is stable. This is much simplified if $H(\cdot)$ is made separable to enable the function to be implemented as the cascade of 3 one dimensional filters operating in orthogonal directions.

$$H(z_1, z_2, z_3) = \frac{1}{(1 - r_1 z_1^{-1})(1 - r_2 z_2^{-1})(1 - r_3 z_1^{sx} z_2^{sy} z_3^{-1})} \quad (3.12)$$

From the above equation therefore, $H(\cdot)$ is stable provided that the three values, r_1, r_2, r_3, all have a magnitude less than 1. To simplify further references to these values they will be called poles, even though strictly speaking, the third system in cascade (involving z_3) is a multidimensional one and so has some sort of pole surface.

Two sequences were used for the experiments. They were generated by driving a 3DAR system with noise until 910 frames of 900 × 900 resolution[3] were

[3]This unusual resolution was forced due to memory constraints in the machine used.

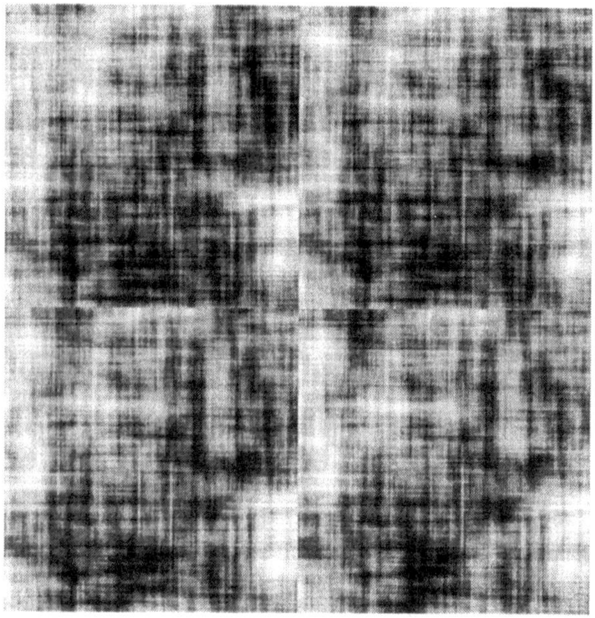

FIGURE 3.2. Top right to bottom left: A sequence of frames similar to AR95 but no motion. The intensities have been scaled to 8 bits.

Model	Coefficients						
	a_1	a_2	a_3	a_4	a_5	a_6	a_7
AR95	-0.95	0.9025	-0.95	-0.95	0.9025	-0.8574	0.9025
AR99	-0.99	0.9801	-0.99	-0.99	0.9801	-0.9703	0.9801

TABLE 3.1. The coefficients used for generating 3DAR signals.

produced. The first 900 were regarded as transient and the experiments used the bottom right hand corner (resolution 32×32) portion of the last 10 frames. The two sequences, AR95 and AR99, had poles at $r_1 = 0.95, r_2 = 0.95, r_3 = 0.95$ and $r_1 = 0.99, r_2 = 0.99, r_3 = 0.99$. The variance of the white noise used was 1.0 and in both cases the motion was $sx = -1.0$, $sy = -1.0$. This small motion was used in order to keep the spatial transients in the generation process to a minimum. The resolution of these frames is too small to include a useful print here, instead, Figure 3.2 shows 4 frames (of resolution 128×128) from a sequence similar to AR95 but with no motion. It is hoped that these prints will give the reader a better appreciation for these synthetic sequences.

Once the poles of the AR system have been chosen and the AR sequence generated, the optimum prediction coefficients, $\mathbf{a}$ can be found by equating the denominator of the system function 3.12 to the expression for the system

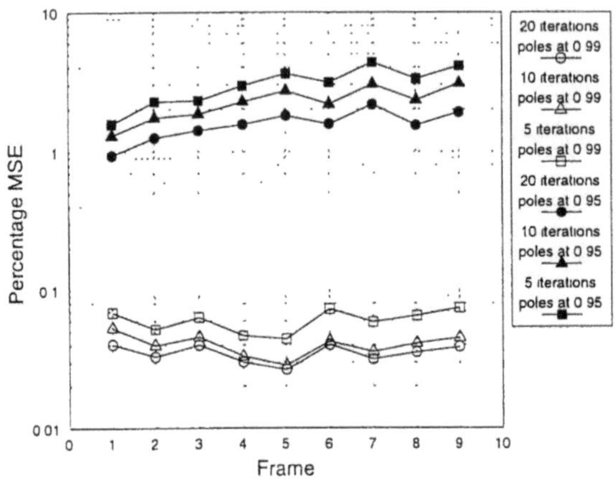

FIGURE 3.3. Comparison of percentage mean square prediction error.

function in terms of the model coefficients, equation 3.10. The coefficients for the two processes examined are given in table 3.1.

3.3.2 The experiments

Having created two sequences for which the model is known, the task is now to examine the behavior of the displacement estimation algorithm. For each of the two sequences, displacements were estimated for 9 out of the 10 frames. To estimate the displacement at a pel, 49 equations were set up using a block of 7×7 pixels centred on the tested pel. The known coefficients for the model were used in setting up the equations. A separate displacement parameter was estimated for each pixel in each frame. To evaluate pixel intensities at fractional positions, bilinear interpolation was used. Two sets of experiments were carried out with μ fixed at 100.0 and 500.0. A border of 5 pixels around the frame was ignored in the results to reduce edge effects. No thresholds were used to test for convergence since the investigations themselves observed the convergence rate with increasing iterations.

3.3.3 Results and observations

With respect to the prediction error, Figure 3.3, shows the Percentage MSE over the 9 frames from both sequences. This measure is the percentage ratio between the Mean Squared prediction error across the entire frame and the Mean Squared intensity across the original frame. It is defined below.

$$\text{PMSE} = 100.0\frac{\sum_{\mathbf{x}} \epsilon_o(\mathbf{x})^2}{I(\mathbf{x})^2} \tag{3.13}$$

The PMSE is about 10 times smaller for the AR99 sequence than for the AR95 sequence.

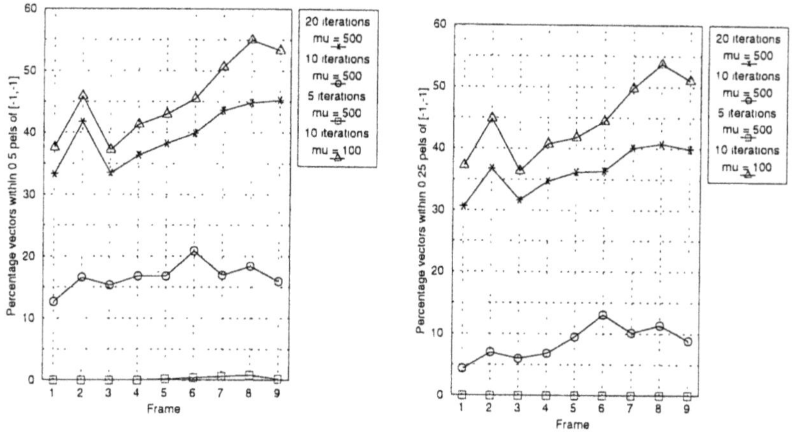

FIGURE 3.4. Accuracy of the algorithm with AR95.

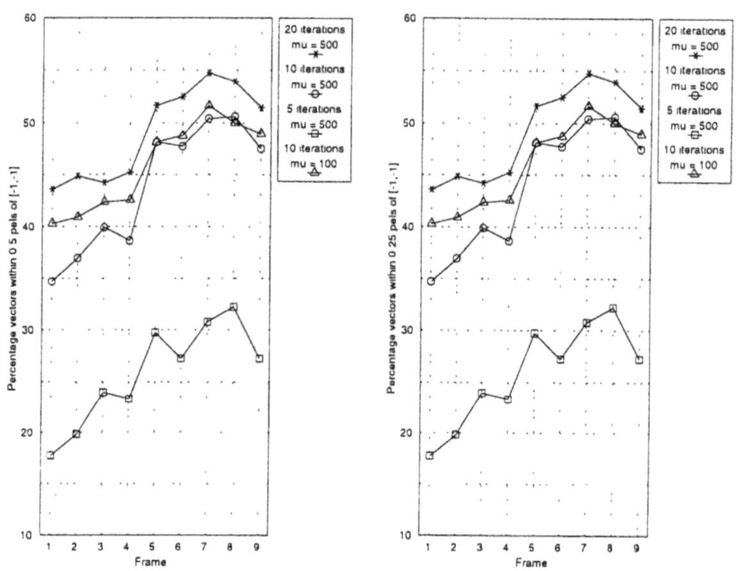

FIGURE 3.5. Accuracy of the algorithm with AR99.

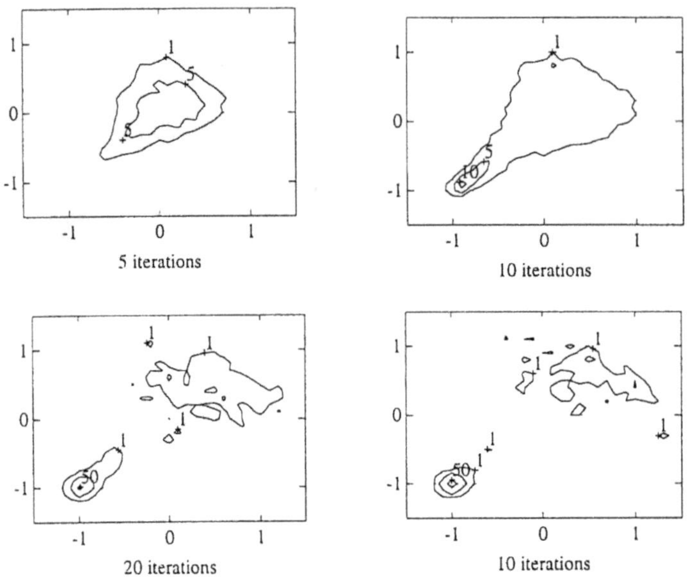

FIGURE 3.6. Distributions of the estimated vectors at 3 stages in the motion estimation process, $\mu = 500$. Poles at 0.95 (AR95). The bottom right plot shows the distribution for $\mu = 100$ at 10 iterations.

Figures 3.4, 3.5 show the percentage number of estimated displacements that were within radii of 0.25 and 0.5 pels of the correct displacement. These results show the different performance between the two sequences with respect to displacement estimation. Figure 3.6 is included to show the typical behavior of the algorithm as the iterations progress. The plots in the figure show averaged 2D histograms of the displacement estimates over the entire frame[4] at a particular iteration. Therefore, as the iterations progress the distribution 'flows' away from the initial cluster at [0.0, 0.0] to eventually converge in a peak at [−1.0, −1.0], the actual displacement. The plots give an overall view of the progression of the method. They also point out that although the number of vectors correctly estimated is just 50.0%, as shown by Figure 3.4, the distribution does show a dominant peak at the correct displacement after 20 iterations. The remaining vector estimates do not converge and are distributed randomly near [0.0, 0.0].

The following observations can be made:

- The motion estimation algorithm is approximately 50% accurate.
- Accuracy is increased with poles closer to the unit circles. The accuracy of the estimation algorithm with sequence AR95 is lower by about 5% at 20 iterations than the estimation with sequence AR99.

[4]The measured distribution over each frame is averaged over all 9 frames to give the distribution that is plotted.

- With increasing iterations there is a corresponding increase in accuracy, but there is an upper bound.
- μ acts as a damper on the estimation system. Small μ yields higher accuracy at smaller numbers of iterations. Compare the right top and bottom plots in Figure 3.11.

3.3.4 Discussion

The observations made above seem to suggest that even with a known system the algorithm is not satisfactory. However attention must be drawn to the fact that even with a given model, a Taylor series approximation is made to linearize the equations. The approximation requires essentially that the higher order terms in the expansion be negligible or can be modelled as a white noise error source. The prints of the sequences shown before indicate that rapid changes in the intensity of each frame are common. This implies that the higher order Taylor series terms can be appreciable. Therefore, contrary to the initial impression, it is understandable that the motion estimator is not quite as good as expected. Of course some of the discrepancy must be due to the random nature of the signal itself. It would be better perhaps to put a bound on the variance of the estimated vectors given that the driving function is random, and observe how close the estimator gets to that bound. Nevertheless, some perspective must be maintained, and it is clear from the prints that this particular driving function does not necessarily result in the type of image features encountered in a real case.

The fact that the performance with poles at 0.99 is better can be attributed to the more 'resonant' nature of the generated signal frequency spectrum. This is a well known phenomenon with regard to generator/estimator experiments like these, and it is considered in [159]. The resulting sequence from the generator is better described by the generator coefficients if the poles are closer to the unit circle.

Finally, the behavior of the estimator with increasing iterations is expected. As the number of iterations is increased more of the estimated displacements can converge on a solution. The degree of this convergence is proportional to the damping factor μ. When μ is high the estimator is heavily damped and necessarily takes more iterations to converge.

3.4 Experiments with a more realistic sequence

To better assess the performance of the 3DAR model, sequences which obey both an AR model and a first order Taylor series expansion must be found. Unfortunately, these requirements conflict. A sequence obeying a first order Taylor series expansion implies a uniformity that can be obtained when the poles of the generating system are far from the unit circle. When the poles are far from the unit circle however, the coefficients of the generator system

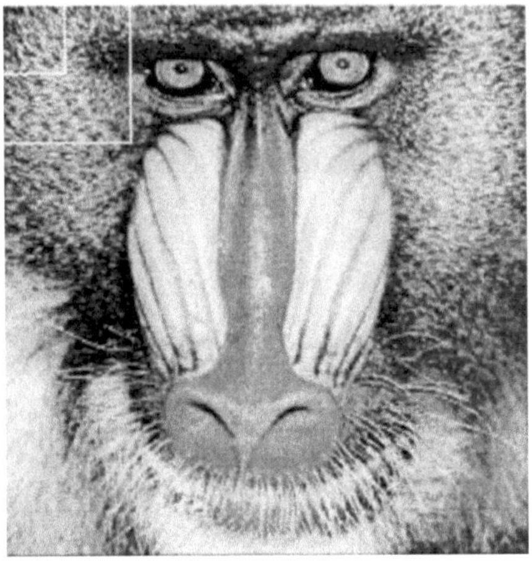

FIGURE 3.7. The Baboon image. The upper left borders delineate two quadrants used for experiments. The larger quadrant (64×64) is used for the multiresolution experiments.

do not describe the generated image well. Given a certain number of trials of generator/estimator experiments, there is a large variance in the estimated coefficients with the same generating system. This does shed some doubt as to the usefulness of experiments such as these and instead this next set of experiments tries to represent a half way point between the real world case and the completely artificial case. Driessen et al [35], in their paper on adaptation of the Biemond motion estimator, used the upper left hand corner of the Baboon image, (see Figure 3.7) to generate a sequence for investigation. Their motivation was that this image was a good test of the estimator since this highly textured image would result in a noisy **G** matrix. This same image was used to generate a test sequence for this algorithm. A window of 32×32 pixels was extracted from the upper left hand corner of the Baboon image to create one frame. This window was displaced by [-2,-2] pixels from frame to frame to create the desired sequence. The sequence was 10 frames long, the first frame of which is delineated by a white frame in Figure 3.7.

In his work on 2D interpolation of images, Veldhuis [184] reported that a much better prediction was gained in real images by using a model as stated in equation 3.14 below.

$$I(i,j) - (p_1 + p_2 i + p_3 j) = \sum_{k=1}^{N} a_k [I(i + q_k(x), j + q_k(y)) - (p_1 + p_2 i + p_3 j)] + \epsilon(i,j) \quad (3.14)$$

The coefficients, p_k, represent a 2D plane. This form dealt effectively with non-stationarity in the image blocks used. To linearize the parameter estimation problem, the plane coefficients were chosen on a least squared basis prior to the AR modelling. Therefore, the coefficients, p_k were chosen to minimize the

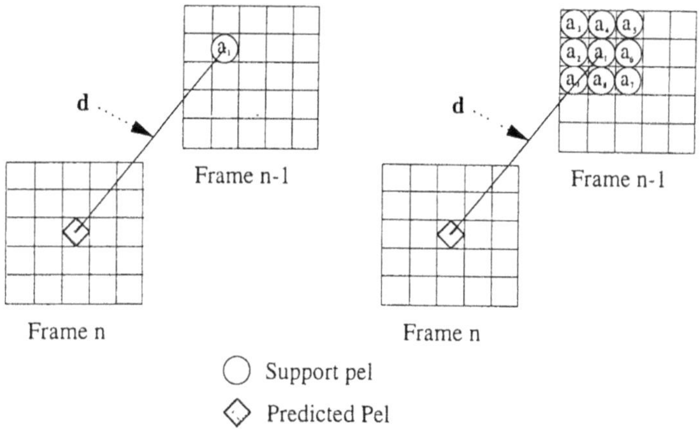

FIGURE 3.8. Two 3DAR models used in experiments. Left: AR1, Right: AR9.

following expression over the chosen image block.

$$[I(i,j) - (p_1 + p_2 i + p_3 j)]^2 \tag{3.15}$$

In this work, an extension of that idea is used for the sequences with real data, and a 3D trend replaces the 2D one of Veldhuis as shown in equation 3.16.

$$\begin{aligned} I(i,j,n) - (p_1 + p_2 i + p_3 j + p_4 k) = \\ \sum_{k=1}^{N} a_k [I(i + q_k(x) + sx_{n,n+q_k(n)}, j + q_k(y) + sy_{n,n+q_x(n)}, n + q_k(n)) \\ - (p_1 + p_2 i + p_3 j + p_4 k)] + \epsilon(i,j,n) \end{aligned} \tag{3.16}$$

This form is equivalent to subtracting a 3D trend from the sequence data, $I(i,j,n)$ prior to the modelling step. The trend acts as a normalizing term and this subtraction is implemented prior to modelling. The normalizing step does not affect the form of any of the equations and the trend is replaced after processing to yield a usable prediction.

3.4.1 Experiments

In working with the Baboon test sequence it is possible to compare the performance of the AR algorithm with the algorithm developed by Biemond et al [17]. However, the AR model for the case of this translational sequence is not known. In order to overcome this problem, the coefficients were determined by solving the Normal equations at the correct, known displacement and then using those coefficients as required at each pixel. The following experiments were then performed:

1. The known coefficients were used in the estimation of the displacement **d**, starting the iterations at $\mathbf{d} = [0.0,\ 0.0]$.

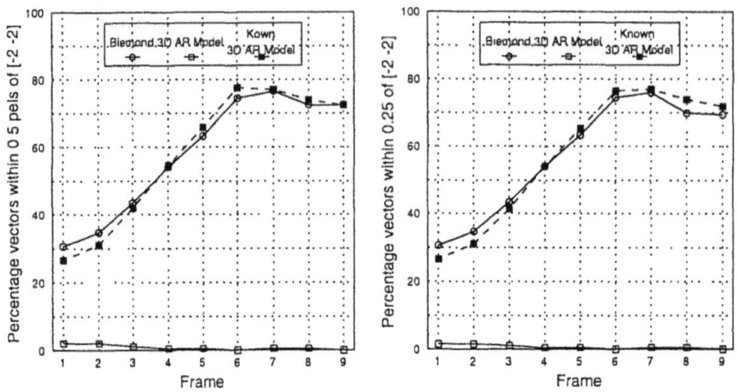

FIGURE 3.9. The accuracy of the motion estimation for various algorithms with respect to the Baboon sequence.

2. The iterations begin with $\mathbf{d} = [0.0,\ 0.0]$ and for each iteration the coefficients are estimated by solving the Normal Equations at the current displacement. Then that displacement is updated by solving for $\mathbf{d}$ using the previously estimated coefficients.

For all experiments, $\mu = 100.0$, blocksize $= 13 \times 13$. The algorithm was run for 20 iterations at all points considered. No threshold was set on the update magnitude nor the mean squared prediction error to halt the process before the maximum number of iterations were reached. The AR9 model, illustrated in Figure 3.8, was used. The PIMSE (defined below) was calculated as well as the percentage of correct vectors estimated. These measures were used to compare the various approaches.

The Percentage Improvement in Mean Squared Error (PIMSE) is defined as follows:

$$\text{PIMSE} = \frac{\text{FDMSE} - \text{DFDMSE}}{\text{FDMSE}} \tag{3.17}$$

FDMSE is the mean squared frame difference without motion compensation and DFDMSE is the mean squared prediction error with motion compensation. For the Biemond algorithm, the DFDMSE is the mean squared displaced frame difference given the estimated vectors. Therefore the PIMSE is a measure of how much better the prediction error is with motion compensation when compared to the difference between frames. 100% PIMSE means that there is no error in predicting the next frame from the last. 0% PIMSE means that the prediction of the next frame using the considered compensation algorithm is just as good as using the previous frame itself as a prediction of the current frame, hence there is no advantage in using the algorithm at all.

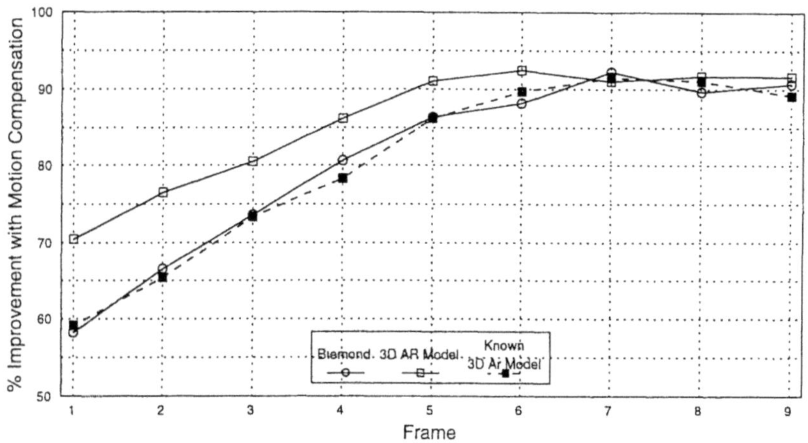

FIGURE 3.10. Comparing the percentage improvement in mean squared prediction error.

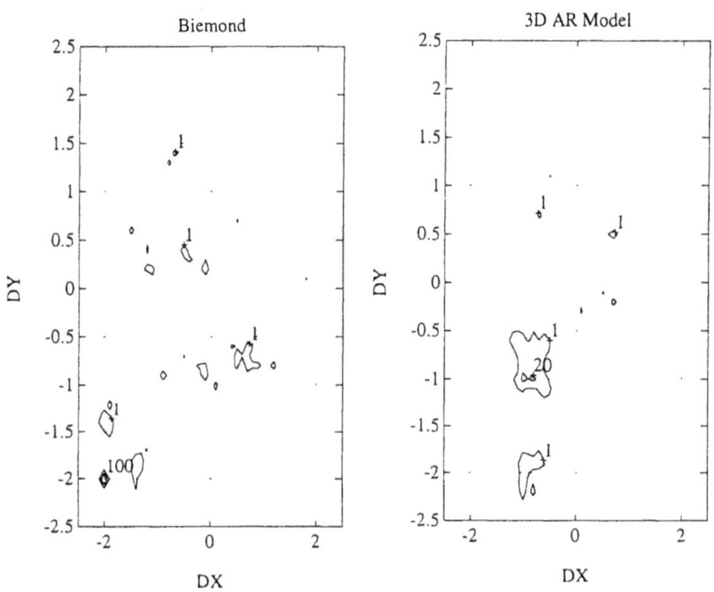

FIGURE 3.11. Comparing the distributions of the estimated vectors for Baboon using the AR9 model.

3.4.2 Discussion

It is clear that for this translational sequence, with integer pixel displacement, the optimal AR model has a one pixel support in the previous frame with associated coefficient value -1.0. This is so in practice. When the 'true' model coefficients are estimated using the Normal Equations at the correct displacement, both the models in Figure 3.8 assign a value of -1.0 to a_1, and the rest are set at 0.0. This will always occur since this assignment of a_1 will always yield the minimum squared prediction error at the correct displacement. Therefore, the Biemond and AR algorithms must perform identically in the experiments given the 'true' AR coefficients. Figures 3.9 and 3.10 illustrate this point in that the curves for both the AR algorithm with known coefficients and the Biemond algorithm are almost identical.

What is surprising is the discrepancy that arises for the case where the AR coefficients are not known and so are estimated at each iteration using the current value for $\mathbf{d}$. This is a more realistic assessment of performance. Figure 3.10, which shows the PIMSE using the AR9 model compared with the Biemond approach, indicates that the AR9 model gives a better prediction than the Biemond approach. However, as far as the accuracy of the motion estimate is concerned, Figure 3.9 indicates the opposite result. The accuracy of the Biemond approach surpasses the AR model result. The experiment highlights the important point that the value for $\mathbf{d}$ given by the AR model does not necessarily reflect the true motion of the scene. This is not unreasonable. The major influence in this observation is the change in the coefficients as the iterative process proceeds.

When the coefficients are estimated using the current estimate for $\mathbf{d}$ the Normal Equations adjust the values to minimize the expected value of the squared prediction error. This implies that the PIMSE is maximized at each iteration. In contrast, the estimation of the motion $\mathbf{d}$, tries to reduce the expected value of the error with respect to the vector $\mathbf{d}$. This is done using the observed image gradients which do not necessarily indicate the correct direction in which to update the current vector. The coefficient adjustment therefore serves as a powerful method of compensating for errors in the vector estimation process. Furthermore, in this case (for AR9), once the iterations have brought $\mathbf{d}$ to within a 1 pixel radius of the correct displacement, the spatial support of the model overlaps the pixel at the actual displacement. The coefficient nearest to that pixel (in this case a_3 can coincide exactly with that location) can take on a value of -1.0 and the others forced toward 0.0 to allow a perfect prediction. Of course this is not exactly the case in practice and Figure 3.11 shows a two dimensional histogram of the estimated vectors estimated using both algorithms. The histograms for each frame were averaged to give this result. The AR case shows a large cluster about $[-1,-1]$ whereas the case for the Biemond estimator shows a large cluster around the actual displacement $[-2,-2]$.

It is interesting to observe the values of the model coefficients after the final iteration. Figure 3.12 was produced by considering the 9 coefficients at each pixel in an 11×11 block at the centre of each 32×32 frame. The distribution

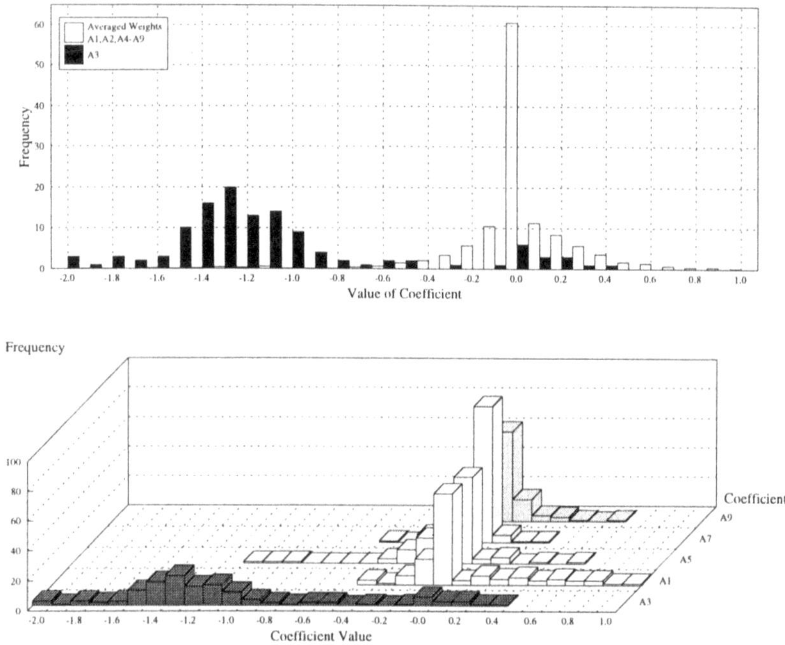

FIGURE 3.12. Distributions of model coefficients after termination of iterative AR process. Top: Comparing the averaged distributions of $a_1, a_2, a_4..a_9$ with a_3. Bottom: Distributions of selected coefficients.

of each coefficient was averaged over all the frames. The three dimensional bar chart in the bottom half of the figure shows the distributions of 5 of the 9 coefficients. Note that a_3 is different from any other coefficient and takes on values between -1.0 and -1.5. The plot at the top of the figure contrasts the distribution of a_3 with that of the averaged distribution of the other 8 coefficients. It is clear that all but a_3 are distributed about 0.0. This reinforces the statements made above in that when $\mathbf{d} = [-1.0 - 1.0]$, the support pel corresponding to a_3 overlaps the correctly displaced position and the corresponding correlation coefficient is high enough to cause $a_3 = 1.0$ and $a_1, a_2, a_4..a_9$ to be near 0.0. The resulting prediction error becomes small and the update vector produced is similarly diminished. The algorithm converges at that point.

It would appear that in order to get a good prediction, estimating the coefficients at each iteration is worthwhile. In order to get accurate motion estimation a model with a small spatial support must be used. To combine these two benefits it would seem reasonable to use an AR1 model and update the coefficient at each iteration. This is not necessarily the case. Consider the small spatial support of model AR1 in Figure 3.8. Depending on the image and the motion involved, there may not be a high correlation between the pixels at a location in the current frame, n, and the corresponding position in frame $n-1$. This is the situation if the iterations start with an initial zero displacement estimate. Of course, this can occur regardless of the support for the model but the

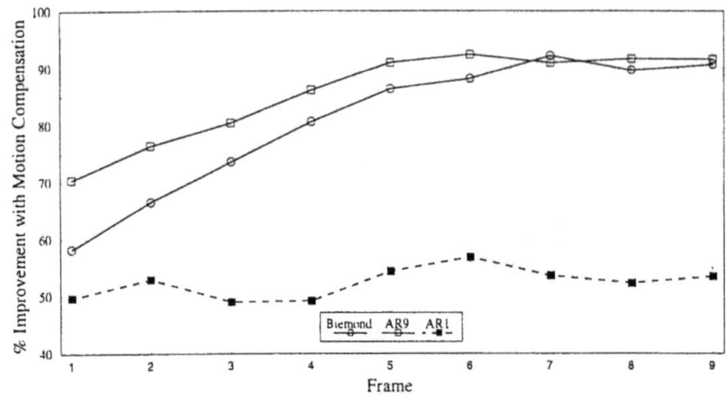

FIGURE 3.13. Comparing performance when using a model with a small spatial support.

effects are exaggerated if the spatial support is limited as in this case. When the Normal equations are solved with this initial value for **d** the low correlation will cause coefficient a_1 to be assigned a very low value. This will weight the elements of the gradient matrix toward zero causing the solution for the motion update to be ill conditioned. The algorithm will not converge on the correct displacement and a good prediction would not be possible. Alternatively, if the image information was such that the initial displacement guess allowed for some useful correlation between frames at the start of the iterations then the situation would be favourable.

Figure 3.13 illustrates this point with respect to the current experiments. It shows that the PIMSE with model AR1, with support as in Figure 3.8, is much worse[5] than with the 9 point AR model. Understandably, the accuracy of the vectors is increased from almost 0% with the AR9 model to 10% with this model (this is not illustrated), but the improvement is insufficient to compensate for the loss of spatial support which in this case yields the low PIMSE. To counteract the ill conditioning caused by a low weighting of the gradient terms, μ can be increased. For this model, increasing μ to 500 results in behaviour which is almost identical to the Biemond estimator, as expected.

The following points make clear the lessons of the preceding paragraphs.

1. The general 3DAR model of the image sequence has a motion estimation accuracy equal to the extent of its spatial support in the previous frames. The model AR9, has a motion estimation error of ± 1.0 pixel. Algorithms involving models with a one pixel support, such as the Biemond algorithm, are potentially arbitrarily accurate since they can only minimize the expected error if the estimated displacement coincides with the actual displacement.

[5] Recall that the Biemond estimator uses a model identical to AR1. The difference in performance is due purely to the process of estimating the coefficient at each iteration. In the Biemond technique, this coefficient is fixed at -1.0.

2. Estimating the coefficients at each iteration based on the current motion estimate is likely to yield useful improvements over other standard gradient based estimators only if the current vector estimate is close to the actual displacement or a large spatial support is used. 'Close' depending on the correlation structure of the image information itself.

These experiments illustrate clearly that the 3DAR model does not necessarily give an accurate estimate for motion. Nevertheless, it can give a better prediction than the Biemond algorithm, and other 'translation only' algorithms. When presented with a textured image sequence the 3DAR model is able to adapt not only to motion but also to the nature of the image contents as well.

3.5 Real sequences

The previous experiments have been able to highlight, in controlled environments, the shortcomings of 3DAR modelling. The best test as always is with real world data. Therefore, the Biemond algorithm and the 3DAR algorithm were compared by observing the improvement in motion compensation over a a few frames from two real sequences. These were as follows:

1. The Miss America Sequence: $256 \times 256 \times 10$ frames.
2. The Cheers sequence: $256 \times 256 \times 10$ frames.

The Cheers sequence is not a standard video sequence. It consists of several objects moving in different directions. It is a short clip from the television series 'Cheers'. It is included in the CD-ROM.

For all three sequences, μ was fixed at 10.0 and the iterations at each pixel considered always began at zero displacement. The motion estimation equations were set up using the 7×7 square at the centre of the block used for estimating the AR model coefficients. The iterative process was halted when one of the following criteria was satisfied,

1. The number of iterations > 10.
2. The mean square prediction error (MSE) < 1.0
3. The size of the update vector < 0.01
4. The MSE at the current iteration $>$ the MSE at the previous iteration.

Because of the size of the images and the heavy computation required for the solution of the Normal equations, a block based method was used for the AR algorithm. One block size of 9×9 was used. The motion estimation equations were set up using the inner 7×7 block of pels and the blocks were overlapped to allow one vector to be estimated for each 7×7 block. The AR estimator was therefore employed to predict tiles of size 7×7. Several different supports for the AR model were tried and they gave the expected behavior, that is, a larger

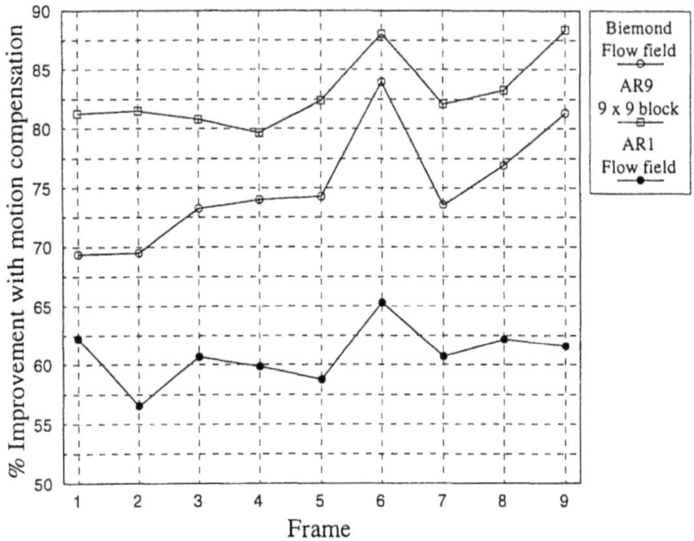

FIGURE 3.14. Comparing performance on the Cheers sequence.

support yielded a lower prediction error. The AR9 model shown in 3.8 gave satisfactory results and was chosen to represent a good compromise between computation and prediction error.

The Biemond algorithm was used to estimate a vector at every pixel in the image. Again a 7×7 block of pixels was used for setting up 49 equations that were then solved for the motion vector.

3.6 Discussion

From both Figures 3.14 and 3.15 it is evident that the AR algorithm performs better than the Biemond technique. This is despite the fact that the latter algorithm estimated a motion flow field (i.e. one vector per pel) for the image whereas the former was a block based estimation process. In both examples, the AR9 estimator improves the PIMSE by about 10% over the Biemond estimator.

For comparison there is also a result using the AR1 model as in figure 3.8. A prediction was generated at every pixel, in the same way as for the Biemond estimator (i.e. not block based processing). There is little difference between these two estimators for the Miss America sequence principally because that sequence is slow moving.

The Cheers sequence shows faster motion therefore the differences are more significant. Recall that the initial start displacement parameter for the AR algorithm is zero. Hence the AR1 model assigns a low value to the one coefficient because the temporal correlation is low at the start of the algorithm. This low temporal correlation biases the motion update towards zero and hence

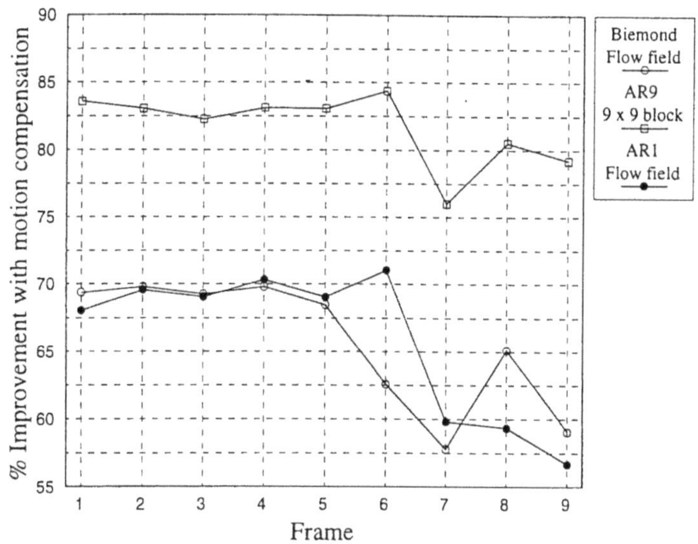

FIGURE 3.15. Comparing performance on the Miss America sequence.

the algorithm does not converge on a useful prediction. This was discussed in section 3.4.2.

Pictures showing the quality of the predictions are not shown at this stage but are left till the algorithm is finalized at the end of this chapter.

3.7 Adaptive estimation

The previous sections have dealt with the basic estimators in order to assess the fundamental differences between the standard methods and this AR approach. The damping factor, μ, has been fixed throughout the investigation. The estimation of the AR coefficients at each iteration is the only adaptive process introduced to the algorithm so far. To propose adaptive schemes that improve the estimator, it is educational to consider the conditions under which the estimator fails. The work by Kearney et al [78] considered gradient based estimators in general. The following sources of error are noteworthy.

1. Measurement error due to noise : Measuring the gradient using simple central differences may yield noisy estimates, especially in a highly textured area. Noise in the image itself will add to this problem. This may cause ill–conditioning in $\mathbf{G}^T\mathbf{G}$.

2. Non–linearity of images : The linear approximation to the image function is not valid everywhere. At edges, for instance, the higher order Taylor series terms are not negligible. This adds to the size of the vector $\mathbf{v}$. Furthermore, this error will be systematic with a definite correlation structure because the underlying image feature producing it is not a ran-

dom one. The white noise assumption for the contents of $\mathbf{v}$ is therefore violated.

3. Ill–conditioning of $\mathbf{G}^T\mathbf{G}$. : The sources of error above may not directly give rise to ill conditioning. In fact, Martinez [117] has shown that this ill conditioning is directly linked to the directions of maximum and minimum spatial gradients for the Biemond estimator. His analysis is repeated for this estimator in Appendix E. That analysis shows that the 3DAR estimator is ill–conditioned if the points used in setting up the motion equations lie along an edge. In this respect it behaves similarly to the Biemond estimator as discussed previously in Chapter 2. The advantage of the 3DAR estimator seems to be more in its robustness to noise since measurements over a small region are weighted to yield one observation that is used in setting up the matrix solution.

Driessen et al [35, 34, 33], Böröczky et al [20, 21] have introduced techniques for adapting the regularizing parameter during the standard Wiener based pel–recursive process. Driessen et al [35] have noted that the techniques which make μ proportional to the ill–conditioning of $\mathbf{G}^T\mathbf{G}$ are useful. Their proposed scheme was introduced in Chapter 2 and is repeated below.

$$\begin{aligned} \mathbf{d} &= [\mathbf{G}^T\mathbf{G} + \mu\mathbf{I}]^{-1}\mathbf{G}^T\mathbf{z} \\ \text{where } \mu &= |\mathbf{z}|\frac{\lambda_{max}}{\lambda_{min}} \end{aligned} \tag{3.18}$$

Martinez [117] recognized that when the solution was ill–conditioned, it was best to align the motion estimate with the direction of maximum spatial gradient. Otherwise, his solution was not regularized.

Although these arguments were introduced for the standard WBME, which in effect uses model AR1 in Figure 3.8, they are equally valid for the general AR based estimator. Therefore, a combined adaptive scheme is now presented which incorporates the adaptive estimate for μ as proposed by Driessen et al and the Martinez 'fallback' mode when the solution is ill–conditioned.

$$\mathbf{d} = \begin{cases} \alpha_{max}\mathbf{e}_{max} & \text{if } \frac{\lambda_{max}}{\lambda_{min}} > \alpha \\ \left[\mathbf{G}_w^T\mathbf{G}_w + \mu\mathbf{I}\right]^{-1}\mathbf{G}_w^T\mathbf{z}_w & \textit{otherwise} \end{cases} \tag{3.19}$$

$$\begin{aligned} \mu &= |\mathbf{z}|\frac{\lambda_{max}}{\lambda_{min}} \;\; \textit{if } \frac{\lambda_{max}}{\lambda_{min}} \leq \alpha \\ \alpha_{max} &= \frac{\mathbf{e}_{max}^T}{\lambda_{max}}\mathbf{G}_w^T\mathbf{z}_w \end{aligned}$$

As before, λ, $\mathbf{e}$ refer to the eigen values and eigen vectors of $\mathbf{G}_w^T\mathbf{G}_w$, and α_{max} is a scalar variable introduced to simplify the final expression. This adaptive algorithm, which involves an estimate of the model coefficients at each iteration prior to the motion estimate being generated, will be referred to as the *adaptive 3DAR* (A3DAR) algorithm for image sequence modelling. Incorporating this scheme into the standard WBME yields an algorithm which will be referred to as the adaptive WBME (AWBME).

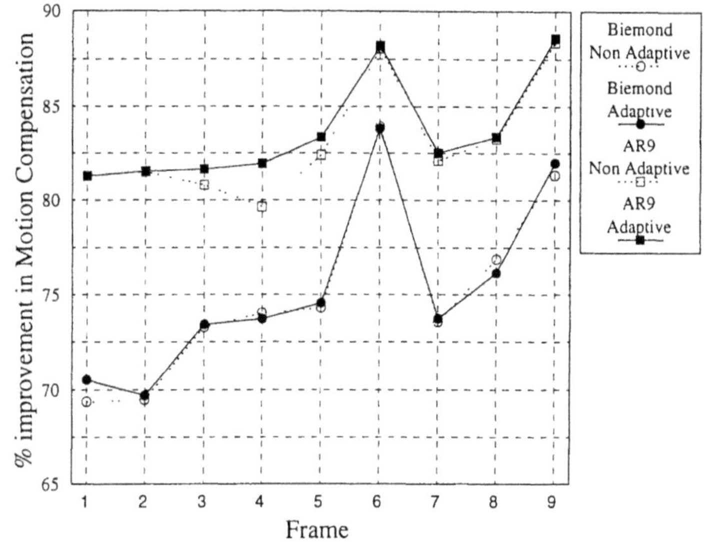

FIGURE 3.16. Comparing adaptive and non-adaptive algorithms with the Cheers sequence.

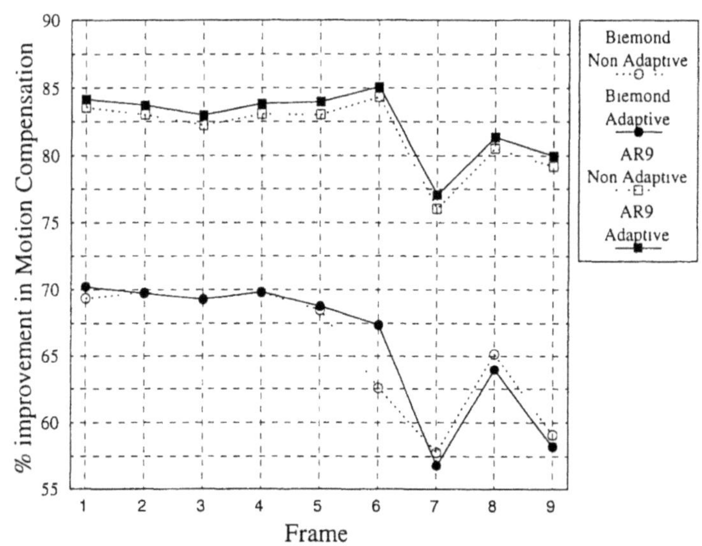

FIGURE 3.17. Comparing adaptive and non-adaptive algorithms with the Miss America sequence.

3.7.1 Experiments

Using the adaptive estimator described in the previous section, two experiments were performed comparing the performance of the AR estimator with and without the adaptation. Two real sequences were used, the Cheers sequence and the Miss America sequence. The results are shown in Figures 3.16, 3.17. In these experiments a block size of 9×9 was used for coefficient estimation and motion estimation. The threshold α was set at 25.0. Four remaining criterion were used to halt the iterative process, when any of the following were satisfied, the process was stopped.

1. The number of iterations > 10
2. The mean square prediction error (MSE) < 1.0
3. The size of the update vector < 0.01
4. The MSE at the current iteration $>$ the MSE at the previous iteration.

Both the Biemond WBME and the 3DAR estimator (employing model AR9 in Figure 3.8) were compared. As before, the Biemond technique was used to generate a motion vector for every pixel in the frame. In contrast, the 3DAR estimator was block based, and so generated a prediction for a block of size 9×9 using one motion estimate for that entire block. The blocks were overlapped so that the final predicted tiles were of size 7×7. The adaptive versions of the estimators, AWBME and A3DAR, used the same adaptive algorithm introduced above.

3.7.2 Discussion

In both estimators there is some improvement using the adaptive estimators. Although this improvement is small, the results indicate that it is safe to use the adaptive estimator to remove the guesswork involved in choosing values for μ. It is notable that the adaptive scheme gives a larger improvement for the 3DAR estimator than the WBME. One can argue that there is still guesswork involved in the choice of the Martinez threshold ($\frac{\lambda_{\text{max}}}{\lambda_{\text{min}}}$), however this threshold operates on the ill-conditioning of the gradient matrix directly. It is therefore further removed from the actual image information than is μ.

3.8 Modelling sequences with large displacements

In real image sequences such as motion pictures, the size of the displacements from frame to frame may be in excess of 10 pixels. Under such conditions, the approximations made to solve for the motion field do not apply and the accuracy of the motion estimates will be severely affected. The main contributing factor to this breakdown is that when there are large displacements the image data at the start point of the iterative gradient based estimator bears little

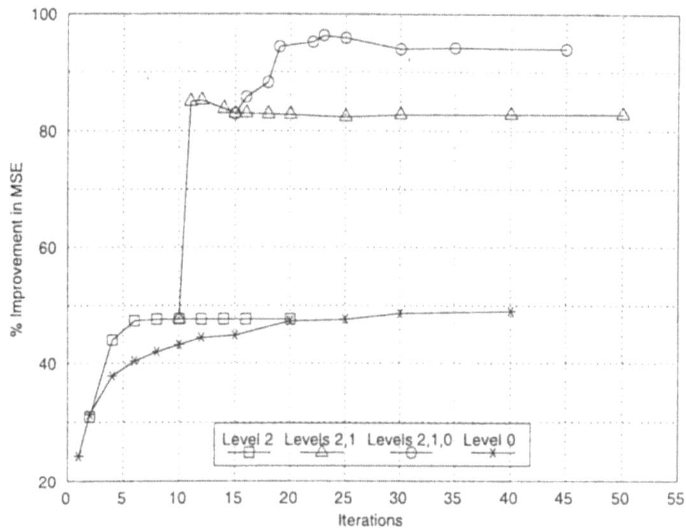

FIGURE 3.18. Multilevel estimation on Baboon using the Biemond estimator.

resemblance to the ideal end point corresponding to the correct motion vector. As stated in the motion estimation review of Chapter 2, a multiresolution scheme is adopted here to solve this problem. The estimator also adopts the motion detection scheme of Bierling [18] to avoid vector haloes.

3.9 Model selection on the pyramid

When considering the performance of the 3DAR predictor for the translating Baboon image earlier in this chapter, it was reported that the spatial extent of the model creates an inherent error in the motion vector estimate. This has extremely serious implications for the multilevel process of modelling a sequence with large displacements.

If a model such as AR9 in Figure 3.8 is used, the motion vectors generated have an error of ± 1.0 pixels. This does not necessarily imply that the prediction will suffer, in fact Figures 3.10 and 3.13 show just the opposite. However, in the case of multilevel estimation, a motion vector error of 1 pixel at the highest level, $L-1$, say, in an L level pyramid, implies an error of 2^{L-1} pixels at the original resolution. If the displacement is large (and one must assume this for real TV sequences), then the prediction at the original level will suffer because of this error. The errors at the higher levels will propagate down the pyramid until the data extracted in compensation will have very little correlation from frame to frame. For this reason, it is important to have accurate motion vector estimation at the higher levels.

'Accurate' motion estimation in this sense means extracting the motion information from the scene that most approximates the true motion of the objects

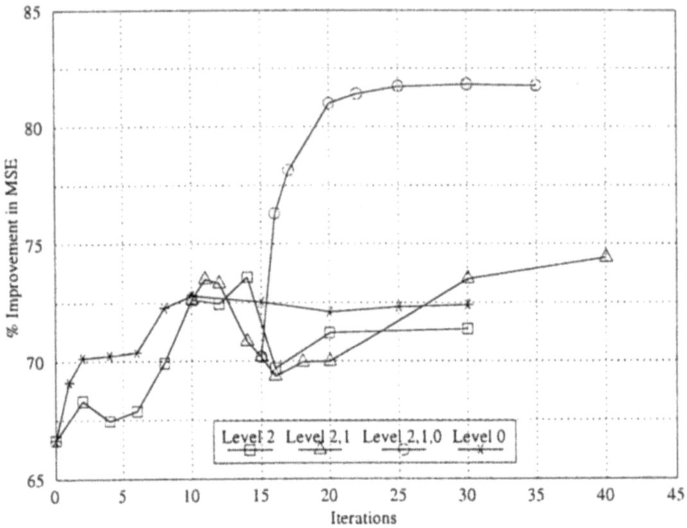

FIGURE 3.19. Multilevel estimation on Baboon using the AR estimator with model AR9.

in that scene. The general A3DAR algorithm presented earlier is therefore inadequate because it involves spatial support larger than 1 pixel. At this point it must be stated that there is no reason for the same model to be used at every level in the multiresolution hierarchy. This presents the solution to the problem.

At the higher pyramid levels, an AR model such as AR1 in figure 3.8 can be used to generate accurate motion estimates. At the lower levels, or just the original level only, a more general AR model may be used, for example AR9 in Figure 3.8. However, at the higher levels there is now less attention to prediction and more attention to motion estimation, therefore, other algorithms such as Block Matching or the Biemond pel recursive estimator may be used. This has the further advantage of reducing computation.

Figures 3.18, 3.19, 3.20, 3.21 show how important accurate motion estimation can be. The upper 64×64 quadrant of the Baboon image was extracted and translated to yield a translation only sequence with displacement $[-5.0, -5.0]$, see Figure 3.7. Multilevel motion estimation was employed using 3 levels and a variance of 1.0 for the Gaussian low pass window. The threshold on ill-conditioning, α, was set at 10.0 but no other thresholds were used. Algorithms were compared on the basis of the PIMSE and figures 3.18, 3.19, 3.20, 3.21 show the progress of the various methods over the second frame in the sequence. The x–axis shows the number of iterations and each differently marked line indicates iterations at a different level in the hierarchy. The lines marked with stars show iterations at the original level only and so this is the progress of the algorithm if no multilevel estimation was employed. The PIMSE was measured at the original level. Therefore the PIMSE for iterations in level 2 only (i.e. none in levels 0,1) was measured by propagating the vectors from level 2 down to level 0 and measuring the PIMSE there. Iterations always begin at level 2 in the pyramid.

Figure 3.18 shows that iterations of the Biemond process at level 2 only can give a PIMSE of about 50% maximum after 20 iterations. But if after 10 iterations, the vectors were refined at level 1 for a further 5 iterations, this improves dramatically to about 80%. Going on to refine vectors at the original resolution results in a PIMSE of about 95%, after a total of 30 iterations. This is much better than the maximum of 50% PIMSE after 40 iterations at the original level alone. This shows that the multilevel approach can provide for useful convergence of the algorithm when displacements are large. Contrast this to the use of the AR9 model at all levels of the pyramid, shown in Figure 3.19. The algorithm does not appear to converge at all for iterations at the first two levels, giving about 72% PIMSE. This is very poor compared to the 72% PIMSE after 10 iterations if no multilevel estimation was employed. Only after refinement at level 0 does the PIMSE converge on a more acceptable result of 82% improvement. This pales in comparison to the much better performance of the Biemond multilevel estimator which requires substantially less computation.

If however, the AR1 estimator is used at levels 2,1 and the AR9 model used at the lowest level 0, the situation is improved. Figure 3.20 shows this result. It takes just 20 iterations in total, 10 at level 2, 5 iteration refinement at level 1, and 5 iteration refinement at level 0, to yield a PIMSE that reflects almost perfect compensation at 95%. Compare this to the 70% PIMSE if the AR9 model was used at level 0 alone. Note also that the algorithm appears to converge rapidly as opposed to the almost random behaviour shown in Figure 3.19.

Using the Biemond estimator at levels 2,1 and the AR9 model at level 0 gives the best result here. The purely Biemond multilevel estimator is compared to the Biemond/AR9 mixture algorithm in figure 3.21. The figure shows that almost perfect compensation (99%) is achieved using the two different algorithms at different levels.

3.9.1 Experiments and discussion

To highlight the improvement gained with multilevel estimation, experiments were performed on the JUGGLER sequence. This is a sequence showing the torso of a juggler whilst juggling. The motion is very fast, > 20 pixels per frame. The juggling balls give numerous problems to motion estimation since there are large areas which are occluded from frame to frame. The rapidly changing shape of the hands pose additional problems. The sequence is of resolution 256×256.

Figure 3.22 shows a comparison between two model based estimators one multilevel and the other single level. The AR1/AR9 multilevel estimator used 4 levels, implementing AR1 on levels 3,2,1 and AR9 on the final level of original resolution. The AR9 estimator used the AR9 model on level 0 alone. The pyramid was generated using $\sigma^2 = 1.0$. A final improvement to the multilevel algorithm was also employed since it is common in multilevel algorithms that

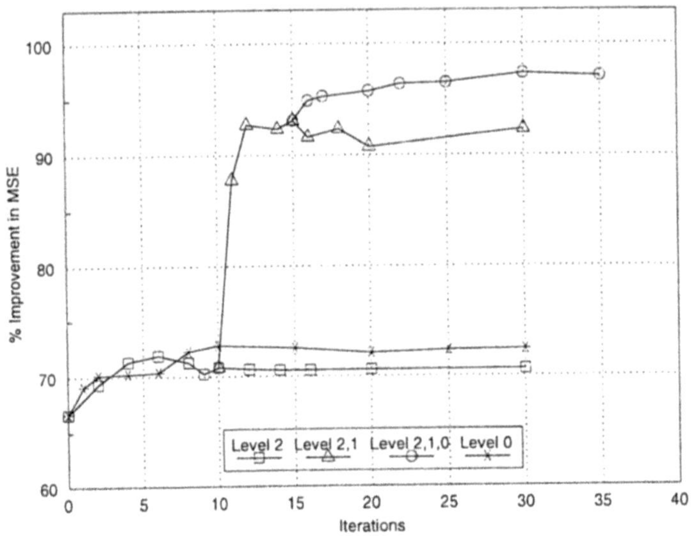

FIGURE 3.20. Multilevel estimation on Baboon using the AR estimator with AR1 at levels 1,2 and AR9 at level 0.

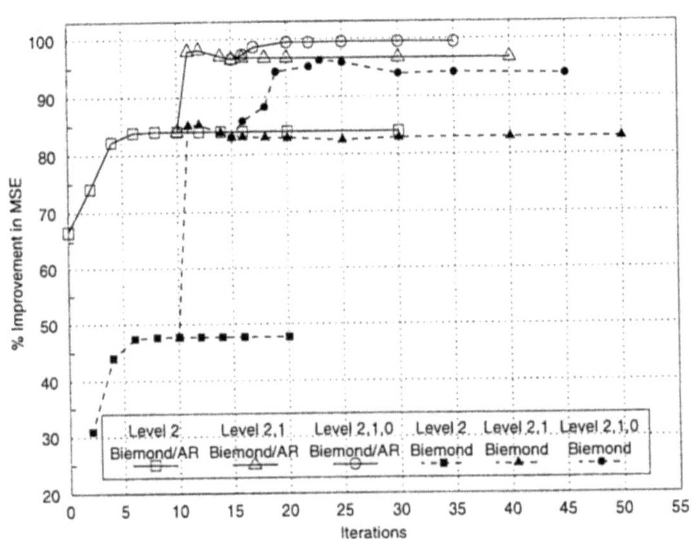

FIGURE 3.21. Comparing multilevel estimation using a mixture of algorithms at each level versus the Biemond estimator at each level.

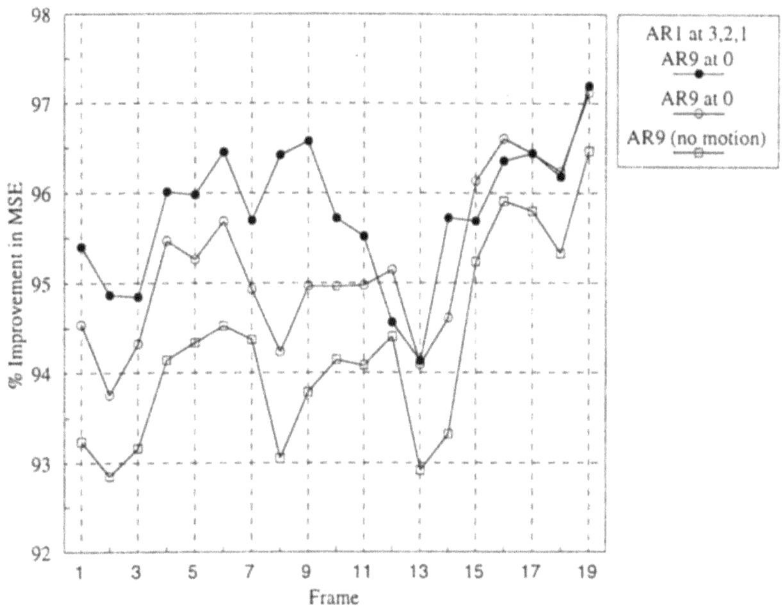

FIGURE 3.22. 3DAR Modelling for JUGGLER using multilevel 3DAR motion estimation.

stationary areas around moving regions are also estimated as being moving. The phenomenon is known as a vector halo effect and was discussed in the review Chapter 2. To reduce this problem a motion detector is applied at the level 0 to verify that a region is moving. This improves the quality of the prediction. The motion detector takes the form of a threshold applied to the non-motion compensated frame difference based on either a MSE or MAE measure, exactly the same as the motion detector applied in standard single level estimation.

The parameters used in the estimation process with the AR1/AR9 system were as follows, (i refers to the iteration index)

1. Modelling block size $= 9 \times 9$, motion estimated using an inner block of 5×5, blocks overlapped to have one motion vector for every 7×7 block.

2. Iterative estimation halted when any one of the following occurred.

$$
\begin{aligned}
\text{MSE} &\leq 40.0 \text{ At all levels} \\
|\hat{\mathbf{u}}_{n,n-1}| &\leq 0.1 \text{ At all levels} \\
i &> \begin{cases} 10 & \text{For level 3} \\ 5 & \text{Otherwise} \end{cases} \\
\mathbf{z}_i^T \mathbf{z}_i &\geq \mathbf{z}_{i-1}^T \mathbf{z}_{i-1}
\end{aligned} \tag{3.20}
$$

The parameters used with the single level AR9 system were the same except there were a maximum of 25 iterations allowed at the single level 0.

The curves do not show an impressive improvement with the multilevel estimator, principally because the motion is so large that any reduction in frame

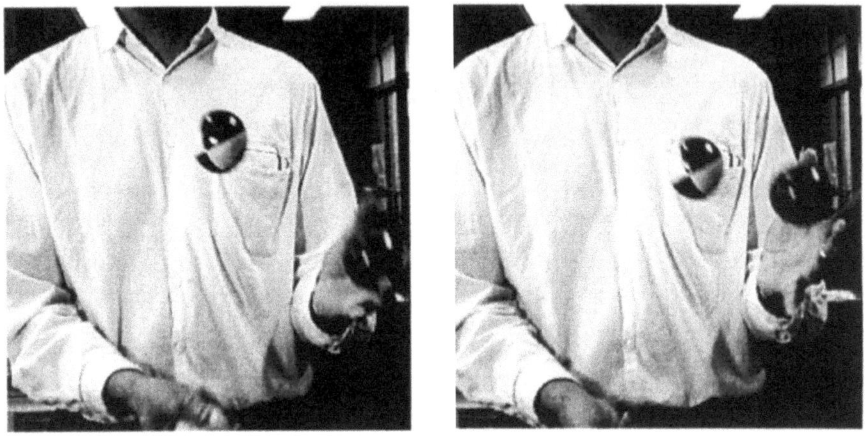

FIGURE 3.23. Frames 9 (left), 10 (right) of original JUGGLER sequence.

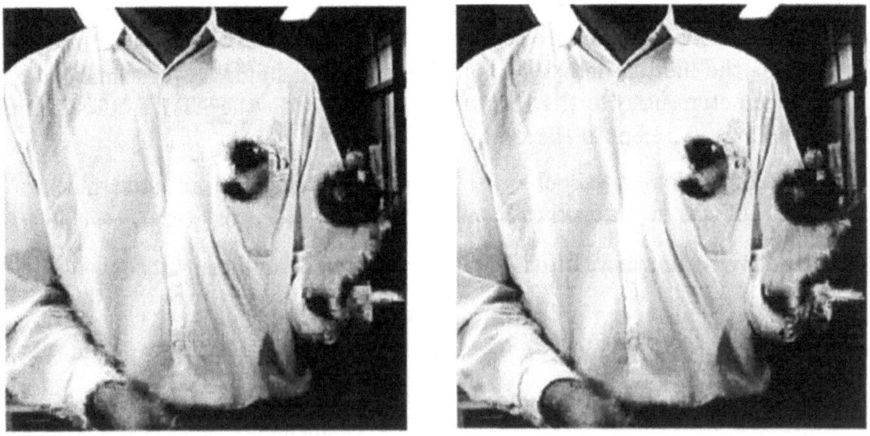

FIGURE 3.24. Predicted frame 10 using single level AR estimator (left) and Multilevel AR1/AR9 estimator (right).

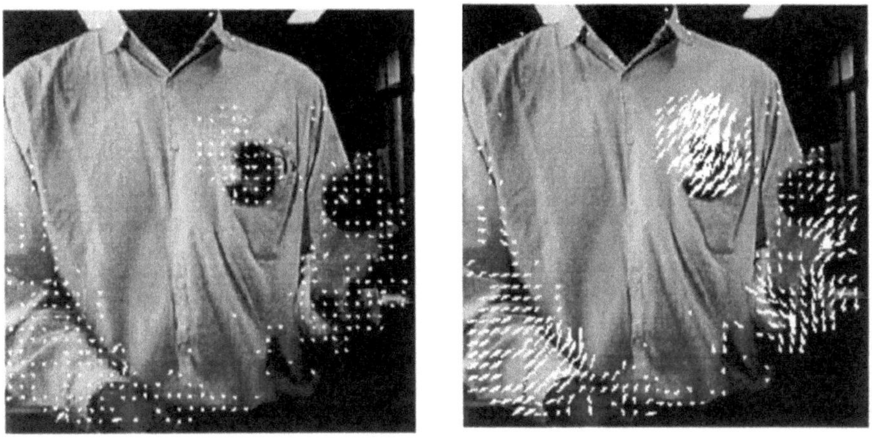

FIGURE 3.25. Estimated vector field for frame 10 using single level AR estimator (left) and Multilevel AR1/AR9 estimator (right).

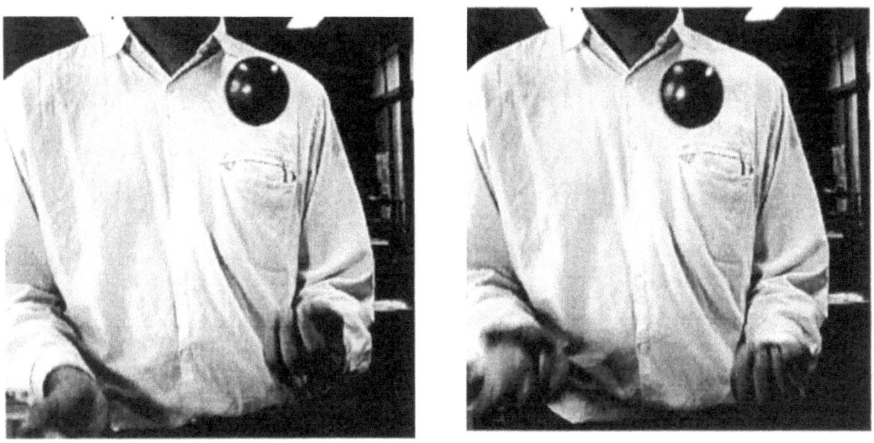

FIGURE 3.26. Frames 14 (left), 15 (right) of original JUGGLER sequence.

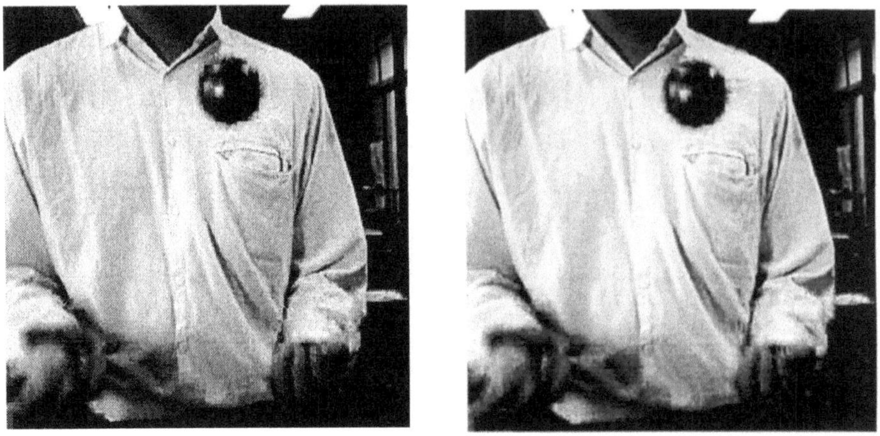

FIGURE 3.27. Predicted frame 15 using single level AR estimator (left) and Multilevel AR1/AR9 estimator (right).

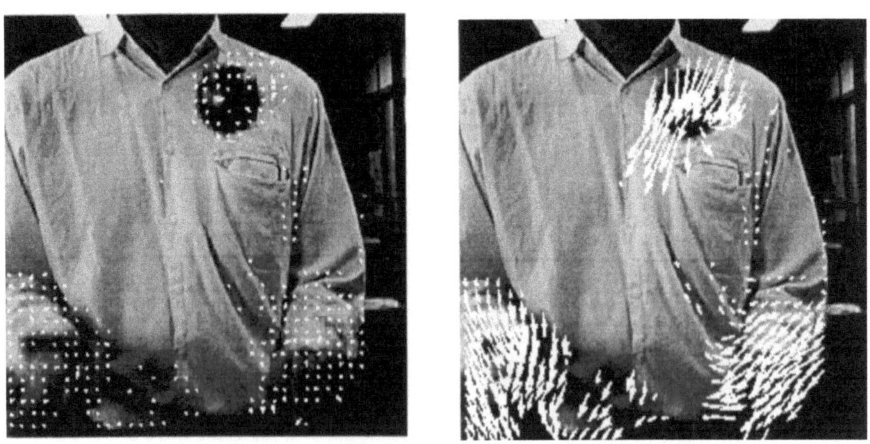

FIGURE 3.28. Estimated vector field for frame 15 using single level AR estimator (left) and Multilevel AR1/AR9 estimator (right).

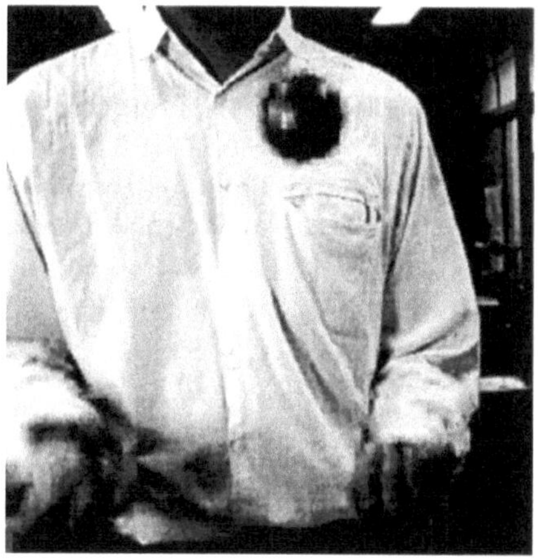

FIGURE 3.29. Predicted frame 15 using single level AR estimator with no motion compensation.

difference is emphasized. This is highlighted with the inclusion of the PIMSE for the AR9 model applied with the displacement parameter set to [0.0, 0.0] everywhere in each frame (AR9 (no motion)).

The reason for this is due to the image material itself. JUGGLER consists of fast moving areas; the hands and the juggling balls, but there are large areas of each frame that are not moving with such high velocity; the shoulders and chest. The fast moving areas contribute to a very large frame difference even though these areas account for about 1/3 of the total image area. This is illustrated in figures 3.23 and 3.26. These figures are photographs of four frames from two different points in the sequence. Frames 9 and 10 were chosen because they represent typical performance. Frames 14 and 15 were chosen to represent bad performance where the PIMSE for the multilevel motion estimator and the single level method cross over (see Figure 3.22). The frames show that the juggling balls rotate as they move and that the hands change shape from frame to frame. The motion estimation algorithm must deal successfully with these problems to be useful.

Because the model can adjust coefficients to compensate for rapid changes in image intensity, the non-compensated model can predict some reasonable grey scale variation in each block even in the fast moving areas. This is shown in figure 3.29. Therefore, the PIMSE is large even for this non-compensated process. The improvement with the single level estimator comes from its partial compensation of the moving areas and uncovered regions. This is well illustrated by frames 9 and 10 in the quality of prediction of the moving ball. The multilevel estimator is able to give a very good estimate for the moving arms and so the resulting prediction is much improved. Figure 3.24 shows this prediction improvement. Figure 3.25 shows the estimated vector fields for each frame using the different estimators. The fields are superimposed on the corresponding predicted frame, and that frame is reduced in intensity slightly to allow the vectors to be seen more easily. It is clear that the multilevel estimator copes with the moving arms much better than the single level estimator. Frames 14 and 15 show the opposite effect, (Figures 3.26 and 3.27). The multilevel estimator does not track the white/black[6] ball well in either frame pairs 9,10 or 14,15. Unfortunately, in 14,15 neither estimators can track the motion of the right arm well because of blurring and shape changing, therefore no improvement is gained. The vector fields in Figures 3.28 verify this observation.

Considering that the AR1/AR9 estimator is computationally lighter and more effective than the AR9 estimator at the single level, it is the better choice for this sequence.

3.10 The motion parameter in the 3DAR model: Using alternative estimators

The previous sections have served to highlight the fact that the modelling process is more effective when the motion of the object is accurately estimated. Therefore, the motion parameter in the model is definitely related to the motion of objects in the scene. It would be advantageous therefore, to consider the modelling process as two processes; a motion estimation process and an image prediction compensation. To this end it is useful to consider the effect of using BM or the WBME to estimate motion and then use this motion parameter in the model framework. The 3DAR model would then act as an interpolator correcting for errors and motion not obeying the pure translational rule. There is also the additional consideration that the adaptive iterative 3DAR process for estimating $\mathbf{d}$ is computationally demanding when compared to either of the two other estimators mentioned.

Figure 3.30 shows the PIMSE for the Juggler sequence using (BBM) Block Matching and WBME to provide the estimate for the motion vector that is then used in the 3DAR modelling process. The adaptive WBME algorithm employed was identical to the previous AR1/AR9 estimator with respect to the number of levels and the parameters used. The MAE criterion was used for the BBM algorithm, with the motion detection threshold set at 10, and the noise ratio[7] threshold, $r = 1.2$ for the level 0 only[8] and 1.0 otherwise. A full search was employed with search space of ± 4 pixels. The block sizes were the same for both algorithms, and the blocks were overlapped to allow one motion vector to be estimated for each inner 7×7 block. There were therefore 36×36 vectors estimated in each frame.

The PIMSE results indicate that the new algorithms perform similarly to the multilevel AR modelling process, with the algorithm employing the adaptive Biemond estimator (Biemond/AR9) performing best overall. However, the figures 3.31 show that the predicted frames 10 and 15 using both the WBME and

[6] Yellow and red in a colourful world.

[7] Boyce ratio; See Chapter 2.

[8] The noise is greatly reduced in the upper levels of the pyramid due to the low-pass filtering that creates the pyramid.

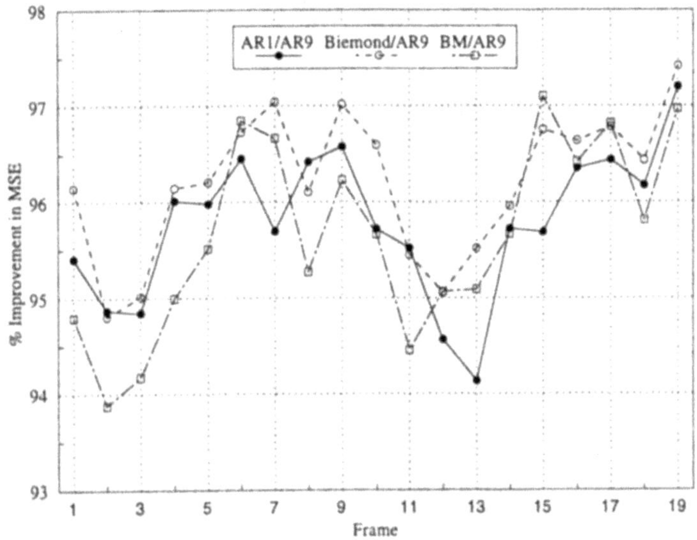

FIGURE 3.30. 3DAR modelling for JUGGLER using alternative motion estimators.

the BM estimator give much sharper rendering of the moving areas, particularly the juggling balls. The vector fields shown in 3.32 verify that this is due to the better performance of these estimators with respect to motion estimation. The use of these motion estimators therefore improves the performance of the modelling process overall.

3.11 Final comments

This chapter has considered the use of the 3DAR model for image sequence modelling. Consideration was given to allowing variation in the model coefficients across the image. The major problem encountered in solving the model equations was the estimation of both the displacement $\mathbf{d}$ and the model coefficients simultaneously.

Two different approaches were investigated. The first was an iterative refinement technique involving estimation of displacement and model coefficients as two separate refinement problems in the iterative scheme. The second approach treated the displacement estimation as a completely separate issue in the modelling process. The displacement estimation was achieved using a standard motion estimation technique and then the modelling was executed using that motion estimate. Both these approaches were implemented finally as part of a multiresolution algorithm for large displacements.

The results show that both approaches can give similar performance with respect to the same sequence when considering PIMSE. Also the modelling approach is superior to standard motion estimation/compensation approaches. This is not surprising. The modelling stage of the proposed algorithms attempts

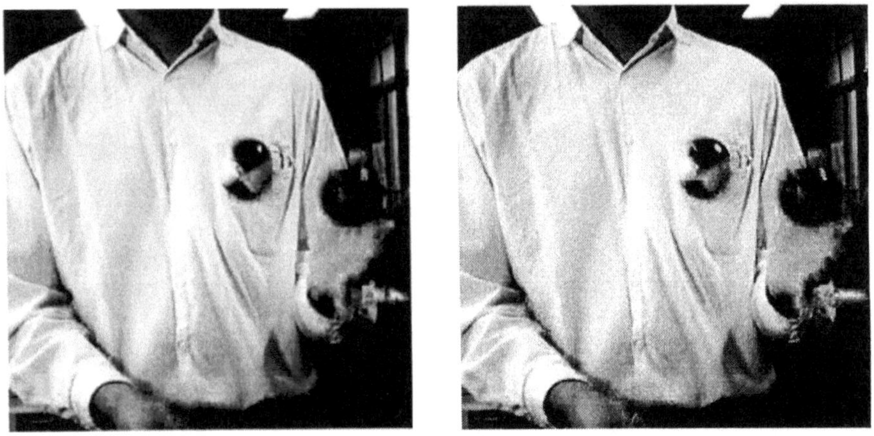

FIGURE 3.31. Predicted frame 10 using different motion estimators, BM (left) and adaptive WBME (right).

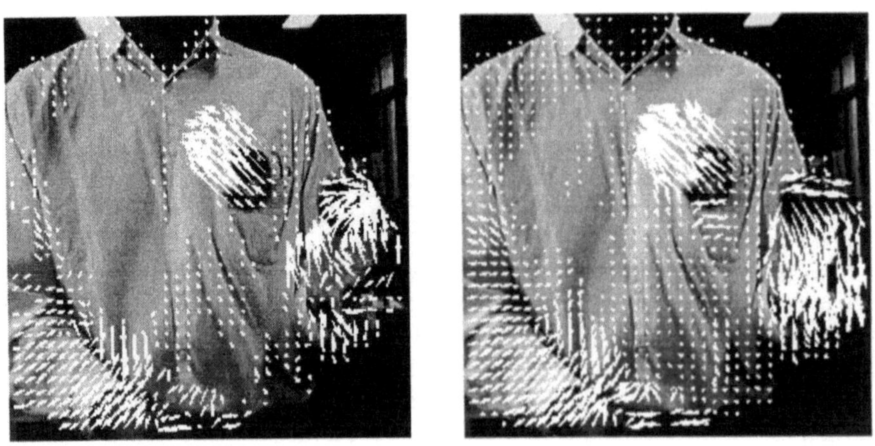

FIGURE 3.32. Estimated vector fields for frame 10 using different motion estimators, BM (left) and adaptive WBME (right).

to minimize the squared prediction error directly, therefore it is not unexpected that the PIMSE is better than standard purely temporal motion estimators.

As a motion estimator, the algorithm employed falls short of standard techniques. This is due in part to the size of the spatial support used in the next frame, and also due to the re–estimation of coefficients (using Least Squares in particular) at each iteration in the motion estimation process. It is conceivable that an initial displacement guess would cause the correlation between the current block and the previous displaced block to be small. This may occur due to large motion or the image information itself. In such a case, the model coefficients estimated would be small and prevent motion updates to be made. The motion estimation would therefore halt at that stage despite the large error.

It was shown that the performance of the modelling process depends on the accuracy of the motion estimate. Multilevel motion estimation using the AR1 model was found to be successful when combined with the AR9 predictor as the last stage. The algorithms employing the BM estimator or the adaptive WBME to estimate the motion used in the modelling stage, can perform better compensation than the 3DAR estimator as far as visual perception of the prediction is concerned. In view of the fact that both of these alternatives are less computationally intensive than A3DAR, they are the better choice for 3DAR modelling frameworks. The restoration work that follows mainly uses the multiresolution AWBME scheme (MWBME), as discussed in this chapter, for motion estimation because of the low computation required and the ability to partially resolve some of the aperture problems.

Chapter 7 re-works the use of the 3DAR image sequence model by incorporating it directly into a scheme for simultaneous motion estimation, model parameter estimation as well as detection and reconstruction of missing data. By employing a Bayesian framework for inference many of the coefficient estimation problems illustrated in this chapter are overcome.

4

Image Sequence Restoration: A Review

Two dimensional image restoration has been an active research topic since the 1960's. Algorithms have been developed to remove blur, noise and impulsive distortion from single images [107, 153, 71]. In contrast, research concerned with image sequence restoration can be traced back only to about the late 1970's [31] when it was mainly associated with image coding. The subject has only become prominent within the last decade [36, 115, 117, 77, 137, 42]. It is during these years that the increasing speed of microprocessors has made the heavy computation practical. That, coupled with the rising importance of video both for communications and entertainment has made the subject grow in importance.

Early work in this field by Dubois and Sabris [36] and Dennis [31], tended to concentrate on the suppression of white noise. The treatment of motion was very simple. The noise suppression filter consisted of some form of weighted frame averaging except where motion was detected. In those areas either little or no averaging was done.

As stated in Chapter 2, it is better to consider the processing of image sequences as a full 3D problem rather than a repetitive 2D one. Although the early work done on Image Sequence Restoration (ISR) could achieve satisfactory results, in cases with much motion or heavy noise levels the performance deteriorated. It was clear that explicit treatment of motion would yield improvements.

Huang [66] in 1981 used a simple correspondence motion estimator to direct the application of a 3 tap temporal FIR filter to suppress white noise. He also employed a 3 tap temporal median filter to suppress line dropout and impulsive distortion. His results showed a marked improvement over techniques employing no motion compensation.

More recently, Martinez [117] has studied this exact problem of ISR. His thesis however, was more concerned with the development of a robust motion estimation technique which was low in computation, to allow rapid sequence processing. Like Huang [66], he implemented 3 tap FIR and Median filters to good effect. His work points out that the performance of the motion estimator drastically affects the performance of the restoration system.

It is only recently with the work of Sezan, Tekalp, Özkan, Lagendijk, Katsagellos, Efstratiadis and others that an attempt has been made to use spatio–temporal restoration techniques combined with a more coherent approach to motion estimation. Their work has concentrated on noise suppression [77, 156, 137] and simultaneous blur and noise removal [42]. The work has generally employed a gradient based approach to motion estimation using both multiple constraint techniques [44] and variants of the WBME [38, 77]. These results encourage the development of more 3D algorithms for noise suppression in image sequences. Note however, that the approach to ISR has traditionally been to treat the motion estimation and noise suppression separately.

In contrast to the work done in noise suppression, there has been little treatment of the problem of suppressing impulsive noise in image sequences. The problem is not a negligible one since TV engineers are quite familiar with motion pictures showing randomly dispersed flashes of light and dark. The effect is called 'Dirt and Sparkle'. The earliest work that can be cited in this area is that of Richard Storey [163, 164] working in the BBC Research Department at Kingswood Warren, England. That early effort resulted in a non-motion compensated process for detection and removal of dirt and sparkle which was incorporated into a machine called Debra. The details of that work are given in Chapter 6. The only other work that can be cited here is that due to Arce, Alp et al [8, 6, 2], and of course the motion compensated efforts of Huang [66] and Martinez [117]. Arce and Alp have presented spatio–temporal Multistage Median filter architectures that are robust to motion but still reject impulsive noise. This class of filters had previously been introduced in [134] and the Median filter itself has a long history as a good tool for impulsive noise suppression in 2D signals. The efforts of all these authors have been combined and extended in this book to introduce new spatio–temporal structures for impulsive noise suppression. New work is also introduced which proposes a model based approach to the problem and unifies all the algorithms developed under the same Bayesian framework.

Aside from Storey, none of the previous authors have considered using a detector for the impulsive noise before engaging the Median operation. The idea is presented in this book (coupled with motion estimation) as an effective way of decreasing computation and increasing output quality. The main philosophy in detection is to restrict the attention of the filtering mechanism only to the distorted areas. The problem is definitely a local one and so must be treated as such.

Having outlined a brief overview of the subject, the review will now continue in three parts. The final two parts discuss the central problems addressed by this book and the first discusses early ISR work which gives some basis to the directions chosen for further work. The work of the author is excluded from this review and is instead incorporated into the discussions in the chapters to follow.

4.1 Early techniques for noise reduction in image sequences

Frame averaging has always been recognized as a cheap, effective way of reducing noise in image sequences. The technique has been used extensively and to good effect for electron microscope imagery. The implementation is usually a recursive estimator of the form

$$\mathbf{i}_n = \frac{1}{n}[(n-1)\mathbf{i}_{n-1} + \mathbf{s}_n] \tag{4.1}$$

Here, $\mathbf{i}_n$ represents a vector of the intensities in the current output image, $\mathbf{i}_{n-1}$ the previous output image and $\mathbf{s}_n$ the vector of current noisy image samples that are input to the system. $\mathbf{i}_n$ can be recognized as the running average of all the past n frames. This is successful in microscopy applications because the image sequences observed represent stationary scenes.

It was only natural, therefore, that the first noise reducers for TV imagery attempted to implement frame averaging. Due to motion in real scenes, it was necessary to limit the averaging effect when motion was detected. The filters implemented were first order recursive of the form

$$\begin{aligned} I(i,j,n) &= I(i,j,n-1) + \alpha(s(i,j,n) - I(i,j,n-1)) \\ &= (1-\alpha)I(i,j,n-1) + \alpha s(i,j,n) \end{aligned} \tag{4.2}$$

When $\alpha = \frac{1}{n}$, the filter shown in 4.1 is the result. The scalar constant α was chosen to respond to the size of the frame difference $|\mathrm{DFD}_0| = (s(i,j,n) - I(i,j,n))$. When this error was large, it implied motion so α was set close to 1.0 to turn off the filtering. A small frame difference implied no motion and so the constant could be set to some small value to allow filtering. A typical example was the strategy used by Dennis [31].

$$\alpha = 1.0 - k_1 e^{-(\frac{|\mathrm{DFD}_0| - k_2}{k_3})^2} \tag{4.3}$$

$$\textit{where } \mathrm{DFD}_0 = s(i,j,n) - I(i,j,n-1) \tag{4.4}$$

k_1, k_2 and k_3 were constants chosen to select the best form of the characteristic.

Although the technique performs well in stationary regions of the image, the final result is not satisfactory because of the following artefacts.

- Moving regions in the output frames would generally have a higher noise level than stationary regions.

- Stationary regions would suddenly become noisy if they began to move.
- There is a smearing effect due to filtering in a direction not close to motion trajectories.

Huang [66] and Dubois et al [36] presented methods for the *motion compensated* temporal filtering of noisy image sequences. Huang [66] employed a 3 tap averaging operation over 3 frames, for use along an estimated motion trajectory. He performed two experiments using two different motion estimators. In one, the motion estimator used a correspondence matching technique that chose the motion estimate to be along the direction that minimizes the variance in pixel intensity. The other estimator used a gradient based approach. The results for both estimators were similar. The filter is interesting in that it uses not only the previous frame but also the next frame as well. However, the filter can still show smearing effects when the motion is not properly tracked.

To adapt the filter algorithm to this effect, Dubois [36] implemented a recursive filter of the type described by equation 4.3. The pel–recursive scheme of Netravali and Robbins [132] was used for motion estimation and the operation of the filter was directed along the estimated motion trajectories. The filter adapted to occlusion effects and errors in motion estimation by varying α according to a piecewise linear characteristic repeated below:

$$\alpha = \begin{cases} \alpha_b & \text{for } |\text{DFD}| \leq e_1 \\ (\frac{\alpha_e - \alpha_b}{e_2 - e_1})(|\text{DFD}| - e_1) + \alpha_b & \text{for } e_1 < |\text{ DFD }| \leq e_2 \\ \alpha_e & \text{for } |\text{DFD}| > e_2 \end{cases} \tag{4.5}$$

Note that DFD is the *displaced* frame difference, i.e. the frame difference along an estimated motion trajectory between the current frame and the previous. The characteristic allows for three regions of operation. When motion is tracked effectively, filtering is performed with $\alpha = \alpha_b$. When motion cannot be tracked the filtering is turned off by setting $\alpha_e = 1.0$. The mid–range linear ramp represents a smooth transition in filtering between the tracked and untracked regions.

This process of adapting the filter in addition to motion compensation enabled better image detail preservation than the strategy used by Huang [66]. In their conclusions, Dubois et al [36] note that improved motion compensation would lead to better motion compensated temporal filtering, and indeed this is so. The lessons from this early work are therefore twofold.

1. Motion compensated filtering of image sequences enables better noise suppression and a better output image quality than non–motion compensated techniques.
2. It is important to adapt the extent of filtering to the accuracy of motion estimation. This would enable a better quality output image by reducing smearing artefacts.

Therefore, although motion compensated filtering is more effective for image sequences, a good algorithm must also be robust to erroneous motion estimation.

An important observation that must be made is that it is often the case when there is rapid motion, this motion cannot be estimated accurately. However this may not affect the subjective quality of the output despite the reduced effectiveness of the filtering operation. This is because the human perception of a fast moving object is much less than a slow moving one. Therefore it is possible to strike a useful compromise between output image quality and the speed of objects in a scene.

The next two sections discuss the development of ISR algorithms that incorporate more spatial information and employ better motion estimators.

4.2 Noise reduction in image sequences

The work by Dubois et al [36] led others [117, 23, 156, 137] to incorporate better motion estimators into a temporal filtering scheme. However workers such as Katsagellos et al [77] and Sezan et al [42, 136, 135] recognized that although the driving force behind purely temporal image sequence filtering was the potential for improved image detail preservation, there was no need to ignore the noise rejection capacity of the spatial operation. This led to the use of 3D operators [77, 73]. In [156, 137, 136], a Wiener filter for the removal of noise and blur has been presented. This operation uses information from several frames at once. The various methods can be considered under the two headings below.

4.2.1 Motion compensated temporal filtering

Purely temporal filtering of image sequences is still popular presumably because of the low computation involved. Boyce [23] has implemented a motion compensated frame averager using BM as the motion estimator. She has made the BM algorithm more robust to noise in the sequence by ensuring that the final match is due to object motion and not to noise. Her BM algorithm was described in Chapter 2. The algorithm used $(N-1)/2$ frames previous to and following from the current frame. Her results showed noise reduction satisfactorily close to the theoretical maximum attenuation of $\frac{\sigma_{nn}^2}{N}$, where σ_{nn}^2 is the noise variance. Using the BBM algorithm implies a robust motion estimator that is able to handle reasonably large motion, therefore the subsequent averaging operation would be better than the approach used by Huang [66]. A multiresolution motion estimation algorithm would be more appropriate for real TV imagery and this is considered in the later chapters of this book.

Of particular interest is the use of the 3D Wiener filter for noise suppression since it can reject more noise given the same number of frames used in a frame averaging operation. Several investigations [115, 117, 156, 137] have employed an N tap Wiener filter temporally to reject noise in sequences. The signal observation model is typically

$$s(i,j,n) = I(i,j,n) + e(i,j,n) \tag{4.6}$$

Here, $s(i,j,n)$ represents the observed, noisy image sequence, $I(i,j,n)$ represents the original, clean sequence and $e(i,j,n)$ the added white Gaussian noise of variance σ_{nn}^2. The single point Wiener filter was used as the estimator, and is stated below.

$$\hat{I}(i,j,n) = \frac{\sigma_{ii}^2}{\sigma_{ii}^2 + \sigma_{nn}^2}(s(i,j,n) - \bar{I}(i,j,n)) + \bar{I}(i,j,n) \tag{4.7}$$

$\hat{I}(i,j,n)$ is the Wiener estimate of $I(i,j,n)$ and $\bar{I}$, σ_{ii}^2 are the mean and variance of the unknown, clean signal, $I(i,j,n)$.

The filter was directed along motion trajectories, and given N frames the following parameter estimates were used, where sx_n and sy_n represent the components of the motion vector mapping the current frame 0 into the frame n.

$$\begin{aligned} \bar{I}(i,j,n) &= \frac{1}{N+1}\sum_{n=-N/2}^{N/2} s(i+sx_n, j+sy_n, n) \qquad (4.8)\\ \sigma_{ss}^2 &= \frac{1}{N+1}\sum_{n=-N/2}^{N/2} [s(i+sx_n, j+sy_n, n) - \bar{I}(i,j,n)]^2 \\ \sigma_{ii}^2 &= \begin{cases} 0 & \text{if } \sigma_{nn}^2 \geq \sigma_{ss}^2 \\ \sigma_{ss}^2 - \sigma_{nn}^2 & \textit{otherwise} \end{cases} \end{aligned}$$

The filter achieves noise reduction by adding to the averaged signal a fraction of the difference between the noisy and averaged signal. This fraction is proportional to the suspected ratio of original to noisy signal variances. The output signal is forced closer to the averaged signal when the observed signal variance is low.

The filter can be seen to behave in the required manner for robust operation when there is inaccurate motion estimation. When there are errors in motion estimation, the estimate for σ_{ss}^2 would be large, and $\sigma_{ii}^2 \gg \sigma_{nn}^2$. Therefore less filtering would be done. Conversely, when the motion estimation is more accurate, the filter would be engaged since the signal variance would then not swamp equation 4.7.

In an earlier work, Martinez and Lim [115] used a cascade of Wiener filters like 4.7 oriented in different directions. No motion estimation was employed so that the cascade of filters would operate as an implicitly compensated filtering operation. The filter directions were not only chosen to be temporal but several directions within the current image plane were also employed. Their results showed good performance in uniform regions but no really improved performance in moving areas. This is because, unfortunately, in many sequences it is unlikely that among a practical number of 1D filters there will be one whose orientation corresponds to the motion of the scene. That work showed that there is no substitute for an explicit treatment of motion.

Martinez [117] went on to implement an explicitly compensated version of the filter introduced in [115]. Only one filter was used in the direction of the motion estimate and some useful results were reported. The motion estimator employed was a non-regularized version of the WBME discussed previously, in a multiresolution framework.

Sezan et al [156, 137] employed a much more complex motion estimator to implement the same filter shown in 4.7. Again only one filter was used, and that was directed along the motion trajectory. The algorithm for motion estimation [44] employed a multiresolution scheme involving a multiple constraint[1] gradient based technique. They compared their results with that using the WBME and the previous Martinez et al algorithm, and found that they were able to achieve superior performance. This demonstrates the importance of good motion estimation in the noise reduction problem.

4.2.2 Motion compensated spatio-temporal filtering.

Purely temporal filtering can provide noise reduction without image detail degradation. Yet the image sequence is a 3D signal and so the spatial information could be used to advantage. The principal flaw in purely temporal filtering is that in cases where the motion is not tracked, the image detail can be destroyed. It is generally expected that incorporating spatial information would yield a system with an improved noise suppression capacity together with the added detail preservation of temporal information when this is available.

Katsagellos et al [77] extended a number of 2D noise filtering algorithms to 3D. Their methods involved spatio-temporal filtering. Parameters for the filters were chosen to optimize subjective image quality. The work of Sezan et al [42, 135, 136] represents the first attempts to 'optimally' filter noise and blur in an image sequence. Their approach used 3D Wiener filtering and they have developed a novel form of the filter. It results in a frequency domain image sequence filter that involves 2D FFTs and not 3D FFTs as is the case with the full 3D frequency domain Wiener filter. The derivation of their filter is somewhat involved, but can be arrived at through an alternative route which is introduced in Appendix G. The motion estimator employed was the one presented by Fogel [44]. The filter was applied to images that showed a global displacement only, aircraft surveillance video in particular.

Their results were encouraging and Chapter 10 in this book presents the use of the Wiener filter for image sequences that show several moving objects. Two forms of the 3D Wiener filter are investigated: the 3D FIR spatiotemporal approach and the fully 3D DFT frequency approach. It is employed for noise reduction in TV imagery.

[1] Oriented smoothness.

4.3 Removing impulsive noise

The problem of impulsive noise suppression in image sequences has only been considered in a few published works. The work of Huang [66] and Martinez [117] represents two attempts to achieve temporal Median filtering along motion trajectories using a 3 tap Median filter.

The Median filter itself was introduced by Tukey [179] in 1977. The technique is one from the generic set of Order Statistic filters [195, 140, 46, 43]. These filters operate on some window of data. The procedure is first to rank the samples in that window and then the output of the filter is chosen to be one of the order statistic values for that ranked distribution. Thus the output can be the Upper Quartile value for instance. The Median filter chooses the median value to be the output. For the case of an odd number of samples, $2N + 1$, in the window, the median value is the Nth value in the ranked array. But when the number of signal samples is even, $2N$, the median value is defined as the mean of the samples at the two middle positions N and $N + 1$.

The median operation has been quite successful in images since it preserves edges well [140]. As it rejects outliers from a distribution, it is effective in removing impulsive noise such as Speckle or Salt and Pepper noise from single images. A typical 2D image filtering application [71, 107, 140] would use a window of size 3×3 and replace the pixel at the centre of the window with the median of the 9 pixels in the window. Every point in the image is treated in this way.

Although the median operation preserves edges, the fact that it rejects isolated impulses implies that it distorts fine detail in the image. Recursive implementation of the filter would reduce image grey scale variation to isolated areas of planar gradations of grey level separated by sharp steps. This is an important shortcoming with respect to median operations and in general it is preferred that the window used for the filter be as small as possible to avoid this distortion. Figures 4.1, 4.2, 4.3, 4.4 show the performance of a 3×3 median operation on an image corrupted by uniformly distributed impulsive distortion both in terms of position and grey scale. Note that the distortion is not additive but exclusive, i.e.

$$s(i,j,n) = \begin{cases} S & \text{with probability } P_s \\ I(i,j,n) & \text{Otherwise} \end{cases} \tag{4.9}$$

Here, S represents a spike of uniformly distributed grey scale which *replaces* the value of the undistorted image $I(i,j,n)$ at that same position. The distortion is essentially a local one and its effect depends on how often the distortion occurs i.e. P_s. In the image shown in Figure 4.2, $P_s = 0.1$, which is a high rate of distortion. The photographs indicate that although the filter rejects the distortion well, it also blurs the image slightly, cf. Figure 4.1 with Figure 4.3.

To allow the median operation to preserve more detail in the image sequence, the multilevel Median filter (MMF) was introduced by Nieminen et al [134]. This class of filters employs a hierarchy of median operations that is able to preserve the fine detail that a simple median operation would remove. In [134]

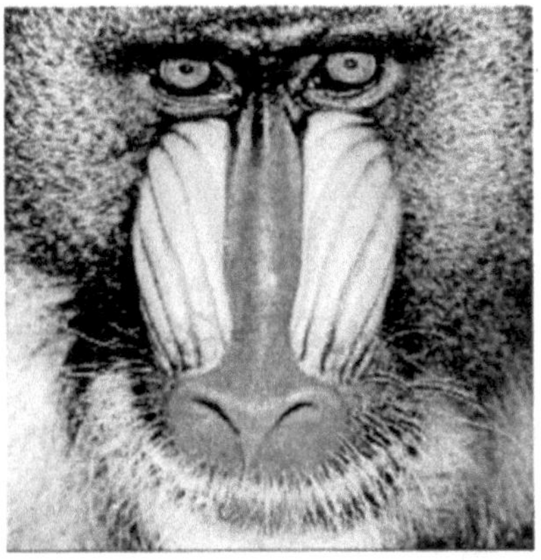

FIGURE 4.1. Original Baboon (8 bit grey scale).

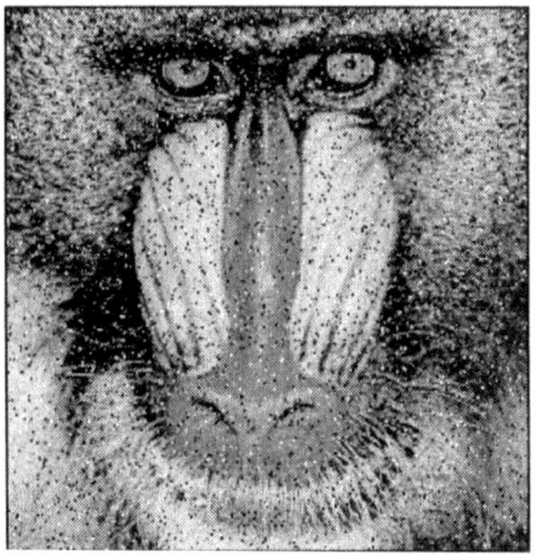

FIGURE 4.2. Baboon with impulsive degradation. $P_s = 0.1$.

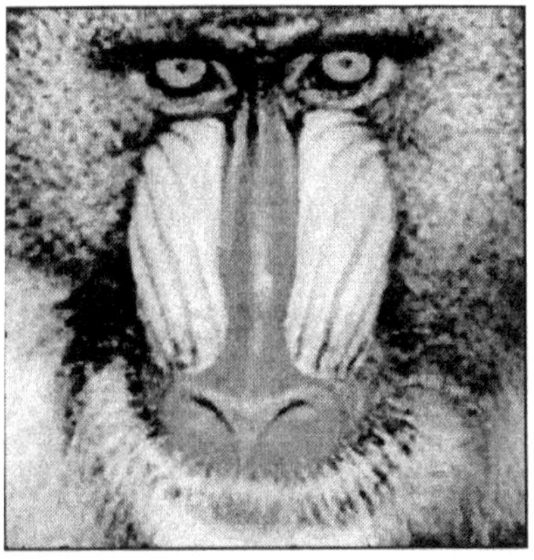

FIGURE 4.3. Standard 3 × 3 Median.

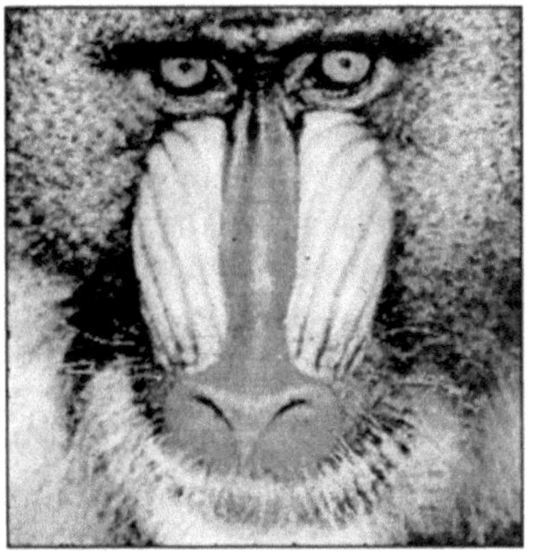

FIGURE 4.4. Multilevel filtering.

the general structure of an MMF was defined. A typical 2D filter structure is shown in Figure 4.5. There it can be seen that the MMF is implemented as the median of the output of Median filters with different topologies. Figure 4.5 illustrates a 2 level median operation, but there is no limit to the number of levels. In the figures, the black squares of each window grid indicate the pixels used in the median operation for that window. The centre pixel in each window is the same pixel[2] in the image and it is that pixel which is replaced with the output of the filter.

The filter is able to preserve more detail because of the different orientations of the sub–filter masks and the inclusion of the original sample into the final median level. A typical filtering result is shown in Figure 4.4. The figure il-

[2]The windows share the same space on the image with respect to the filtered location.

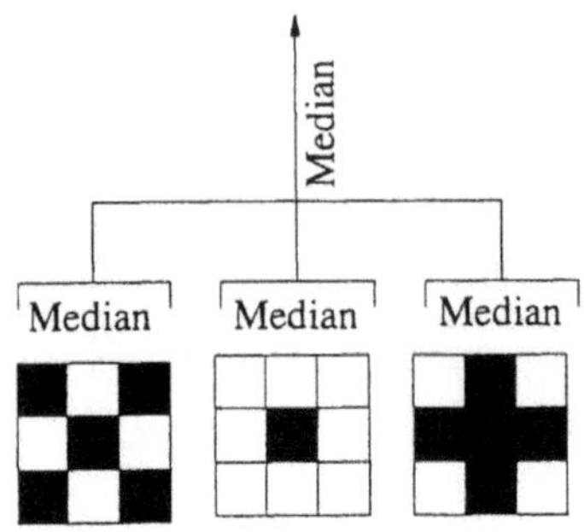

FIGURE 4.5. An example of a 2D multilevel Median filter using 3 sub filter windows.

lustrates the compromise that must be struck between detail preservation and impulsive noise rejection. The whiskers on the Baboon image are better preserved with the MMF than with the standard 3×3 median operation, at the expense of slightly less impulse rejection capacity. Nieminen [134] also introduced the FIR/Median hybrid filter which incorporates FIR filters in the lower hierarchy together with median filters. This class of filters is not considered in this book principally because of the additional computation involved.

In 1989 Arce et al [8] introduced the use of the min/max MMF for image sequence processing. The filters were studied in more detail in [6] and were examined through their ability to reject constant impulse Speckle and contaminated Gaussian noise. One of the filters presented can be defined with the help of the filter masks in Figure 4.6. The bi-directional filter was defined as follows.

$$\begin{aligned} z_l &= \text{median}[W_l] \;\forall\; 1 \leq l \leq 4 \\ z_{max} &= \text{MAX}_{1 \leq l \leq 4}[z_l] \\ z_{min} &= \text{MIN}_{1 \leq l \leq 4}[z_l] \\ \text{Bi-directional Filter output} &= \text{median}[z_{max}, z_{min}, C] \end{aligned} \tag{4.10}$$

The centre window in each 3 window set refers to a window in the current image. The filters presented used data from three frames to achieve filtering. The Uni-directional filter that was also presented, was the same as the Bi-directional filter except that it used 5 windows, only one of which included information from the surrounding two frames, window 7 in Figure 4.6. The other windows were the centre frame masks of windows 1–4.

The motivation for the form of the filters was to incorporate more 3D information into the median operation for image sequences. These filters would be better than the purely temporal implementation in [66, 117] because of the better impulsive noise rejection. The filters were introduced by Arce to be robust to motion, because the computational requirement of motion estimation was seen to be too large. Therefore the filters were implemented in a direction orthogonal to the frames. When motion occurs, the filters reduce to a purely spatial operation because the pixels in the two surrounding frames would be unlikely to bear any relation to the pixels in the current position.

A similar set of MMF's was introduced by Alp et al. The windows are also indicated in Figure 4.6. They utilize median operations only. These were also

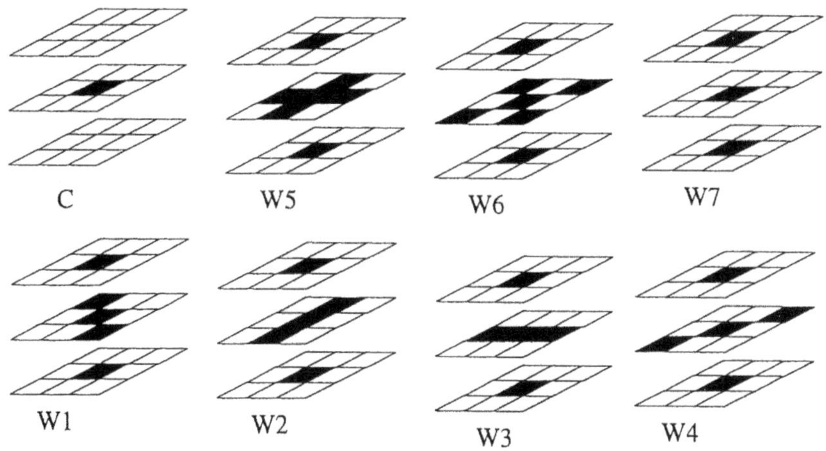

FIGURE 4.6. Sub–filter masks used for the MMFs presented by Arce and Alp et al.

not motion compensated implementations. However, because they incorporate more spatial information in each window, their impulse noise rejection capacity is better than those proposed by Arce. Their detail preservation capacity is reduced from those of Arce's for the same reason. The ML3D MMF introduced by Alp et al is defined below.

$$\begin{aligned} z_l &= \text{median}[W_l] \;\; 5 \leq l \leq 6 \\ \text{ML3D Filter output} &= \text{median}[z_5, z_6, C] \end{aligned} \tag{4.11}$$

The form of the filters show that when motion occurs, the MMF again reduces to a purely spatial operation in the current frame. It would be an improvement of course to implement these filters along a motion trajectory, and that is pursued later in the book. Further, the topology of the MMF's introduced means that impulsive distortion greater in size than 3×3 cannot be completely removed after one pass. Further passes (i.e. a recursive implementation) would remove such distortion but adversely affect the output image quality. 'Dirt and Sparkle' is typically larger than this and so this book also spends some effort in introducing new MMF's that can solve this problem.

It is possible to quantify the performance of Median filters if suitable simplifying assumptions are made [7, 6, 43, 46, 134]. However these analyses are of limited use in practice. Nevertheless there is active interest in optimal adaptive order statistic filtering which quantifies the effects of this general class of filters [108, 29]. This book does not investigate the prospect of these optimal schemes, preferring instead to concentrate on low complexity heuristics and a probabilistic model based approach which is outlined in the next section.

4.3.1 Model based missing data reconstruction

There has been little work in the area of impulsive noise suppression in images that does not involve Median filtering, the notable exception being the work of Geman [48] and Morris [119]. For the 1D (audio) case two authors, Veldhuis

[184] and Vaseghi [182], have presented practical alternatives which have been quite successful. Recently, Godsill [54] has been improving on these techniques. They consider the problem of scratch and click removal in degraded audio signals such as archived gramophone recordings. The methods are similar so only the ideas of Vaseghi will be considered.

Vaseghi argues that given an appropriate model for the audio signal (he used an AR model), the signal error between the prediction and the observed signal value can give useful information. At an impulsive distortion, the error signal has an amplitude which is virtually unchanged from the observed, distorted signal, but otherwise the prediction error is quite small. Essentially, the impulsive distortion cannot be handled by the model. Therefore, the error signal can be used as a detector for impulsive noise. He goes on to show that the AR model can be used to interpolate the missing information at the detected location by minimizing the squared error of the predicted samples that would be used to fill missing gap. His algorithm shows impressive performance when used to correct degraded audio.

Veldhuis also developed a 2DAR version of the audio 'click' interpolator which is based on the same ideas as used by Vaseghi. The important point about these ideas is that by detecting the impulsive distortion, attention may be concentrated only at the degraded signal portions. This would isolate any artefacts that may be introduced in a global operation. This has important implications for Median filtering in particular since, as discussed earlier, the global median operation invariably blurs fine texture in images.

Chapters 3 and 7 investigate further the 3DAR work of Efstratiadis et al [37] discussed in Chapter 2. The work of Veldhuis and Vaseghi is then extended to deal with the problem of detection and interpolation of Dirt and Sparkle in motion pictures. In Chapter 7 the entire model based approach to missing data detection is reworked and generalized to treat the detection and motion estimation problem simultaneously.

4.4 Summary

Most of the work done in ISR to date involves white noise suppression and deblurring. Within the last 5 years Özkan et al. [135] have introduced techniques for the optimal 3D filtering of image sequences. These techniques are superior to the purely temporal operations that have been implemented previously.

Very little attention has been paid to the problem of missing data (impulsive distortion) removal in image sequences. Arce [8, 6] and Alp et al [2] represent two works that have developed 3D median filters for impulsive noise suppression in image sequences. Those filters were implemented without motion compensation and so the results did not realize the full potential of these structures. Further, the median operation, although quite successful in the suppression of impulsive noise in images, invariably introduces distortion. This distortion primarily takes the form of blurring fine image detail. The methods of Veldhuis and Vaseghi can be extended to deal with this difficulty. The basis of the ideas of Vaseghi, that of detecting the distortion, can be used to introduce some heuristics for control of the median operation. This is considered in Chapter 6. Building on the work of Efstratiadis et. al [37], Strobach [165], Vaseghi, Godsill, and Rayner [183, 51, 54], it becomes possible to treat this 3D problem within a model based framework that represents a superior alternative to Median filtering. (See Chapters 6, 7, 8.)

5

Line Registration for Jittered Video

Video signals must contain synchronization information to allow the video display device to properly locate lines and frames relative to each other in space and time. Noise in the video signal, or degradation of the storage medium on which the signal is stored (video tape) can cause the synchronization signals to be corrupted. This can cause the loss of 'lock' in video digitizing and playback apparatus. The loss of line synchronization pulses will prevent the video manipulation device from locating the actual start and end of each line thus yielding random line displacements (line jitter) in the observed video images. In addition, the lack of accurate frame timing would cause a vertical rolling effect in the digitized video. In practice, the line *jitter* effect is more common. The most visible effect of line jitter or loss of line synchronization is that edges which should be vertical in the image appear to be jagged. It is accepted that generally speaking, digital video tape recorders and players are less susceptible to the problem of loss of synchronization information. However the problem is still important in those cases where images must be acquired in harsh environments e.g. archived analogue video tape.

Whatever the source of the degradations, the line jitter effect is disturbing to the viewer. The jitter can range from ± 1 pixel to more than ± 5 pixels in severe cases. Figures 5.6 and 5.8 illustrate typical distortion. It is possible to employ machines called 'Time Base Correctors' which process the analogue video signal and recover the line synchronization pulses by trying to remove the noise on the non-picture parts of the video signal. This chapter considers an alternative solution to the problem which relies only on image data [89, 94]. The data can then be 'de-jittered' even if the original analogue recording is unavailable.

Possible simple solutions to the problem involve either searching for the shift between lines that maximizes the correlation between them or averaging lines.

Correlation matching tends to cause a bias towards vertical lines in the estimated shifts since it has a limited 'memory'. That is to say, such a method involves only two lines in the image and good matches across these lines may not reflect the overall image structure. Averaging can remove much of the perceived jitter, provided the jitter is within $\pm$ 1 pel, but this is at the expense of half of the vertical resolution. It is shown to be very unsuccessful in Figure 5.5. What is needed therefore, is a method capable of retaining longer range vertical structure in the restored image without a loss of resolution. Such a method is developed here through a model based approach for estimating the jitter displacements. Although a variety of image models exist, it is thought that a spatial linear autoregressive structure is adequate for this problem, especially since it is able to capture the notion of local image smoothness.

The method introduced in this chapter to provide an estimate for the relative shifts between lines is closely related to the 3DAR pel-recursive modelling framework discussed in Chapter 2. The method presented here uses a non-homogeneous 2-D AR model in a similar way that Efstratiadis et al [37] use a homogeneous model for the considered pixel. The chapter begins with a model description and then explains the registration algorithm.

5.1 The model

The assumption is made that the original image[1] obeys a non–homogeneous 2DAR model. This model is in exactly the same form as the 3DAR model described in Chapter 2, with the important exception that it uses support pixels only in the same frame. The model used is also explicitly non–homogeneous in that it is acknowledged that different areas of the image are modelled by different AR coefficients.

The situation in the original image can be defined following some of the notation introduced in Chapter 2. The model support is defined by a set of N spatial vectors $\mathbf{q_k} = [q_k^h,\ q_k^v]$. The corresponding 2DAR coefficient at a particular location is given by $a_k(h,l)$, where h,l are respectively the horizontal location in pixels along a line and the line index.

Therefore the predicted grey level $\hat{I}(h,l)$ at the location (h,l) is given in terms of N surrounding pixel values $I(h+q_k^h,\ l+q_k^v)$ as follows. The image is assumed to be zero mean.

$$\hat{I}(h,l) = \sum_{k=1}^{N} a_k(h,l) I(h+q_k^h, l+q_k^v) \tag{5.1}$$

---

[1]The image prior to the jitter distortion.

The corresponding error between the predicted value and the actual grey level of the pixel at (h, l) is then given by

$$e(h,l) = I(h,l) - \sum_{k=1}^{N} a_k(h,l) I(h + q_k^h, l + q_k^v) \qquad (5.2)$$

Changing the sign of the coefficients and defining $a_0(h,l) = 1.0$, allows a more useful expression for the prediction error (or excitation).

$$e(h,l) = \sum_{k=0}^{N} a_k(h,l) I(h + q_k^h, l + q_k^v) \qquad (5.3)$$

After the image frame is jittered so that the lines are displaced relative to each other, the prediction equations must incorporate this displacement. In order to predict the intensity of the hth pixel at line l in the *degraded* image, the offset vectors $\mathbf{q}_k$ must be modified by the displacement between the current line and the line on which the support point is to be found. Thus the model of the degraded image can be written using equation 5.2 as

$$I(h,l) = \sum_{k=1}^{N} a_k(h,l) I(h + q_k^h, l + q_k^v + s_l - s_{l+q_k^v}) + e(h,l) \qquad (5.4)$$

where s_l is the absolute displacement of line l away from its true location. An example model arrangement as regards the jittered lines is shown diagrammatically in Figure 5.1.

The object is now to estimate the displacements s_l in such a way that by displacing the line l by $-s_l$, the original image is the result. Non-homogeneous AR coefficients must be employed (i.e. AR models whose coefficients vary with position) since the statistical nature of the image can change drastically along a line. In practice, it is sufficient to divide the line into horizontally non-overlapping blocks and to employ a single model for each block.

5.2 Displacement estimation

As has been stated, to achieve a restoration of the jittered image, the relative displacement between lines must be found. Unfortunately, using the model structure above requires knowledge of the model coefficients for the unknown original image. Therefore, the solution proposed uses an iterative process which estimates a set of coefficients given a current estimate for the displacement and then estimates an update to the current displacement based on the estimated coefficients and so on.

Each line or group of lines is considered separately. It is assumed that displacement estimates for all lines are available except the estimate for line l (or group of lines) and the lines involved in the immediate model neighbourhood. It is simpler to consider the absolute displacement of each line bearing in mind

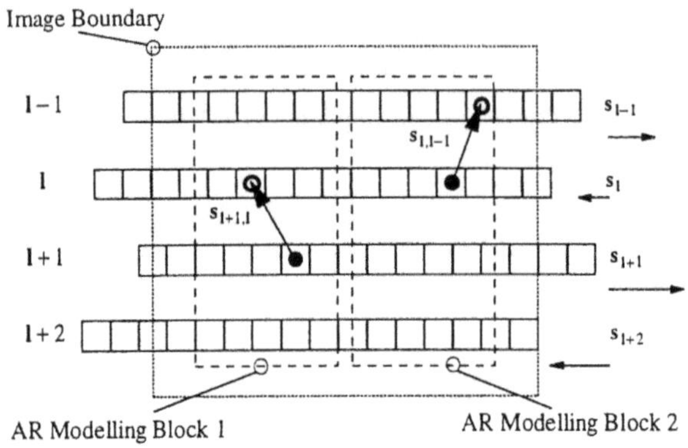

FIGURE 5.1. A simple (1 tap) 2DAR model applied to an image of 4 jittered lines showing two separate data blocks (– –) which employ different sets of AR coefficients. Example predicted locations are shown as • with their corresponding support pixel as ∘, suitably compensated for relative displacement. The relationship between absolute (s_l) and relative ($s_{l,l-1}$) shifts is also shown. The line indices are shown on the left hand side. Note that the shifts represented are only a guide and are meant to illustrate that non-integer jitter is possible.

that the displacement of the top or bottom line in the image can be arbitrarily assigned. Suppose that some initial displacement estimate s_l^0 for line l is already obtained (this could be zero). The task is then to estimate an update displacement u_l such that $s_l = s_l^0 + u_l$. Assuming the relative displacement defined by $s_{l,l+j} = s_l - s_{l+j}$, is small, and that the AR coefficients are known, the Taylor series expansion of equation 5.4 around the current displacement estimate s_l^0 leads to an explicit function in s_l as follows:

$$I(h,l) = \sum_{k=1}^{N} a_k(h,l) \Big[\quad I(h + q_k^h, l + q_k^v + s_l^0 - s_{l+q_k^v}^0) + (u_l - u_{l+q_k^v}) \frac{\partial}{\partial x} I(h + q_k^h, l + q_k^v + s_l^0 - s_{l+q_k^v}^0) + O(h, l + q_k^v) \Big] + e(h,l) \tag{5.5}$$

where s_n^0 is an initial estimate of s_n which may be zero, $O(h, l + q_k^v)$ represents the higher order terms in the Taylor expansion and $\frac{\partial}{\partial x} I(\cdot)$ the horizontal gradient of the image given the current estimate for displacement. This approach [89] was employed in a similar manner for motion estimation in image sequences [17, 37]. It is assumed that the effect of $O(h, l + q_k^v)$ is equivalent to additive Gaussian noise.

The linearized equation may be rearranged as follows:

$$I(h,l) - \sum_{k=1}^{N} a_k(h,l) I(h+q_k^h, l+q_k^v + s_l^0 - s_{l+q_k^v}^0) =$$
$$\sum_{k=1}^{N} a_k(h,l) \left[(u_l - u_{l+q_k^v}) \frac{\partial}{\partial x} I(h+q_k^h, l+q_k^v + s_l^0 - s_{l+q_k^v}^0) + O(h, l+q_k^v) \right]$$
$$+ e(h,l) \quad (5.6)$$

To simplify the arguments to follow, consider the model shown in Figure 5.1. The equation above becomes

$$I(h,l) - a_1(h,l) I(h + s_l^0 - s_{l-1}^0, l-1) =$$
$$(u_l - u_{l-1}) a_1(h,l) \frac{\partial}{\partial x} I(h + s_l^0 - s_{l-1}^0, l-1) + a_1(h,l) O(h, l-1) + e(h,l) \quad (5.7)$$

It is now easier to recognize that the left hand quantity is the current measured prediction error given the current estimate for s_n, s_n^0 and the coefficient $a_1(h,l)$. The right hand side consists of a weighted gradient term, an excitation term $e(h,l)$ and a truncation error term $a_1(h,l)O(h,l-1)$. Defining the current prediction error as $z(h,l)$ and assuming that the effect of the other error and excitation term is the same as that of added noise, $v(h,l) \sim \mathcal{N}(0, \sigma_{vv}^2)$, the equation may be simplified as follows:

$$z(h,l) = (u_l - u_{l-1}) a_1(h,l) \frac{\partial}{\partial x} I(h + s_l^0 - s_{l,l-1}^0, l-1) + v(h,l) \quad (5.8)$$

Generally, observations of the prediction error and horizontal gradient may then be collected together horizontally along a particular line, giving a set of equations involving the various displacement updates $u_l, u_{l+q_k^v}$ i.e.

$$\mathbf{z} = \mathbf{G}\mathbf{u} + \mathbf{v} \quad (5.9)$$

where the composition of these matrices are defined implicitly in equation 5.8. In addition, such equations can be assembled for more than one line. In fact, note that for the model shown in Figure 5.1, the line l is involved in prediction error equations set up both in that line *and* in the next. Therefore to get a complete description of the local model behaviour it is best to set up the equations for several lines at once.

To see how the elements of the equation 5.9 can be arranged for multiple lines, it is necessary first of all to adopt the simpler notation that $u_{n,k} = u_n - u_k$ is the update for the relative displacement between lines. The equations for $M = 3$ points (positions x_1, x_2, x_3) along each line in a block of $N_l = 3$ lines for the model shown in Figure 5.1 then become

$$\begin{bmatrix} z(x_1,l) \\ z(x_2,l) \\ z(x_3,l) \\ z(x_1,l+1) \\ z(x_2,l+1) \\ z(x_3,l+1) \\ z(x_1,l+2) \\ z(x_2,l+2) \\ z(x_3,l+2) \end{bmatrix} = \begin{bmatrix} a_1(x_1,l-1)\mathcal{G}(x_1,l-1) & 0 & 0 \\ a_1(x_2,l-1)\mathcal{G}(x_2,l-1) & 0 & 0 \\ a_1(x_3,l-1)\mathcal{G}(x_3,l-1) & 0 & 0 \\ 0 & a_1(x_1,l)\mathcal{G}(x_1,l) & 0 \\ 0 & a_1(x_2,l)\mathcal{G}(x_2,l) & 0 \\ 0 & a_1(x_3,l)\mathcal{G}(x_3,l) & 0 \\ 0 & 0 & a_1(x_1,l+1)\mathcal{G}(x_1,l+1) \\ 0 & 0 & a_1(x_2,l+1)\mathcal{G}(x_2,l+1) \\ 0 & 0 & a_1(x_3,l+1)\mathcal{G}(x_3,l+1) \end{bmatrix} \begin{bmatrix} u_{l,l-1} \\ u_{l+1,l} \\ u_{l+2,l+1} \end{bmatrix} + \begin{bmatrix} v(x_1,l) \\ v(x_2,l) \\ v(x_3,l) \\ v(x_1,l+1) \\ v(x_2,l+1) \\ v(x_3,l+1) \\ v(x_1,l+2) \\ v(x_2,l+2) \\ v(x_3,l+2) \end{bmatrix} \tag{5.10}$$

where $\mathcal{G}(h,k)$ is the horizontal gradient of the current de-jittered version of the image at pixel h on line k. It is assumed that the image lines have already been compensated by the current estimate for displacement. Therefore in this case three updates can be considered at once, bearing in mind that the top line of the block (line $l-1$) can be fixed at its current location. When the model involves more than one line of memory it is necessary to use the relation

$$s_{l,l+n} = s_{l,l+1} + \sum_{k=0}^{n-2} s_{l+1+k,l+1+k+1} \tag{5.11}$$

in order to express the relative shifts between pixels several lines apart (lines l and n in the equation above) in terms of the sum of relative displacements between each intervening line.

Following Efstratiadis et al [37] and Biemond et al [17], a Wiener solution may be derived for an estimate of u defined $\hat{u}$. The estimated update is assumed to be uncorrelated with the error signals and has variance σ^2_{uu}.

$$\hat{\mathbf{u}} = [\mathbf{G}^T\mathbf{R}_{vv}^{-1}\mathbf{G} + \frac{1}{\sigma^2_{uu}}]^{-1}\mathbf{G}^T\mathbf{R}_{vv}^{-1}\mathbf{z} \tag{5.12}$$

where $\mathbf{R}_{vv}$ is the correlation matrix of the random vector $\mathbf{v}$. This solution is similar to that considered in Appendix C. Care must be taken in applying the relative shifts to compensate for the jitter at each iteration. The actual displacement applied to compensate for the shift at each line may require a cumulative tally to be kept from the top of the line in each block, depending on how the user chooses to compensate for the shifts.

This pel-recursive solution is virtually identical to the motion estimation solution of Efstratiadis et al [37] and Biemond et al. [17]. Nevertheless, there are two important differences.

1. Within a frame an image is highly non-homogeneous. Hence it is necessary to use a different model at different positions along the line. Therefore the correlation matrix $\mathbf{R}_{vv}$ cannot be represented by a scaled identity matrix as was the case in [37, 17]. When equations for one line are set up, this matrix is diagonal if the model error $e(h,l)$ is assumed to be white noise of variance σ^2_{ee}. The matrix can then be described as below

$$\mathbf{R}_{vv} = \begin{bmatrix} \sigma^2_{ee}(0,l) + \sigma^2_{OO}(0,l)a_1^2(0,l) & \dots & 0 \\ \vdots & \ddots & \vdots \\ 0 & \dots & \sigma^2_{ee}(M,l) + \sigma^2_{OO}(M,l)a_1^2(M,l) \end{bmatrix}$$

   where there are M observations of $v(x,l)$ made along a line, and σ^2_{oo} is the variance of the truncation error term $O(h, l + q_k^v)$. In practice this decomposition of σ^2_{vv} into its component terms involving $\sigma^2_{ee}, \sigma^2_{oo}$ is not employed. In general, for several lines, under these assumptions, $\mathbf{R}_{vv}$ is *block* diagonal.

2. The model is two-dimensional and the displacement parameter required is a scalar for each line.

Of course it is not necessary to use a different set of coefficients at every pixel. In practice the line is divided into blocks which are then assigned different model coefficients.

It is acknowledged that it may be possible to use the 2DAR model itself to treat the line l as missing and so interpolate some meaningful information[2]. However the form of this distortion *is known* to be the relative displacement between the lines, therefore it is thought that a method that explicitly attacks this problem would be superior.

---

[2]See Chapters 7 and 9

5.2.1 *Increasing stability*

Solving for many line updates at once does help increase the stability and convergence of the iterative process, but much depends on the assumption about the statistics of $\mathbf{v}$. The assumption of whiteness in $\mathbf{v}$ is improved by employing only purely causal AR models. However, this causes an accumulation of error in the displacement estimation process [89]. Instead, it is possible to propose a much more stable estimation process by employing both causal *and* anti-causal models at the same time. The anti-causal model contributes another set of equations to the Wiener solution in equation 5.12. Defining the shape of the anti-causal model support to be the exact mirror of the causal model, it is then possible to redefine $\mathbf{z}$ and $\mathbf{G}$ as follows

$$\mathbf{z} = [\mathbf{z}_c^T \; \mathbf{z}_{ac}^T]^T \tag{5.14}$$
$$\mathbf{G} = [\mathbf{G}_c^T \; \mathbf{G}_{ac}^T]^T \tag{5.15}$$

where $\mathbf{z}_{ac}, \mathbf{G}_{ac}, \mathbf{z}_c, \mathbf{G}_c$ are the anti-causal and causal observations respectively.

5.3 Implementation

First of all the support used in the model is restricted by the assumption made about the correlation structure of the model error $e(h, l)$. If equation 5.10 is set up only for one line at a time, the correlation terms used in $\mathbf{R}_{vv}$ all concern horizontal offsets. The assumption about the error implies that all the pure horizontal offset correlation coefficients are zero. These terms are only zero if the support of the model is either causal or, if not, does not include taps in the current line. This is analogous to the situation with regard to 3DAR modelling in Chapter 2 and is due to the fact that the error from a non-causal AR model is not white. (See Appendix B). These statements are justified by considering the correlation function of the residual (or error) from a non-causal AR process stated below

$$r_{ee}(\mathbf{x}, \mathbf{q}_n) = \begin{cases} \sigma_{ee}^2 & \text{for } \mathbf{q}_n = [0, \; 0] \\ a_n \sigma_{ee}^2 & \forall n = 1 \ldots N \end{cases} \tag{5.16}$$

It is assumed that the other correlation terms not defined by the above equation are zero. The terms in the correlation matrix $\mathbf{R}_{vv}$ that refer to the model error are of the form $r_{ee}(\mathbf{x}, \mathbf{x}+[k, \; 0])$. They are terms that apply to the horizontal lags only if the observations are assembled from line l alone. Therefore, as long as the model does not contain taps in this line, there will be no contribution from the term $a_n \sigma_{ee}^2$ in the expression 5.16 and so $\mathbf{R}_{vv}$ is diagonal as required. However, when many lines are de-jittered in the same step, i.e. employing equation 5.10, any non-causality in the support of the 2DAR model will necessarily lead to a non-diagonal $\mathbf{R}_{vv}$, thus violating the assumption that $\mathbf{R}_{vv}$ is block diagonal. To avoid any of these complications, it is best to employ purely causal supports for the AR process. Note that the only reason for preferring diagonal $\mathbf{R}_{vv}$ is to allow

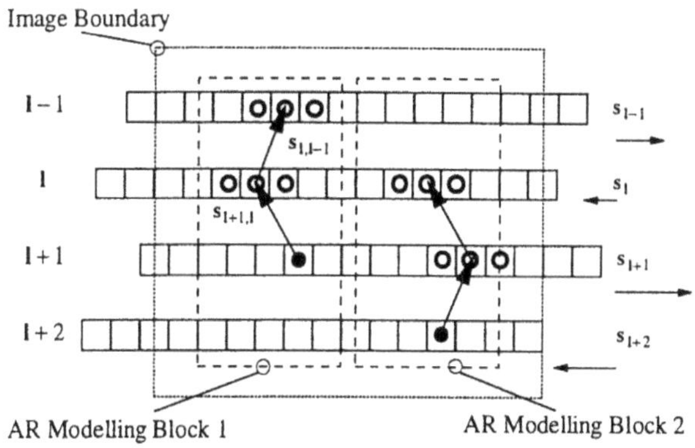

FIGURE 5.2. The semi-causal 6 tap 2DAR model used in the de-jittering results shown. • represents a support pixel and ∘ a predicted pixel. The support for the model is shown suitably compensated for relative shifts. Two separate data blocks employing different sets of AR coefficients are shown. Note that they do not overlap horizontally.

the fast application of the Wiener solution by avoiding the need to calculate $\mathbf{R}_{vv}^{-1}$ when this matrix is not diagonal. In practice it is found that there is little difference in performance whether causal (see Figure 5.1) or semi-causal (see Figure 5.2) supports are used.

5.3.1 Model parameter estimation and overlapped blocks

The AR coefficients are re-estimated in each image block at each iteration using the Normal equations and the current estimates for the displacements. In practice it is found to be more robust to employ a vertically median filtered image (at each iteration) to yield estimates for the AR coefficients. In the results shown later, a 5×1 vertical window was used for the median operation. Furthermore, assuming the major term in $\mathbf{v}$ is the variance of the current prediction error, $\mathbf{R}_{vv}$ can be measured at each iteration.

It is computationally intractable to estimate the displacements of all the lines in the image at once. As discussed previously, it is preferable to de-jitter some subset of lines, L, at a time. The stability of the process is further improved by overlapping the model estimation areas. For the results illustrated in Figure 5.3 for instance, a single 2DAR coefficient set is used for a 32×32 block of pixels. These blocks are tiled across the horizontal width of the image without overlap, but vertically overlapped 2:1. An estimation area of 64 lines over 4 rows of overlapped blocks was used for generating Figure 5.3.

5.3.2 Interpolation

During the iterative procedure, it is necessary to evaluate pixel intensities at fractional displacements. There are a variety of techniques available to do this. Bilinear interpolation is found to cause too much blurring and instead windowed SINC interpolation is used. Polynomial interpolation is a good alternative [71, 107].

5.3.3 Multiresolution

To improve the quality of the small displacement assumption it becomes advisable to successively refine estimates on a hierarchical basis. The image is low pass filtered and subsampled horizontally to create $\mathcal{L}$ horizontally "compressed" image levels (including the original level 0). Displacement estimation begins at level $\mathcal{L} - 1$. Then estimates are refined at each successive level until the original resolution level has refined the final estimates.

5.3.4 Enhancing vertical features

Some amount of horizontal high frequency emphasis helps to increase the contrast of the jitter phenomenon and improve the displacement estimation process. In the algorithm to follow, this emphasis was similar to unsharp masking. To further restrict attention to more significant vertical features, only pixels which were at locations having a horizontal gradient > 5 grey levels per pixel were used to set up the equations.

In the pictures shown later, the actual emphasis process was as follows.

1. Generate a horizontally low pass filtered version of the image using a Gaussian blur with a shape of variance 1.5 and truncated to a 5 tap FIR filter. Generate a difference image ORG $-$ LO where ORG is the original image and LO the low pass filtered version. The difference image is a high pass filtered image.

2. Sites are marked where the horizontal gradient in LO (the low pass filtered image) is greater than 5.0 (arbitrarily chosen).

3. At each of the sites marked above, add a proportion of the difference image back into ORG. The proportion used was 0.5 in the de-jittering of the images shown, and the Gaussian FIR filter was normalized so that the sum squared coefficients was 1.

It was found that the algorithm was not sensitive to the actual emphasis process used, but some emphasis was always worthwhile.

5.4 Algorithm flow

A horizontal pyramid of image resolutions is set up, of $\mathcal{L}$ levels including the original resolution level at level 1. The pyramid is created using a Gaussian low-pass filter and 2:1 downsampling at each level. In the same manner as hierarchical motion estimation, the jitter estimation begins at the coarsest level, $\mathcal{L}$. After estimation is terminated at level $\mathcal{L}$, the jitter displacement vector is scaled by a factor of 2 and used as the initial estimate at level $\mathcal{L} - 1$. This estimation/propagation/initialization process continues until the jitter vector updates terminate after several iterations of updating at level 1, the original resolution. The jitter update procedure is defined as follows. Block sizes may change at each multiresolution level.

1. Define $B \times B$ to be the block size (in pixels) over which the image is assumed homogeneous and N_l to be the number of lines to be de-jittered at the same time.

2. Each block of $B \times B$ pixels in the area to be de-jittered is to employ a separate set of AR coefficients. Associate with each block a surrounding block from which to estimate the model parameters: size $\mathcal{B} \times \mathcal{B}$.

3. Set the current estimate of displacement (s_l^0) for all lines to be 0 pixels.

4. Estimate the model parameters for each $B \times B$ block using the Normal equations, measuring correlation coefficients over the $\mathcal{B} \times \mathcal{B}$ modelling block. Employ a vertically median filtered image for coefficient estimation. In the pictures shown later, a 5 tap median filter was used.

5. Set up $\mathbf{z}, \mathbf{G}, \mathbf{R}_{vv}$. The gradients are calculated using simple difference techniques, as indicated in Chapter 2. An estimate for the unknown model error variance can be gained by measuring the variance of the prediction error over the blocks associated with each line segment. A fixed value (1.0) for σ_{uu}^2 is assumed. There is no need for the points used on each line to occupy consecutive locations along the line. For practical purposes, the blocks of $B \times B$ pixels need not overlap horizontally, although they should overlap vertically as indicated in the previous discussions.

6. Solve for the update using equation 5.12.

7. If $\sum |\mathbf{z}| \leq z_t$ where z_t is some previously set threshold, then halt the process. The correct displacement has been found. Alternatively the MSE may be used. The Sum Absolute Error was used in the experiments because of the simpler computation.

8. If $|u| \leq |u|_t$ then assume that the algorithm has converged and halt the process.

9. Update the displacement $\hat{s}_{l,l+1}^{i+1} = \hat{s}_{l,l+1}^{i} + u_{l,l+1}$ where $\hat{s}_{l,l+1}^{i} + u_{l,l+1}$ is the current estimate of relative displacement between lines l and $l+1$ and $\hat{s}_{l,l+1}^{i+1}$ is the updated displacement.

10. If the number of iterations has exceeded a maximum allowed limit then halt the process.

11. Goto 4.

After the algorithm has terminated, line $l+1$ is shifted by $-\hat{s}_{l,l+1}$. Then the next line is considered until all the lines in the image have been treated. In practice it is better to limit the maximum estimated displacement and perform several passes over the entire frame. This will displace all the lines by small amounts with each pass, eventually converging on a smooth image. Limiting this displacement prevents the reconstructed image showing gross lateral drift over parts of the image.

In the pictures to follow, de-jittering was achieved using only two stopping criteria for the iterations, a maximum number of allowed iterations and a minimum size of update displacement.

5.5 Interframe processing

For de-jittering a sequence of images it is possible to employ the spatial process described above on each frame. However, the resulting frames are then misregistered with respect to each other, and objects may appear to warp from frame to frame. To avoid this problem, a second stage process can be introduced which assumes that one or more stationary reference regions that cover the whole frame in a vertical sense have been identified. In these regions the line shifts can be estimated using some direct matching criterion by using the first de-jittered frame as a template for registering the lines in subsequent frames. In [94] a *phase correlation* method is applied which determines phase shifts in the Fourier domain :

$$\begin{aligned} S[f] &= \frac{F_1[f]F_2[f]^*}{|F_1[f]F_2[f]^*|} \\ &= exp(j\theta_1[f])exp(-j\theta_2[f]) \\ &= exp(j(\theta_1[f]-\theta_2[f])) \end{aligned} \tag{5.17}$$

where $j=\sqrt{-1}$, $F_i[f]$ are the Fourier transforms of the reference line segment and its corresponding line segment in the frame currently being processed, and $\theta_i[f]$ the phase. The location of the maximum of S^{-1} gives the relative shift between the lines. As magnitude information is discarded this method is relatively insensitive to low contrast. Also assuming the noise to be uncorrelated and Gaussian, the noise is spread evenly over the spectrum, making the method more noise robust.

5.6 Performance

The algorithm relies heavily on the performance of the spatial displacement estimation process to produce a successful result after the temporal phase correlation is applied to a set of images. Therefore the results concentrate on the performance of the spatial part of the algorithm.

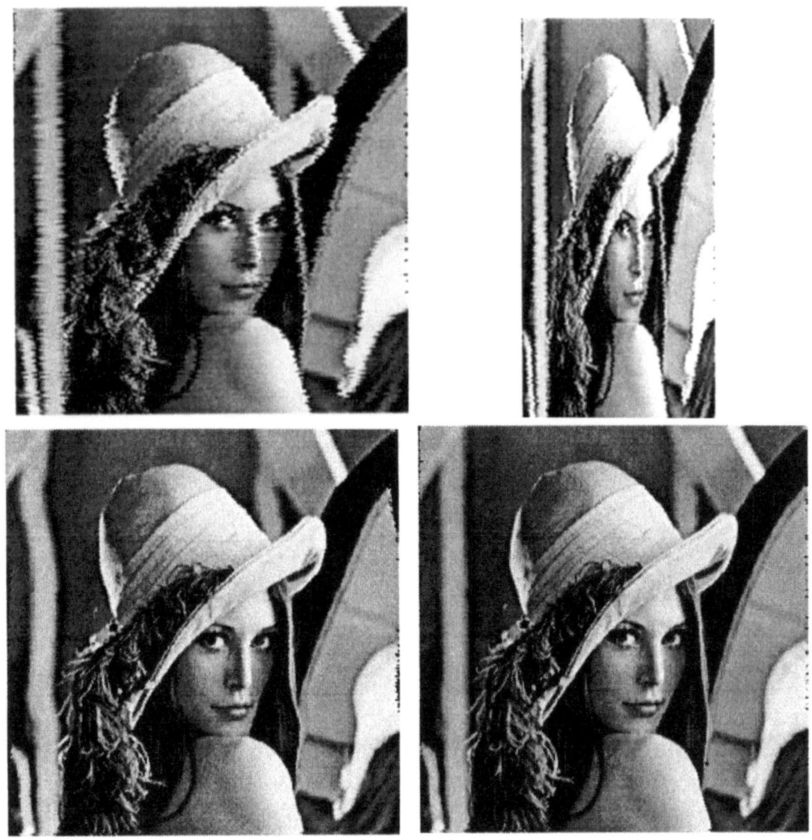

FIGURE 5.3. Top Row: Artificially Jittered Lenna, Multiresolution levels 1 (left) and 2 (right); sizes 256 × 256 and 256 × 128 respectively. Bottom Row: De-jittered Lenna *before* drift compensation (left) and *after* drift extraction (right).

Figure 5.3 shows a portion of the Lenna image severely artificially degraded by displacing each line using $s_l \sim \mathcal{N}(0, 1.0)$. The resulting *actual displacements* are graphed in Figure 5.4 for a subset of 128 lines. All the estimated data is included on the CD-ROM. The top row of Figure 5.3 shows the two levels of the pyramid that were used for displacement estimation. These have been emphasized as previously described. The estimation algorithm used 5 iterations at each level and a modelling block size of 32 × 32. The support for the causal and anti-causal models employed 6 locations (see Figure 5.2) immediately above (or below, as in the case of the anti-causal model) the predicted pixel in the lines $l-1$, $l-2$, with 3 points of support in each line centred on the predicted location. 64 lines were de-jittered at a time as explained previously.

The bottom left image in Figure 5.3 shows the result of compensating for the displacements using the *raw* estimates from application of the de-jittering algorithm described. It is assumed that the image is zero outside the given image grid (256 × 256). Thus when lines are shifted, the edge of the shifted

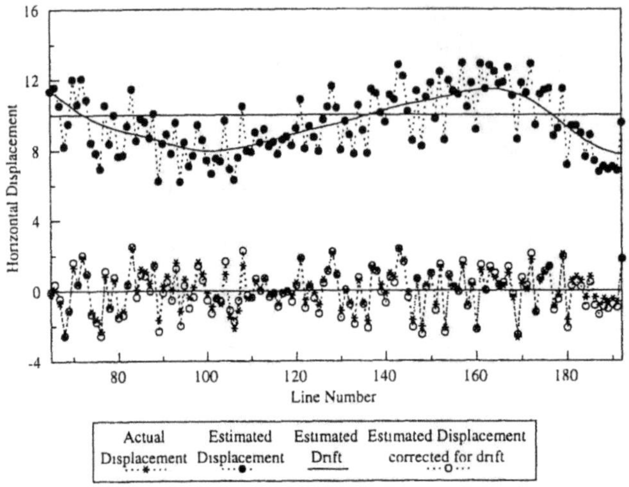

FIGURE 5.4. Actual and estimated displacements.

lines can be seen at the left or right hand edge of the image. These *estimated displacements* are also shown (offset for clarity) in Figure 5.4. The estimation shows a substantial low frequency component which is erroneous[3]. This can be seen both from the plot of displacements *and* the fact that the image looks warped if these estimates are used to compensate for jitter. This drift can be removed by *whitening* the displacement signal which results in the *corrected estimated displacements* indicated after subtracting the *estimated drift*. The remaining small error can be compensated after 1 or 2 iterations of the algorithm at the original resolution. The whitening may be performed using an estimated 1-D AR process, but it is simpler to estimate the low frequency component using a low-pass filter. A 31 tap hamming window was used as an FIR filter. The resulting de-jittered image using the corrected displacements (after 2 further iterations) is shown in the bottom right hand image in Figure 5.3. The displacements are now properly compensated and there is no visible artefact, this is also seen in the accuracy of the displacement estimates as compared to the true displacements in Figure 5.4. Note that the low frequency overestimation phenomenon is worse when only a causal model is used. However, it can generally be corrected in this algorithm using the whitening approach.

This result is illustrative of the ill-posedness of the jittering problem. The AR process has no knowledge of overall image structure, it cannot differentiate between smooth, wavy edges which are true features and such edges which are errors in the estimation process. The only way to solve the problem at this low level of image understanding[4], is to insert more prior knowledge about the line jitter characteristics. The whitening process implicitly forces the assumption

[3]This phenomenon was observed in [96]

[4]Through the use of local pixel interactions quantified by the 2DAR process.

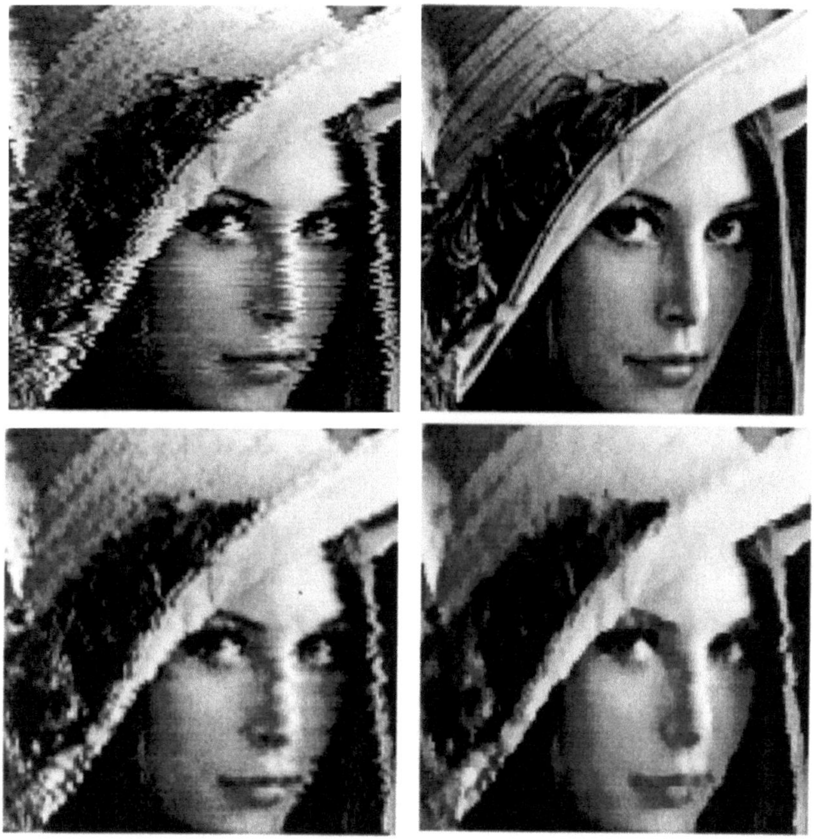

FIGURE 5.5. Clockwise from top left: Artificially jittered Lenna 128×128 portion, De-jittered using algorithm described, De-jittered using 5×2 median filter, De-jittered using vertical averaging.

that the jitter source is random, white noise, thus effecting a good correction in Figure 5.3.

The results for Lenna are compared to two alternative techniques in Figure 5.5, where zooms on several reconstructions are shown. One involves vertically weighted averaging of pels in the image so that the intensity on line l, $I(l)$, is replaced by the intensity $0.25I(l-1)+0.5I(l)+0.25I(l+1)$. The other uses a vertical median filtering operation using a window of 5×2 pixels (5 rows and 2 columns). The CD-ROM contains full versions of these images as well as a further comparison with a direct matching technique using correlation. In all cases the de-jittered attempts are inferior to the model based technique introduced here. The averaging method causes too much blurring and the median filtered version removes too much detail.

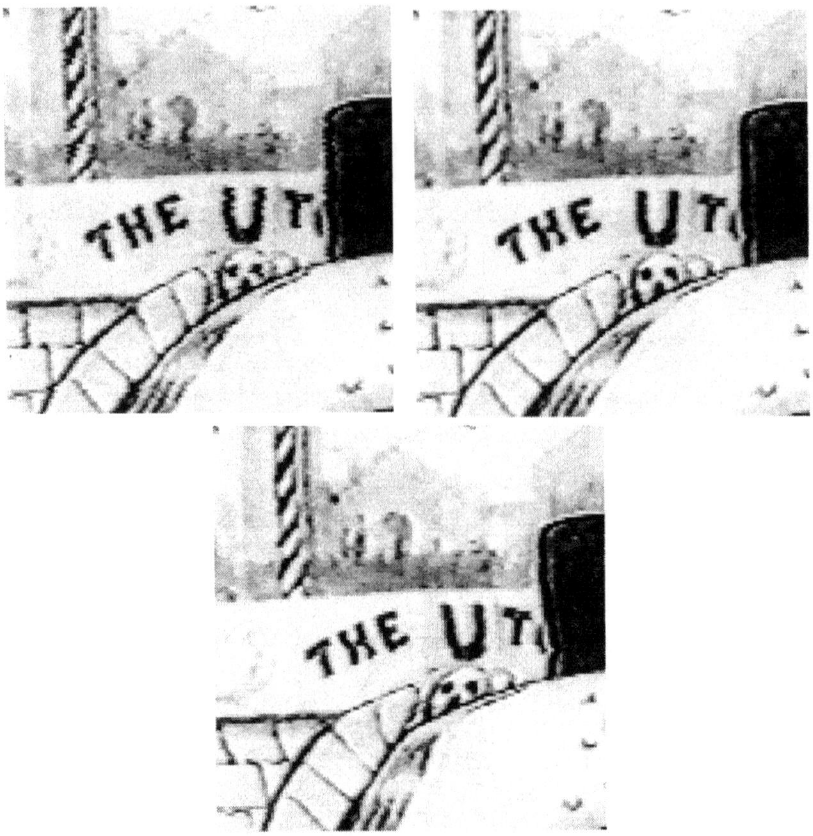

FIGURE 5.6. Top Row: Real jittered cartoon (left) UTUBE, Restored (right). Bottom image shows de-jittered result *before* drift compensation.

5.7 Real jitter

Figure 5.6 shows an example of relatively mild jitter occurring as a result of normal digitization of a video signal from a VHS tape, which was copied from some original. The degraded sequence can be seen in the top left hand image and shows clear 'jaggies' on the vertical side of the chimney, the lettering " THE UT", and the bottom of the barber's pole. De-jittering used the same model and parameters as above for Lenna, only one level of multiresolution (i.e. no subsampled levels), 10 iterations and a minimum update threshold of 0.01. The de-jittered image (after drift compensation) is shown on the right of the top row in Figure 5.6. It shows definite improvement as there are no more visible 'jaggies'. Figure 5.7 also shows the raw and compensated displacements, illustrating the effectiveness of the drift compensation in a real case. The lower image in Figure 5.7 shows the use of the *raw* displacements for de-jittering. This *raw* compensation shows distortion in the circular shapes near the bottom of the image, and the 'kink' in the chimney at its base.

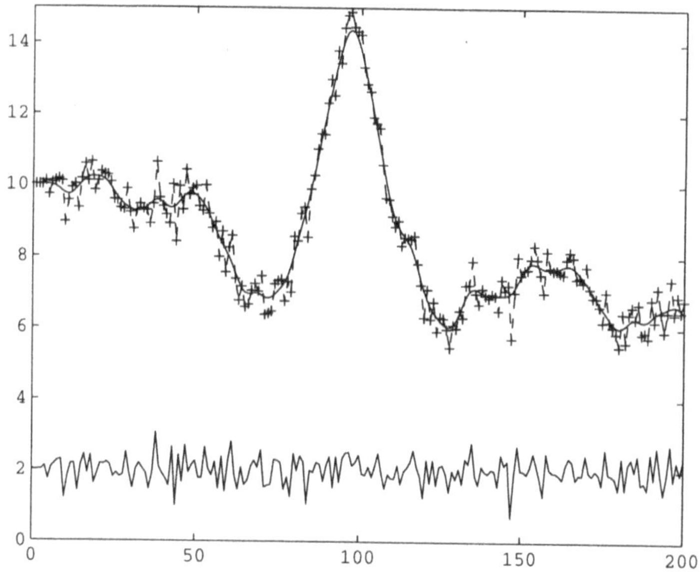

FIGURE 5.7. Estimated displacements for 200 lines of UTUBE. Lines are offset for clarity. Top lines (offset by 10) show estimation of drift (solid line) by filtering raw displacements (dashed line) with FIR filter using 11 tap Hamming function. Bottom line shows the estimated displacements *after* drift compensation (offset by 2).

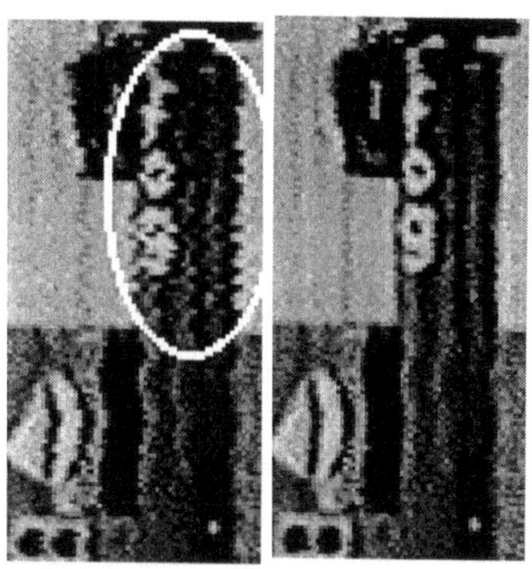

FIGURE 5.8. Left: Portion of original frame 1 of real degraded sequence. Right: Unjittered frame 1 using spatial process.

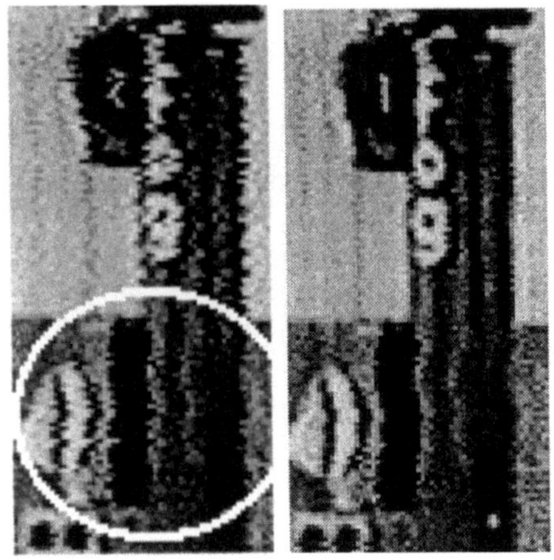

FIGURE 5.9. Left: Portion of original frame 2 of real degraded sequence. Right : Unjittered frame 2 using matching with processed frame 1.

Figure 5.8 (left) shows a portion of a frame from a real degraded sequence showing quite marked jitter. An area where the jitter can be easily seen is highlighted. The right hand portion shows the result of the spatial registration process using blocks of 32 × 32 pixels with an overlap of half the block size vertically and none horizontally. The AR model and parameters used were the same as for Lenna with 2 pyramid levels. Figure 5.9 shows the next degraded frame in the sequence and the registered image generated using the second stage matching process. These pictures give the typical performance of the algorithm and show good improvement in the observed quality of the reconstructed image. A region is highlighted in Figure 5.9 which shows that the jitter displacements here contain in addition a low frequency component which causes vertical structures to look 'wobbly'. This is discussed in the next section.

Figure 5.10 shows a very severe case of jitter. The line jitter displacements are not constant along each line. In fact it appears that each line is stretched rather than translated. To complicate matters, there are many lines in which there is dropout. Applying the exact algorithm described here does improve the situation as shown in the figure (right). The left edge of the hair shows marked improvement. However, there is some residual 'stretch'. There is a sudden shift of the fingers of the hand because of dropout on one line. The stretch may be accounted for by modelling each jitter displacement not as a translational component only, but also with an additional affine stretch component. Thus the jitter at position h on line l becomes $s_l + m_l h$ instead of s_l alone. The same model parameters and multiresolution levels as for Lenna were used.

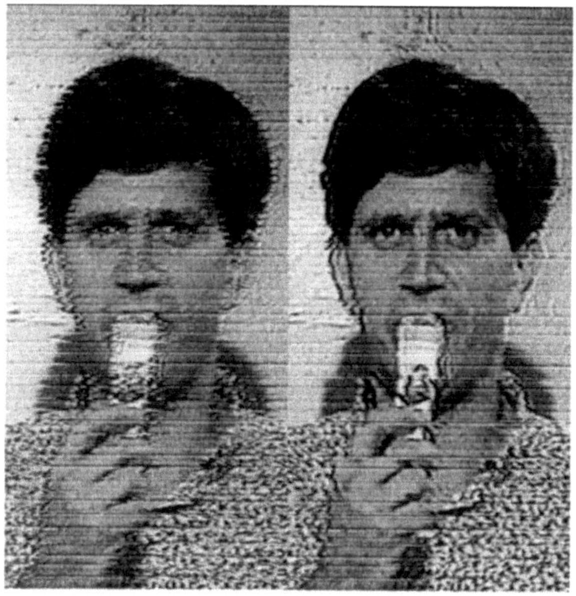

FIGURE 5.10. Left: Original, real degraded image showing *severe* line synchronization problems. Right: De-jittered using model based algorithm.

5.8 Final Comments

The line de-jittering algorithm has been demonstrated to be quite effective in removing the disturbing inter-line jitter that is often observed in captured frames. The success of the method lies primarily with the relatively good modelling capacity of the non-homogeneous 2DAR process and that the jitter between lines causes only relative horizontal shifts and no other distortion.

Nevertheless, the problem of de-jittering an image based on spatial *image* information alone is difficult. The essence of the proposed algorithm is to shift lines so that the vertical image gradient (manipulated through the AR framework) is small over the whole image: thus removing jagged vertical edges and in so doing compensating for jitter. This is only effective when the corrupting jitter is primarily random, or some high frequency corruption. There is little that a low-level image processing algorithm can do to distinguish a smooth diagonal feature from a low frequency jitter corruption, which can occur (see Figures 5.8, and highlighted area in 5.9). It may be possible to improve this algorithm by incorporating knowledge of a generating process for the line jitter, such as a low frequency component plus white noise. Then it may become feasible for a purely image based algorithm to distinguish between true smooth diagonal features and those that have been caused by the low frequency part of the jitter generating source. Although rare, severe jitter manifesting as non-linear line transformations e.g. Figure 5.10, provides a fruitful avenue for further work.

The second stage of the process requires the user to identify stationary regions in the scene in order to 'lock' lines in subsequent frames into position with reference to the first dejittered frame. The location of stationary areas may be done automatically and this is a subject for further work.

6

Heuristics for Reconstructing Missing Data

One of the most common and striking distortions in archived motion picture film is the presence of scratches and blotches. The blotches may take the form of regions of high contrast which appear at random positions in the frame as in Figures 6.1 and 6.14. They rarely occur in the same location in consecutive frames. Line scratches are a separate problem which is considered in Chapter 9. This form of distortion tends to be correlated from frame to frame.

The Oxford English Dictionary explains the term 'heuristic' as a *method of solving problems by inductive reasoning, by evaluating past experience and moving by trial and error to a solution.* As the title therefore suggests, this chapter discusses simple techniques for automatic removal of the random blotches (also called Dirt and Sparkle). The techniques draw upon experience gained from observing the nature of blotches as they appear in degraded image sequences[1]. The processes discussed here were designed by directly exploiting characteristic features of blotches. The model based approach to system design is more subtle than this and relies instead on the ability of the designer to discover (or propose) the underlying processes generating the observed signals. Systems designed through a model based approach are described in the next chapter.

An important consideration in the treatment of any real distortion is an appreciation of the agents that cause the problem and the manifestation of the degradation. Dirt and Sparkle on film is caused by the physical interaction of the film material with the projecting equipment as it is transported through the mechanism. These distortions will occur also in Telecine equipment since the mechanism of film transport is the same. Dirt and Sparkle would then

[1]One could also say that the techniques draw upon experience gained from watching many dirty movies.

occur on the video sequences as well. Abrasion is not a normal side effect of the equipment, it can be caused by foreign particles caught in the mechanism. Because the film material is abraded, bright flashes of light (in the case of accumulation of particles, flashes of dark) are seen. The distortion is clearly a local one and there are large areas of each frame which are not affected at all.

Dirt and Sparkle can be effectively modelled as randomly distributed impulsive distortion which *replaces* the image information in selected regions. These distortions are not limited to single pixel impulses but can also occur as variably sized patches of uniform grey level which are highly contrasted with the surrounding image. These patches represent regions of missing information in the frame. They are referred to as *blotches* in this text. The chapter begins with a discussion of the missing data problem and its implications for the standard applications of spatiotemporal median filtering. It then goes on to present a number of detectors of missing data and finally introduces the notion that motion should also be seen as part of the missing data which is to be estimated.

6.1 Two step removal

It is usually assumed that the missing data problem is a problem of impulsive noise corruption, in that the process of corruption introduces outliers into the observed image sequence. The usual methodology for treating impulsive corruption in images is to apply a median filtering process everywhere [71, 107, 140]. As is illustrated by the actual degradation shown in Figure 6.14, the missing data problem here is clearly a local one. In applying the median filter globally to the image the expectation is that the filter would reject outliers where they occur and otherwise leave the image untouched. In practice the global filtering operation does affect uncorrupted areas and the usual artefact is a fading of image sharpness. It does not make sense to treat the entire image when only just a small proportion of the image needs to be reconstructed. Therefore the approach followed here is first to *detect* the corrupted regions and then engage a correcting mechanism, for instance a median filter [91, 90, 86, 87].

An important point is the size of the area of missing data being considered. Unlike typical impulsive noise suppression applications, it is possible for blotches on motion picture film to be larger than 20×20 pixels. A spatial median filtering operation thus becomes less effective in the centre of such distortion primarily because it is then considering many missing pixels in its input. Of course one could design a median filter which uses more intra-frame information, and this is illustrated in the section on 3D Multilevel filters.

In addressing the issue of data reconstruction for image sequences, it is necessary to recognize that a fully 3D operation would hold much more potential for higher image fidelity than a 2D operation. Of course the problem then arises of estimating motion, and it becomes important to acknowledge the errors that will occur in this estimation process. With respect to reconstruction, a good algorithm would take advantage of both spatial and temporal informa-

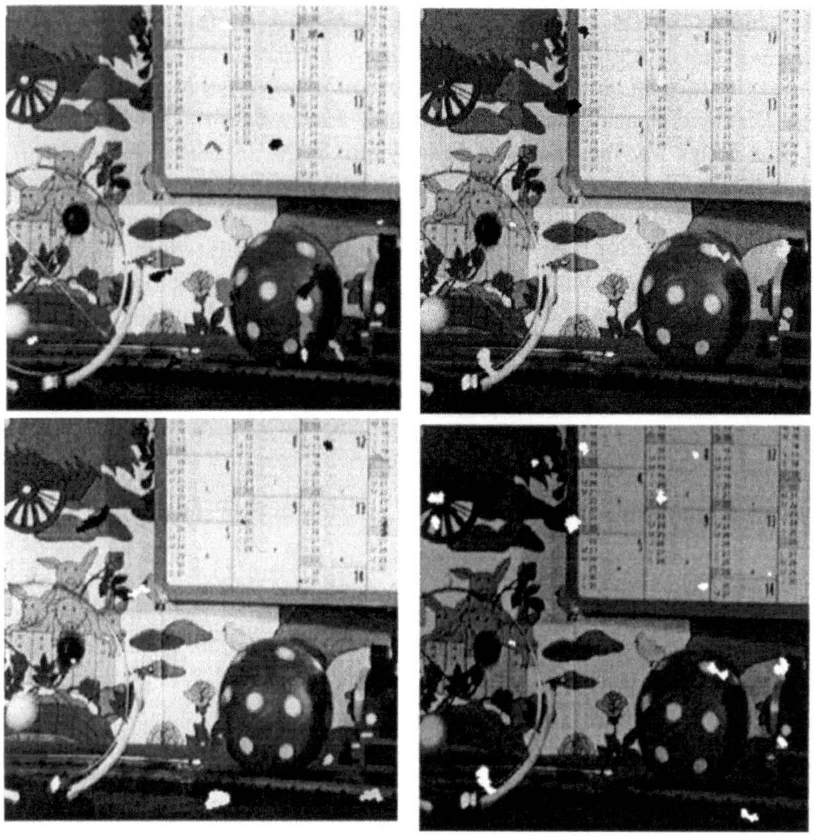

FIGURE 6.1. Clockwise from top left: Frames 8,9 of the degraded MOBCAL sequence, frame 9 with location of distortion overlayed in white, degraded frame 10.

tion, and would be able to emphasize one or the other in spatially or temporally inhomogeneous[2] regions of the sequence.

A typical reconstruction system would involve first motion estimation, then detection of the missing regions, and finally reconstruction of the detected missing regions. The basic feature used for detection of Dirt and Sparkle in particular, is that they represent regions of temporal discontinuity in intensity along motion trajectories. The same applies to other forms of missing data, such as those occurring due to digital dropout. It is this feature which is used in the heuristics for detection that are described next.

---

[2]Inhomogeneous due to either non-trivial motion or erroneous motion estimation.

6.2 Heuristics for detecting missing data

Perhaps the earliest work on designing an automatic system to 'electronically' detect Dirt and Sparkle was undertaken by Richard Storey at the BBC [164, 163]. The design was incorporated directly into hardware which was subsequently used in-house for video restoration before broadcast. In his work he describes detection mechanisms which were not motion compensated. This handicap (lack of knowledge of motion) led to the problem of separating temporal discontinuities in the image sequence which were caused by missing data from those caused by motion. Now that motion estimation is a better understood process it is straightforward to propose motion compensated versions of his detector. His exact non-motion compensated system is not discussed further.

6.2.1 The SDI detectors

Assume the simplest model for the clean, original image sequence as follows,

$$I_n(\mathbf{x}) = I_{n-1}(\mathbf{x} + \mathbf{d}_{n,n-1}(\mathbf{x})) + e(\mathbf{x}) \tag{6.1}$$

where $e(\cdot) \sim \mathcal{N}(0, \sigma_e^2)$ and the terms have their usual meaning. Then it follows that provided σ_e^2, the noise in the sequence is small, then the Displaced Pixel Difference (DPD), $I_n(\mathbf{x}) - I_{n-1}(\mathbf{x} + \mathbf{d}_{n,n-1}(\mathbf{x}))$ should generally also be small when the motion is correctly estimated. When a location is corrupted, however, this is not so and the DPD takes on a high magnitude since the corruption violates the underlying simple model with some outlying signal value. Unfortunately this also occurs when motion discontinuities, e.g. occlusion and uncovering occur. This problem can be overcome by recognizing that in cases of occlusion and uncovering, the temporal discontinuity occurs only in one direction along the motion trajectory. For example, regions covered in the next frame do have a match in the previous frame. Thus the simplest detector for missing data can be designed by noting that missing data will cause high DPD in both temporal directions at a pixel site. Defining directional DPD's as follows

$$\begin{aligned} E_b &= I_n(\mathbf{x}) - I_{n-1}(\mathbf{x} + \mathbf{d}_{n,n-1}(\mathbf{x})) \\ E_f &= I_n(\mathbf{x}) - I_{n+1}(\mathbf{x} + \mathbf{d}_{n,n+1}(\mathbf{x})) \end{aligned} \tag{6.2}$$

allows the simplest detector, the SDIa (Spike Detection Index - a) (see Kokaram et al [86]) to be defined as follows

$$b_{\text{SDIa}}(\mathbf{x}) = \begin{cases} 1 & \text{for } (|E_b| > E_t) \text{ AND } (|E_f| > E_t) \\ 0 & \text{otherwise} \end{cases} \tag{6.3}$$

where E_t is a user defined threshold. The field $b_{\text{SDIa}}(\mathbf{x})$ is set to 1 wherever a pixel is detected as corrupted and is 0 otherwise.

Storey observed another interesting phenomenon. When blotches did occur, their intensity was well outside the intensity of the surrounding region. Given that the simple image model holds, this implies that not only are the DPDs large, but also that the sign of E_b was the same as that of E_f. This of course reduces the ability of the detector to spot all missing data, but does allow it to detect the most visible corruptions. This alternative to the SDIa, named for this text as the SDIp, is therefore defined as

$$b_{\text{SDIp}}(\mathbf{x}) = \begin{cases} 1 & \text{for } (|E_b| > E_t) \text{ AND } (|E_f| > E_t) \\ & \qquad \text{AND } \operatorname{sign}(E_f) = \operatorname{sign}(E_b) \\ 0 & \text{otherwise} \end{cases} \tag{6.4}$$

Independent of this, Kokaram et al [96, 91, 90] introduced another detector, called the SDI (Spike Detection Index) for detecting blotches. It is related to SDIa and Storey's SDIp. Define $p = I_n(\mathbf{x})$ to be the intensity at the current pel, $f = I_{n+1}(\mathbf{x} + d_{n,n+1}(\mathbf{x}))$ (for forward) to be the intensity at the next pel along the motion trajectory from the present pel into the next frame, and $b = I_{n-1}(\mathbf{x} + d_{n,n-1}(\mathbf{x}))$ (for backward) the intensity at the previous pel along the motion trajectory from the present pel into the previous frame. Then the SDI is defined as follows,

$$\begin{aligned} d_1 &= |p - f| \\ d_2 &= |p - b| \\ SDI &= 1 - \left| \frac{d_1 - d_2}{d_1 + d_2} \right| \quad \text{for } d_1 > t_1 \text{ or } d_2 > t_1 \\ SDI &= 0 \quad \text{otherwise} \end{aligned} \tag{6.5}$$

where t_1 is a low threshold which overcomes problems when d_1 and d_2 tend to zero. The SDI is limited to values between 0 and 1 and the decision that a blotch is present is taken when the SDI at the tested pel is greater than some predefined threshold.

To understand how this approach works, assume that the motion is purely translational. Now consider the following points,

- Occlusion: $|p-f|$ will be large and $|p-b|$ will be zero. Therefore $SDI = 0$.
- Uncovering: $|p-f|$ will be zero and $|p-b|$ will be large. Therefore $SDI = 0$.
- Normal Motion: Both $|p - f|$ and $|p - b|$ will be zero. As both $p - f$ and $p - b$ tend to 0 the SDI is not well behaved. However, when this happens, it means that the motion estimator has found a good match in both directions hence the current pel is not likely to be a scratch. Therefore in this case the SDI is set to zero.
- A blotch at the current pel but in the position of an object showing normal motion: Both $|p - f|$ and $|p - b|$ will be large and the same and so $SDI = 1$. They would be the same since f, b would both be the same pels on the object at different times thus having the same intensity, provided the assumption of pure translation holds.

- A blotch at the current pel but in a position on an object image showing occlusion or uncovering: It is difficult to say how the SDI behaves here. The SDI will take some undefined value, not necessarily zero or one. This value would depend on the actual intensities of the occluding regions.

- A blotch at the current pel but f and/or b represent pels at which blotches have also occurred. Again the SDI is not defined, but if the blotches are fairly constant valued the index tends toward 0.

The general rule is that when the SDI is 0 the current pel is uncorrupted. When it is 1 the current pel is corrupted. In order to allow for the cases where occlusion and multiple corruptions along the motion trajectory are possible, there must be some threshold to make the decision. The threshold also allows some tolerance in the case of real sequences where motion is not purely translational and one has to deal with slight lighting changes not due to motion.

For real sequences there must be some lower threshold, t_1, for the forward and backward differences which will indicate that the match found is sufficiently low so that the current pel is uncorrupted. This is necessary because in real sequences the motion is not translational and due to lighting effects the intensities of corresponding areas do not necessarily match. Further, there will be errors from the motion estimator.

6.2.2 *The ROD detector*

Nadenau and Mitra [123] have introduced another heuristic detector for blotches in image sequences. It is the Rank Order Detector (ROD). It is generally more robust to motion estimation errors than any of the SDI detectors although it requires the setting of three thresholds. It uses some spatial information in making its decision.

Define a list of pixels as

$$\begin{aligned}
p_1 &= I_{n-1}(\mathbf{x} + \mathbf{d}_{n,n-1}(\mathbf{x}) + [0\ 0]) \\
p_2 &= I_{n-1}(\mathbf{x} + \mathbf{d}_{n,n-1}(\mathbf{x}) + [0\ 1]) \\
p_3 &= I_{n-1}(\mathbf{x} + \mathbf{d}_{n,n-1}(\mathbf{x}) + [0\ -1]) \\
p_4 &= I_{n+1}(\mathbf{x} + \mathbf{d}_{n,n+1}(\mathbf{x}) + [0\ 0]) \\
p_5 &= I_{n+1}(\mathbf{x} + \mathbf{d}_{n,n+1}(\mathbf{x}) + [0\ 1]) \\
p_6 &= I_{n+1}(\mathbf{x} + \mathbf{d}_{n,n+1}(\mathbf{x}) + [0\ -1]) \\
I_c &= I_n(\mathbf{x})
\end{aligned} \tag{6.6}$$

where I_c is the pixel to be tested.

Sort p_1 to p_6 into the list $[r_1, r_2, r_3, \ldots, r_6]$ where r_1 is minimum. The median of these pixels is then calculated as $M = (r_3 + r_4)/2$. Then three motion compensated difference values are calculated as follows:

If $I_c > M$

$$\begin{aligned} e_1 &= I_c - r_6 \\ e_2 &= I_c - r_5 \\ e_3 &= I_c - r_4 \end{aligned} \tag{6.7}$$

If $I_c \leq M$

$$\begin{aligned} e_1 &= r_1 - I_c \\ e_2 &= r_2 - I_c \\ e_3 &= r_3 - I_c \end{aligned} \tag{6.8}$$

Three thresholds are then selected for comparison. If any of the differences exceeds these thresholds, then a blotch is flagged as follows

$$b_{\text{ROD}}(\mathbf{x}) = \begin{cases} 1 & \text{if } (e_1 > t_1) \text{ OR } (e_2 > t_2) \text{ OR } (e_3 > t_3) \\ 0 & \text{otherwise} \end{cases}$$

where $t_3 \geq t_2 \geq t_1$. The choice of t_1 is the most important. The procedure used to generate the results discussed later (section 6.7) varied t_1 only and left the other thresholds at nominal values of $t_2 = 39.0$ and $t_3 = 55.0$. The detector works by measuring the 'outlierness' of the current pixel when compared to a set of others chosen from other frames. The choice of the shape of the region from which the other pixels were chosen is arbitrary.

6.2.3 Pre-processors

It is clear that noise in the observations of the various inter-frame differences (DPDs) will degrade the performance of the above detectors. This is particularly true of the SDIa and SDIp. It is possible to reduce the effect of noise by filtering the DPDs before processing. A typical filtering path is the one employed by Storey [164]. The idea is to filter the DPDs with a separable low pass filter before any further processing is done. In addition, some action must be taken to prevent large DPDs (due to motion or blotches) from causing a halo effect in detection in the surrounding region. The design of these filtering strategies depends to a great extent on the size and contrast of the missing data to be detected, since these define the usable window size of the filter. This book does not consider the various forms of pre-processing possible. It is noted that this is a practical step which may yield advantages depending on the noise in the input image sequence and the contrast of the target blotches.

6.2.4 Adaptive detectors

In each of the detectors above the decision to flag a blotch is taken based on some DPD values. It is possible to employ an adaptive process to estimate the required threshold locally from the surrounding image data. The detector performance is then expected to improve because this process would allow for the variation of movement and image information across the image. Instead of thresholding the SDI, for instance, with a constant value across the image, it is better to use as the threshold some multiple of the local mean of the SDI. In effect, this makes the thresholding operation adapt to the local spatial information, thus improving upon the purely temporal nature of the SDI itself.

To avoid instances where large distortions cause the local mean to be biased away from a value which will give a good detection rate, it is necessary to filter the local means of the SDI to remove impulses due to large size or frequently occurring distortion. The local means are calculated for each block of some fixed size and then filtered with a multistage median filter. A typical filter to be used is the ML3D filter introduced by Alp et al. [2] (see Chapter 4) without the temporal taps. This topology would preserve directional information in the local means due to motion edges or spatial image edges.

Unfortunately, in practice the success of this kind of adaptivity is also heavily dependent on the size of distortion to be detected [96]. It may be possible to tailor a particular scheme for certain situations such as digital dropouts in video, in which the expected size of a block of missing data is known. This could result in very useful improvements for real time use. However, as the next chapter introduces ideas which are implicitly more robust to this kind of problem, these methods are not explored here. See [96, 91, 90] for an example of an adaptive SDI detection system.

6.3 Motion estimation

All the detectors rely on good motion estimation to be effective, otherwise uncorrupted, moving regions would be erroneously flagged. The motion estimator adopted for use here is the Multiresolution Adaptive Wiener Based Motion Estimator (MWBME) as discussed in Chapters 2 and 3. It is gradient based and as such it is sensitive to the kind of distortion considered here. This is because the gradient based motion estimator cannot track motion if there is no gradient information in the local area searched in the previous or next frame. Therefore, if a blotch occurs in the previous frame at the same location showing motion in the current frame, the motion estimator will not be able to correctly track the motion (since the blotch grey scale profile is predominantly flat) and so may yield a false alarm. A similar argument holds at the edge of blotches where the gradient information may not yield the correct search direction. Block matching, on the other hand, attempts to minimize the DFD using a direct search. It will therefore be immune to these problems to some extent. An in-depth investigation into the robustness of motion estimation schemes to missing data is outside of the scope of this book, but it is educational to consider some of the phenomena.

To illustrate the effect of missing data on two basic types of motion estimator, frames from an artificially degraded MOBCAL sequence are employed. They

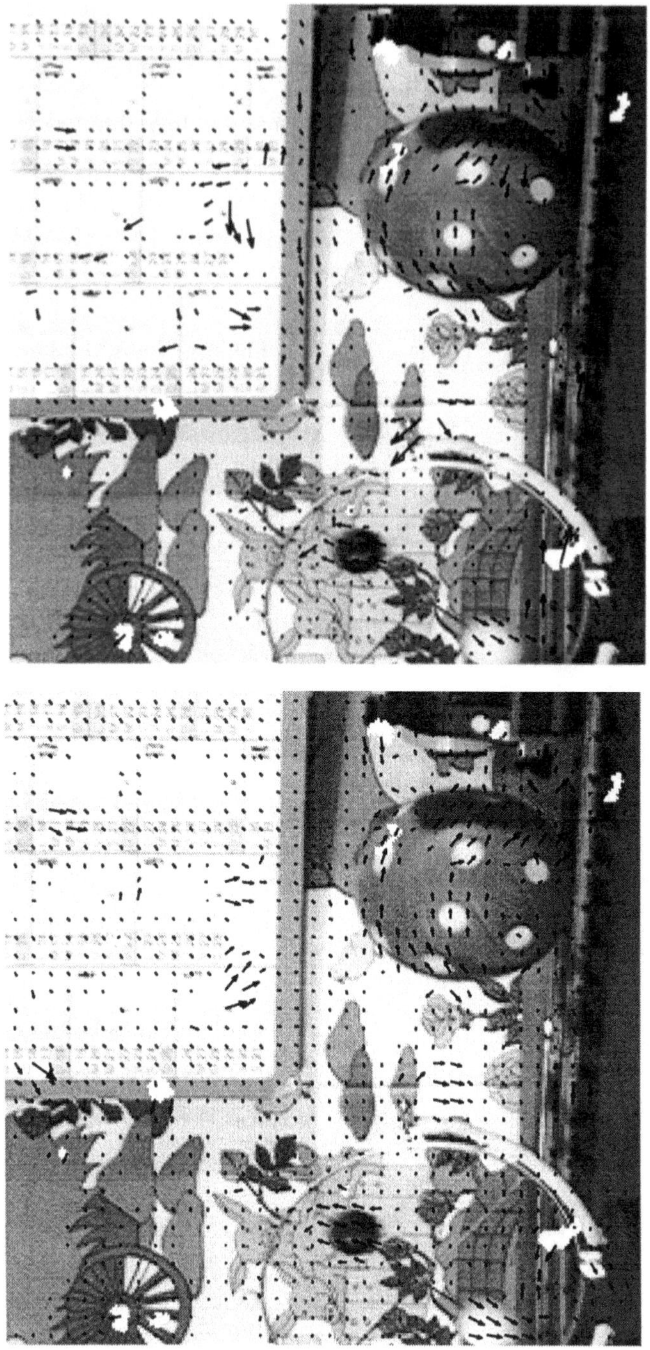

FIGURE 6.2. Motion Estimation fields from frame 9 into frame 8 using Multiresolution Wiener Based Motion Estimation (left) and Multiresolution Integer Accurate Block Matching with Exhaustive Search (right).

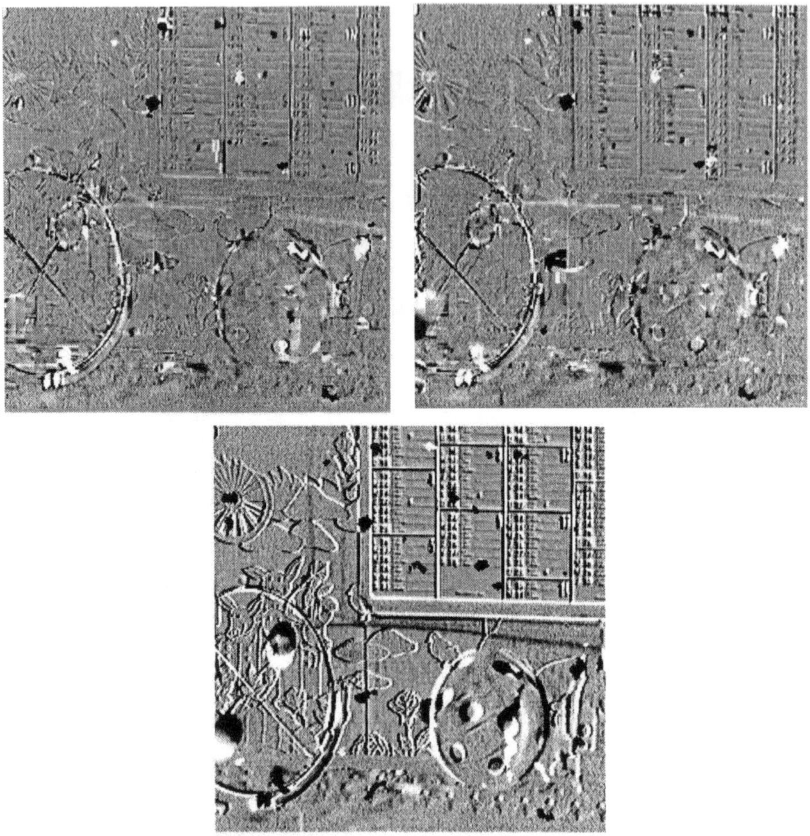

FIGURE 6.3. DFD from motion estimation from frame 9 into frame 8. Left Top: Multiresolution Wiener Based Motion Estimation, Right Top: Multiresolution Integer Accurate Block Matching with Exhaustive Search, Bottom: Non motion compensated DFD

are shown in Figure 6.1. The locations of blotches in the second frame are superimposed as bright white pixels on a slightly darkened image. Section 6.4 describes the sequence further. Figure 6.2 shows the motion field resulting when the MWBME and integer accurate hierarchical block matching is used to estimate motion in frame 9 of the *degraded* sequence. Three levels of the pyramid as well as a block size of 9×9 with an overlap of 2 pixels in each direction were used for both motion estimators. Both algorithms are as described in previous chapters. The block matching algorithm used a full search of ± 4 pixels displacement at each level with a noise ratio r_b set at 1.2 for the first level and 1.0 otherwise. The gradient based algorithm used bilinear interpolation to evaluate fractional displacements and an ill-conditioning threshold of 50.0. A motion detector was used in both cases to turn on motion estimation in each block only when motion was detected. The threshold MAE (Mean Absolute Error) was 5.0 for the block matching algorithm. Threshold MSE was 25.0 for the MWBME. Only the motion field mapping frame 9 into frame 10 is shown. Corresponding DFD values evaluated using one vector per 7×7 inner block are shown in figure 6.3.

The gradient based motion estimator is seen to give a better estimate of 'true' motion. A good example of this is at the lower edge of the calendar where block matching yields a variety of motion directions whereas the gradient based technique correctly identifies consistent or spatially smooth motion. In general the MWBME does better at estimating true motion wherever the data is uncorrupted. Another good example is the bottom right hand sector of the rotating ball. As expected, at sites of missing data (most blotches in this frame happen to be coloured approximately white) both motion estimators are detrimentally affected as far as estimation of 'true' motion is concerned. Note that the MWBME tends to have a larger halo of incorrect vectors around blotches than does the block matching process. However this does not necessarily affect the DFD as shown in Figure 6.3. In regions which are corrupted in other frames, both estimation processes also suffer. Refer to Figure 6.1 to identify such regions. However, because block matching directly minimizes the DFD, it is able on average to generate a better match for the uncorrupted data which just happens to coincide with locations of corrupted data in the previous or next frame. This can be observed in the lower right hand sector of the moving ball and near to the the lower left hand corner of the calendar (see Figure 6.3). In both these regions, corruptions in the previous frame cause higher DFD with the gradient based motion estimator than with block matching. Since the detectors introduced tend to rely on DFD for detection, these phenomena imply that the absolute performance of the detector is influenced by the choice of motion estimator. Figure 6.3 also shows the non-motion-compensated DFD for the two frames considered, in order to put the performance of both types of motion estimator into perspective.

Despite these remarks about DFD, gradient based motion estimation is adopted for use in the subsequent chapters which deal with missing data because it tends to give a better motion field in uncorrupted regions. These can then be used more effectively to infer correct motion, or motion in subsequent frames since there is much more temporal consistency in the estimated motion field. This is important in Chapter 8 in particular. Furthermore, in later discussions (in this and the next chapter), a technique is presented for correcting motion at sites of detected missing data and this recovers robustness lost due to initial motion estimation errors. An additional factor in choosing the MWBME is that it is fast, and when dealing with frames of sizes 720×576 for television and 2880×2048 for film, speed is paramount.

Finally note that the motion estimators used in this book are all initially block based, although Chapters 7 and 8 propose refinement strategies for pixel resolution motion fields. The detectors presented above require DFD values at each pixel. These are generated by using one vector per block of pixels. *On average* this performs better than interpolating the motion vector field independent of image data using bilinear interpolation, for instance, as was initially proposed in [96, 86]. The reason is simply that at the edges of moving objects, interpolating the motion vectors from block to pixel resolution may yield a worse DFD than accepting the block based motion information because of the 'blurred' motion field which would be interpolated at the edge boundary.

6.4 Detector performance

The performance of the detectors described above is illustrated using an artificially corrupted image sequence. Real degradation is considered separately later. The sequence MOBCAL was degraded with realistic blotches synthesized according to the models described in the next chapter (see equation 7.10). The CD-ROM contains the binary files of the corrupted (BCALA) and original sequence (GC256). Only a 256×256 portion of the full frame size was used in creating the sequence. Figure 6.1 shows an interesting set of three consecutive frames. There is also one darkened image in which the position of the inserted blotches in the second frame is shown in white. Each blotch has a flat intensity profile which is set at a random grey scale that is uniformly distributed. This type of corruption is a compromise between the kind of distortion caused by digital dropout and Dirt and Sparkle which tends to have slightly softer edges depending on the resolution of the digitized image. The artificial blotches inserted here are deliberately of a fairly large size, although rarely larger than 20×20 (approximately), in order to better simulate somewhat heavier corruption than encountered in practice.

Note that in the generation of the artificial sequence, no attention was paid to suppressing the location of blotches at the same sites in consecutive frames. This does occur in the sequence and will cause a worse result with respect to detector performance than would occur in practice. In addition some blotches may be coloured only marginally differently from the original clean image at particular sites. The blotch contrast would then be so small that it would be visually the same as the original sequence. No attempt was made to filter out these blotches from the ROC measurements.

It is possible to generate many different comparisons at this stage which examine detector behaviour with respect to different constraints, such as constant threshold or constant correct detection rate etc. A useful summary of performance for detectors in general is the Receiver Operating Characteristic (ROC). At each possible threshold for each detector, the false alarm rate and correct detection rate of the detector is measured. Given an image of size $N_1 \times N_2$ pixels, the actual number of corrupt pixels as N_c, the number of pixels correctly flagged as corrupted as C_b and those incorrectly flagged F_b, then the required rates of correct detection r_c and false alarm r_f are

$$
\begin{aligned}
r_c &= \frac{C_b}{N_c} \\
r_f &= \frac{F_b}{N_1 N_2 - N_c}
\end{aligned}
\tag{6.9}
$$

The ROC is a plot of r_c vs. r_f at all possible detector thresholds. This line or characteristic gives an indication of the overall behaviour of the detector. The simplest ROC to generate will only show detection and false alarm rates with respect to *pixels* that were detected, and not with respect to the *blotches* detected. Generating a ROC which gives some idea of the portion of blotches detected is somewhat more involved and is not attempted here. It is easier to compare the detectors visually with performance on a particular frame in order to augment the description provided by the ROC for each detector.

The ROD detector requires three thresholds and the higher two were set at 30.0 and 40.0. The lowest threshold was stepped to create the ROC. The thresholds were varied through the range 2 : 2 : 50 (matlab notation), giving 24 points for each ROC for each frame. The ROCs for each frame were then averaged over 20 frames in the sequence, excepting the first since detection cannot be performed on that frame using these detectors.

Figure 6.4 shows the ROC for each of the detectors discussed above using the MWBME and hierarchical block matching. Detectors which give curves that are closer to the upper left hand corner of the plot are better than those which do not (low false alarm and high correct detection rate). High thresholds give points on the characteristic close to the bottom left hand corner of the graph, and the lowest threshold corresponds to the point at the top right hand end of each characteristic. Figure 6.5 shows a zoom on a portion of frame 9 showing in bright white the pixels flagged by three detectors SDIa, SDIp, ROD using the MWBME. Figure 6.6 shows the ROC for each of these detectors in this frame only. The thresholds were chosen to give roughly the same false alarm rate of 0.01. The detection accuracy was $0.857, 0.837, 0.836$ for the ROD, SDIp and SDIa respectively.

The ROD and SDIp are seen to be the best detectors and the basic SDI the worst. The reason is that the ROD and SDIp detector is more robust to motion estimation errors caused when the motion model breaks down, for example in the region of the spinning mobile. In the case of the ROD this is a direct consequence of its use of a spatial neighbourhood in the surrounding frames. The SDIp gains this robustness because it uses the additional assumption about the sign of the DFD. The SDIa detector uses no such neighbourhood and so is sensitive to the large DFD in regions where motion is not properly compensated. This can be seen in Figure 6.5. For SDIa and SDIp, the pixels constituting the constant false alarm rate tend to be clustered close to the objects showing difficult motion such as below the bottom left hand sector of the black counterweight in the mobile. There is slightly less clustering in the SDIp in general for this sequence. The detection field for the ROD is more randomly distributed. This is good behaviour in a detector since distortion of motion in these regions due to possibly poor reconstruction can yield a visibly worse quality sequence. Note that all detectors detect the corrupted regions with good accuracy. In a real situation, this detection field would be post processed to improve performance, see section 6.7 on real degradation.

The SDIp is interesting in that its performance can follow that of the SDIa (see figure 6.4), but for a given threshold, its false alarm rate is lower, albeit

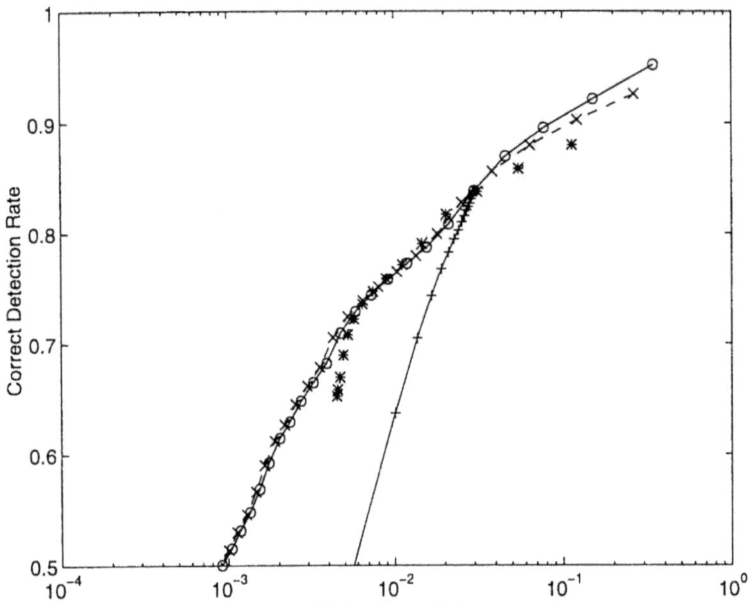

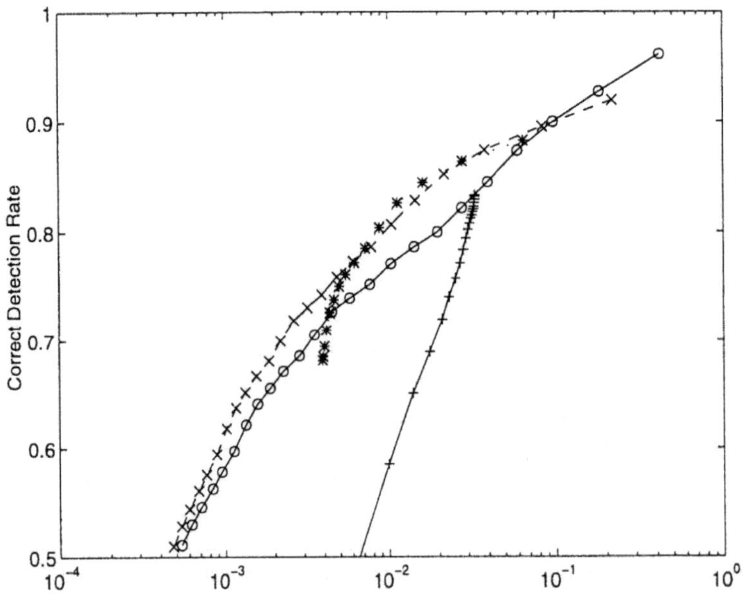

FIGURE 6.4. ROC for various detectors using MWBME (top) and integer accurate block matching (bottom). SDIa $-\circ$, SDIp $--\times$, ROD $\cdots *$, SDI $-+$.

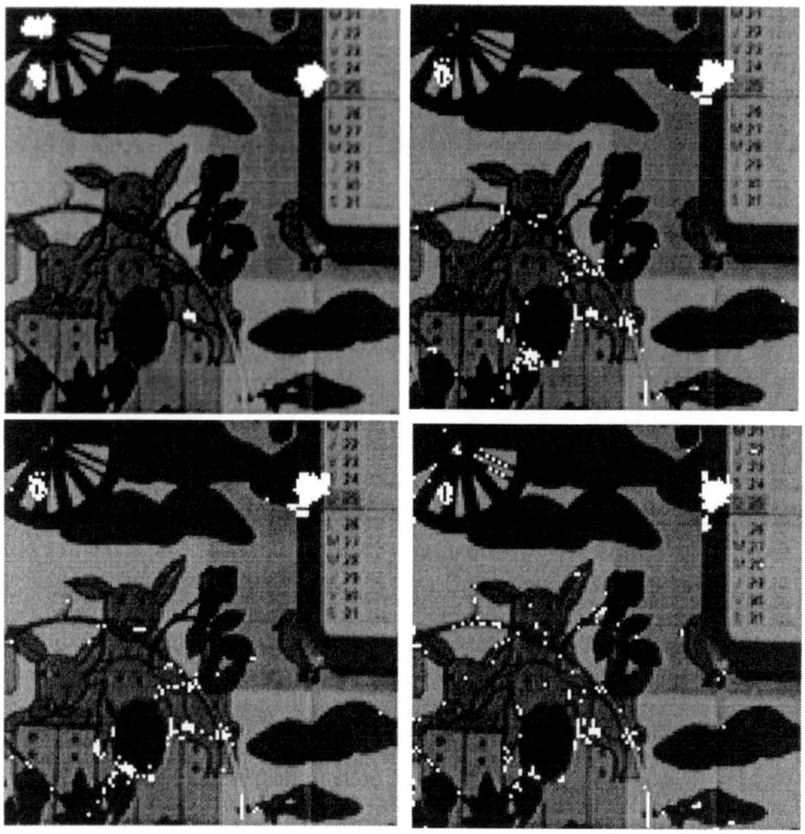

FIGURE 6.5. Clockwise from top left: Frame 9 of the degraded MOBCAL sequence (true blotch location overlayed on darkened image), SDIa detection, ROD detection, SDIp detection.

with a loss of detection accuracy. This is because on average, the sign of the DFD is the same in both forward and backward directions, when a blotch occurs. Thus on average, the use of this extra discriminatory feature is more useful than was initially expected (see [86]) although it must be accepted that it is not possible to achieve as high a correct detection rate as with the SDIa.

The SDI performs poorly compared to any of the other detectors, simply because the the range of its characteristic is severely limited. This is primarily due to the use of the low threshold t_1 used to calculate the index. In effect this limits the detector performance of the SDI and makes its use similar to a soft thresholding operation on the SDIa. As t_1 is changed, the top of the SDI curve follows the SDIa characteristic. This phenomenon is not illustrated in the figures as only $t_1 = 10.0$ was used.

The comparison between the plots in Figure 6.4 is interesting. Note first of all that a good rule of thumb is that an operating point of correct detection rate 0.8 and corresponding false alarm rate of 0.01 is just acceptable. Generally ≥ 0.8 at

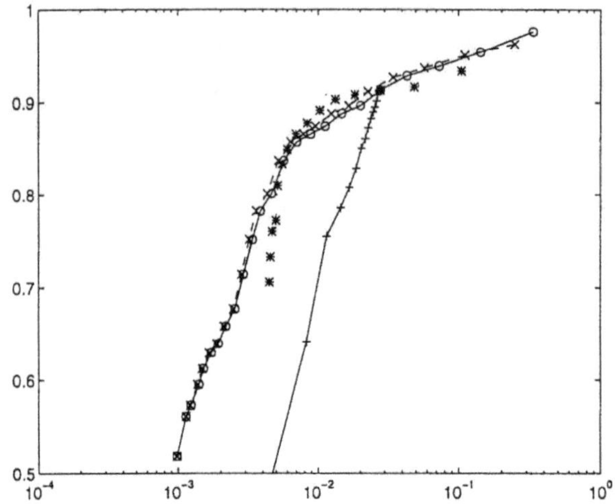

FIGURE 6.6. ROC for various detectors on frame 9 using MWBME. SDIa –o, SDIp - -×, ROD ···∗, SDI –+.

0.001 false alarm rate is desirable. These ad-hoc observations have been gained by experience with different detection/reconstruction schemes. The more robust the reconstruction scheme is to false alarms, the less important it is to have a low false alarm rate. The detectors generally perform better using integer accurate block matching than they do using gradient based motion estimation. This improvement however is not as great as the graphs initially imply given the fact that operating points with > 0.01 false alarm rate are not useful. Below this rate the improvement is about 10% on average. Note that for the MOBCAL sequence there is a considerable spread in performance characteristic for each of these detectors from frame to frame. This is most likely due to the acceleration of the mobile which makes the objects more difficult to track in some frames. Overall the differences between the detectors is emphasized when changing motion estimator. It is interesting that for the SDIa the improvement is only noticeable at very low false alarm rate or very high threshold values. At the lower threshold values the SDIa has worse performance with block matching. This is due in part to the integer accuracy of the block matching scheme used and also due to the fact that the SDIa relies solely on the magnitude of the DFD as opposed to SDIp and ROD. Also note that an important contributing factor to the differences in performance when using different motion estimators is that a large portion of the MOBCAL sequence shows simple, translational motion.

Nevertheless the reader is encouraged to note that the relative performance of the detectors remains the same regardless of the scheme used for motion estimation. It is not intended that this discourse should lead to a full investigation of the effect of many motion estimation schemes. It is sufficient for the reader to appreciate the tradeoff between performance and speed in choice of motion estimation scheme, and the relative performance of these detectors. Note that

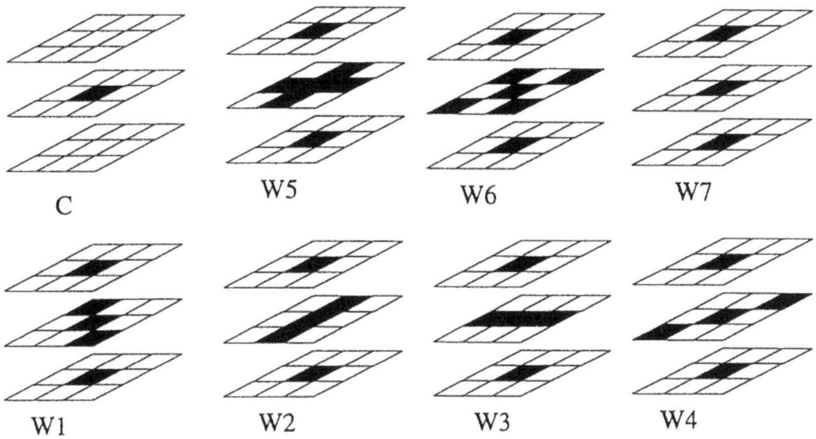

FIGURE 6.7. Sub–filter masks used for the MMFs presented by Arce and Alp et al.

in practice, given the postprocessing steps which are discussed later, the effect of the motion estimation scheme on performance is diluted.

6.5 Motion compensated median filtering

Having detected corruption the idea now is to remove it using a median filter. Chapter 4 has already outlined the work of Alp [2] and Arce [8, 6] who have both previously introduced 3D median filter structures for removing impulsive noise. The structures were introduced without a motion compensated implementation. Both Huang [66] and Martinez [117] implemented a 3 tap motion compensated median operation with good results.

6.5.1 The filters

For convenience, the filters that were reviewed in Chapter 4 are defined again here with the help of the windows in Figure 6.7. Arce's bi–directional filter (BI) was defined as follows.

$$\begin{aligned} z_l &= \text{median}[W_l] \ \forall \ 1 \leq l \leq 4 \\ z_{max} &= \text{MAX}_{1 \leq l \leq 4}[z_l] \\ z_{min} &= \text{MIN}_{1 \leq l \leq 4}[z_l] \\ \text{Bi–directional Filter output} &= \text{median}[z_{max}, z_{min}, C] \end{aligned} \tag{6.10}$$

The centre window in each 3 window set refers to a window in the current image. The filters presented used data from three frames to achieve filtering.

The ML3D MMF introduced by Alp et al is defined below:

$$\begin{aligned} z_l &= \text{median}[W_l] \ \ 5 \leq l \leq 6 \\ \text{ML3D Filter output} &= \text{median}[z_5, z_6, C] \end{aligned} \tag{6.11}$$

The filter used by Huang and Martinez was a motion compensated three tap filter (LIN), and is the median of the samples shown in window 7 of Figure 6.7.

A fourth filter, the CUBE, is also considered. Its output is the median of all the pixels in a $3 \times 3 \times 3$ volume centred on the pixel to be considered. The filter is used to compare a simple full 3D median operation to the multilevel filters, and hence to demonstrate that the 3D multilevel filters are more effective than the simple median operation even in a motion compensated application.

The motion compensated application of these filters involves selecting the pixels required for each window as directed by the motion between the frames. Hence if the pixel at $\mathbf{x} = (i, j)$ in frame n is to be filtered using the LIN filter, the output is defined as below,

$$\hat{I}(\mathbf{x}, n) = \text{median}[\ G_n(\mathbf{x}),\ G_{n-1}(\mathbf{x} + \mathbf{d}_{n,n-1}),\ G_{n+1}(\mathbf{x} + \mathbf{d}_{n,n+1})] \tag{6.12}$$

where $\mathbf{d}_{n,n-1}$, $\mathbf{d}_{n,n+1}$ are the motion vectors mapping information at $\mathbf{x}$ in the current frame into the previous and next frames. Similarly for the MMF's that were defined above. The filters are applied to every pixel in the image.

6.6 An improved spatio–temporal MMF

The MMF's introduced so far cannot remove an impulsive distortion which is larger than the size of the spatial median window in the current frame. Consider the ML3D filter in a situation where it operates on some occurrence of Dirt or Sparkle which is of size 3×3 pixels in the current frame. Assume for simplicity that the image itself is just a uniform grey level covering a large area. When the filter is applied to the centre pixel in the distorted region, the portion of the median subfilter windows in the current frame would be filled with 'blotched' pixels. In each of the two spatio–temporal windows used in ML3D, there are 7 pixels involved, 5 of which are in the current frame. The output of the filter is the 4th sample in the ranked distribution of these 7 pixels. If 5 are corrupted, then it is impossible to yield a median output which is not one of these corrupted pixels. Since Dirt and Sparkle usually occur as a region of high contrast which is uniform, the output pixel would be of the same intensity.

However, this would only occur inside the distortion. At the edges of the degradation, where less of the current frame window is occupied by distorted pixels, the output is more likely to be a suitable image value. Therefore, both BI and ML3D will partially remove large sized distortions. It would take a recursive[3] implementation of these filters to completely remove artefacts such as these. Such a recursive implementation is likely to affect the detail in the rest of the frame adversely. Incorporating more information from the surrounding frames would be more successful in removing distortion provided the Blotch

[3] Apply the filter again to the output image.

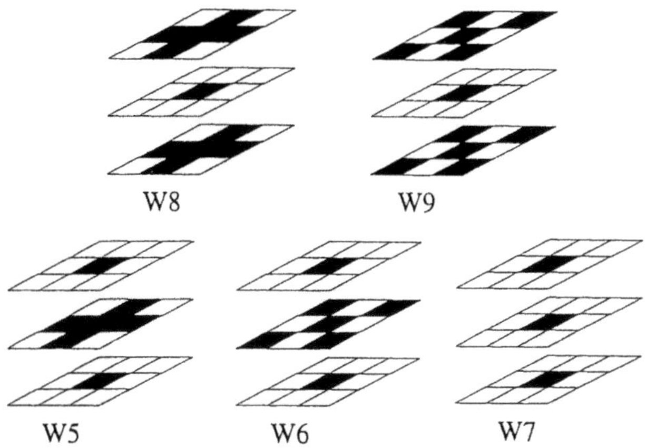

FIGURE 6.8. Sub-filter masks used for the new MMF: ML3Dex.

does not occur in the same place in successive frames. In general this is true for Dirt and Sparkle.

The proposed filter structure is therefore defined with the help of the sub-filter windows shown in Figure 6.8. The MMF is similar to that of Alp et al [2] and is called ML3Dex for Extended ML3D. The output of the filter is defined below.

$$\begin{aligned} z_l &= \text{median}[W_l] \quad 5 \leq l \leq 9 \\ \text{ML3Dex Filter output} &= \text{median}[z_5, z_6, z_7, z_8, z_9] \end{aligned} \tag{6.13}$$

Two additional windows have been incorporated which contain minimal information from the current frame and extensive information from the outer frames. Consider once more the situation with a large 'Blotch'. It can be seen that although the windows $W5$ and $W6$ would output Blotch values inside the degraded area, the three windows $W7$, $W8$, $W9$ would still be able to yield a correction using image data. Therefore, provided that the Blotch does not occur at the same position in the three frames, the 5 value 2nd level median is not dominated by scratch data. Furthermore, the additional information improves the scratch rejection capacity of the filter. This is accompanied by a subsequent loss of detail preservation.

6.6.1 *Motion interpolation*

At sites of missing data there are *two* pieces of missing information, the image data *and* the motion field. Figure 6.2 shows clearly that at sites of missing data the motion field is likely to be completely wrong. The previous discussions have indicated that this does not affect the detectability of the missing data. Nevertheless, if the corrupted motion field was used in a motion compensated application of any of the filters discussed above for reconstruction of the missing data, the result is likely to be useless. This effect is particularly severe with large corrupted regions. A good example is shown in Figure 6.10. It shows a large blotch in frame 13 of the corrupted MOBCAL sequence being corrected using the ML3Dex filter *without* attempting to correct the erroneous motion field in that area. The filter was only used at the true sites of corruption to separate the detection performance from interpolation performance. The corrupted motion field derived from the degraded sequence is overlayed on the reconstruction. It is clear that the result is useless and in some ways worse than the degraded image itself. Therefore it is necessary to design a mechanism for interpolating the motion vector field at the location of the blotch or equivalently, to correct the motion estimate there. Also important is to ensure that the interpolation process is robust enough to ignore or to *de-emphasize* data collected using an incorrect motion vector. A robust interpolation process is necessary even when the blotch is not substantial since 100% confident motion estimates cannot be guaranteed. At this point, the author asks forgiveness for deviating a little from the 'heuristic' nature of this chapter to introduce a coherent solution to the motion interpolation problem. This solution was first introduced in [83].

It is assumed that the image sequence follows the usual translational model as discussed in Chapter 2 and restated in equation 6.1. The conditional probability of the vector $\mathbf{d}_{n,n-1}(\mathbf{x})$ given the frames I_n and I_{n-1} and some neighbourhood subset of motion vectors, $\mathbf{S}_n(\mathbf{x})$, can be written using Bayes theorem as

$$p(\mathbf{d}_{n,n-1}(\mathbf{x})|I_n, I_{n-1}, \mathbf{S}_n(\mathbf{x})) = \frac{p(I_n, I_{n-1}|\mathbf{d}_{n,n-1}(\mathbf{x}), \mathbf{S}_n(\mathbf{x}))p(\mathbf{d}_{n,n-1}(\mathbf{x})|\mathbf{S}_n(\mathbf{x}))}{p(I_n, I_{n-1}|\mathbf{S}_n(\mathbf{x}))} \tag{6.14}$$

The situation between frames $n, n+1$ can be written similarly. The most obvious choice for the likelihood, in view of equation 6.1, is a zero mean Gaussian distribution of DFDs, with variance σ_e^2. However, for interpolation just one vector per block containing the blotch is required. This implies that the vector field considered is defined on the block lattice and not the pixel lattice. Therefore the likelihood should contain a contribution from every pixel in the block centered on location $\mathbf{x}$ as follows:

$$p(I_n, I_{n-1}|\mathbf{d}_{n,n-1}(\mathbf{x}), \mathbf{S}_n(\mathbf{x})) = \frac{1}{\sqrt{2\pi\sigma_e^2}^{N_w}} \exp - \left(\frac{1}{2\sigma_e^2} \sum_{\mathbf{x}\in B} w^2(\mathbf{x})[I_n(\mathbf{x}) - I_{n-1}(\mathbf{x} + \mathbf{d}_{n,n-1}(\mathbf{x}))]^2 \right) \tag{6.15}$$

where B is the set of all locations in the block. A weight $w(\mathbf{x})$ is associated with each DFD measurement. This weight is set to 0 wherever the pixel site is flagged as corrupted by the detector and set to 1 otherwise. N_w is the sum of these weights over the block.

It is assumed that the motion vectors in frame n constitute a Markov Random Field (MRF)[4]. In the manner of e.g. [98, 162, 118] a Gibbs Energy prior is used

[4]It is assumed that the reader is familiar with the concept of Markov Random Fields. See [118, 172, 47, 98] and the discussion in Chapter 2.

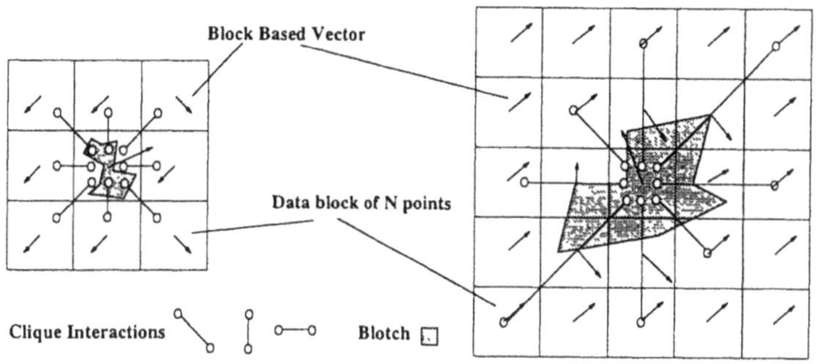

FIGURE 6.9. Left: Neighbourhood and cliques used for $p(\mathbf{d}_{n,n-1}(\mathbf{x}))$. Right: Altered neighbourhood used with a large blotch.

for $p(\mathbf{d}_{n,n-1}(\mathbf{x}))$ as follows

$$p(\mathbf{d}_{n,n-1}(\mathbf{x})|\mathbf{S}_n(\mathbf{x})) \propto \exp - \left(\sum_{\mathbf{v}\in\mathbf{S}_n(\mathbf{x})} \lambda(\mathbf{v})[\mathbf{d}_{n,n-1}(\mathbf{x}) - \mathbf{v}]^2 \right) \tag{6.16}$$

where $\mathbf{v}$ is each vector in the neighbourhood represented by $\mathbf{S}_n(\mathbf{x})$, and $\lambda(\mathbf{v})$ is the weight associated with each clique. The situation is illustrated in the left hand portion of Figure 6.9. Note that the cliques employed here assume first order interactions even though the eight connected neighbourhood can involve some second order cliques [118].

In order to discourage 'smoothness' over too large a range, $\lambda(\mathbf{v})$ is defined as follows

$$\lambda(\mathbf{v}) = \Lambda \frac{N_w}{|\mathbf{X}(\mathbf{v}) - \mathbf{x}|} \tag{6.17}$$

where $\mathbf{X}(\mathbf{v})$ is the location of the block providing the neighbourhood vector $\mathbf{v}$ and Λ is a constant which scales the strength of the interaction. This location is measured in terms of block lengths. N_w is the number of uncorrupted pixels in the block. Note that there is a trade off between the values of Λ and σ_e^2. Large Λ encourages motion vector smoothness, and small σ_e^2 encourages vectors which minimize the DFD.

It is required to choose the correct motion vector at each block site given that some detector has already flagged sites of suspect distortion. This implies manipulating expression 6.14. It is true to say that equations 6.14, 6.15, 6.16 are sufficient to estimate the motion field itself [118, 98]. This is explored in the next chapter. A simple correction strategy is employed here.

In practice, after the use of a typical motion estimator and the blotch detector, it is already possible to make a very confident assessment of the locations of corruption. In which case, there is no further interest in the uncorrupted regions. Furthermore, in areas which are not corrupted, it is found that the initial motion estimates are quite accurate enough to be used for subsequent processing. Therefore, in order to generate a fast MAP estimate for the vector at the missing block, it is assumed that the vectors in the neighbourhood blocks which contain less than 10% corruption are confident estimates of the flow field there.

Fast solution

Rather than relax the vector field around the corrupted location using, for example, the Gibbs Sampler [118], in practice it is found sufficient to reduce the solution search space[5] to the vectors in the neighbourhood of the blotch. Each vector in turn is tested as a candidate solution to the corrected vector by substitution in equation 6.14. The candidate which maximizes expression 6.14 is selected as a working approximation to the MAP estimate. Note that the denominator of equation 6.14 is constant and can be ignored. An estimate for σ_e^2 is made from the measured weighted DFD for each vector candidate. In the case where the DFD does not vary much with different candidate vectors (possibly because much of the current area is corrupted by a blotch) the operation reduces to a type of weighted vector median [10].

A problem arises when the blotch is large enough to engulf several neighbouring blocks. In this case the candidate vector set is chosen from an altered neighbourhood arrangement so that the vectors in the neighbourhood are associated with blocks which contain less than 20% corruption. This 'acceptable' corruption rate was chosen by experience gained from processing several image sequences. This strategy is illustrated in Figure 6.9.

The steps for motion interpolation at a site of missing data are therefore

1. Delineate the contour of the missing region detected. Use some kind of boundary tracking algorithm.

2. Centre a bounding box on the missing region containing at most 20% corrupted data. The percentage corruption that can be tolerated depends on the motion estimator. The safest assumption is not to use any candidate vector from blocks in which there is any suspect corruption of image data.

3. Generate candidates for corrected motion in that box from the eight nearest block based motion vector neighbours which had *uncorrupted* (or almost so) blocks. Generate four more candidates by perturbing the current (erroneous) candidate by ± 1 pixel.

4. Choose the interpolated motion vector to be that candidate which maximizes the expression 6.14 above.

5. This process may be applied iteratively over the image.

Both the forward and backward motion fields are treated in this way.

This fast approximate MAP estimate for the corrected vector is found to be extremely effective. The resulting improved interpolation (using ML3Dex

[5] A similar simplification was made by Stiller [162] for motion field smoothing.

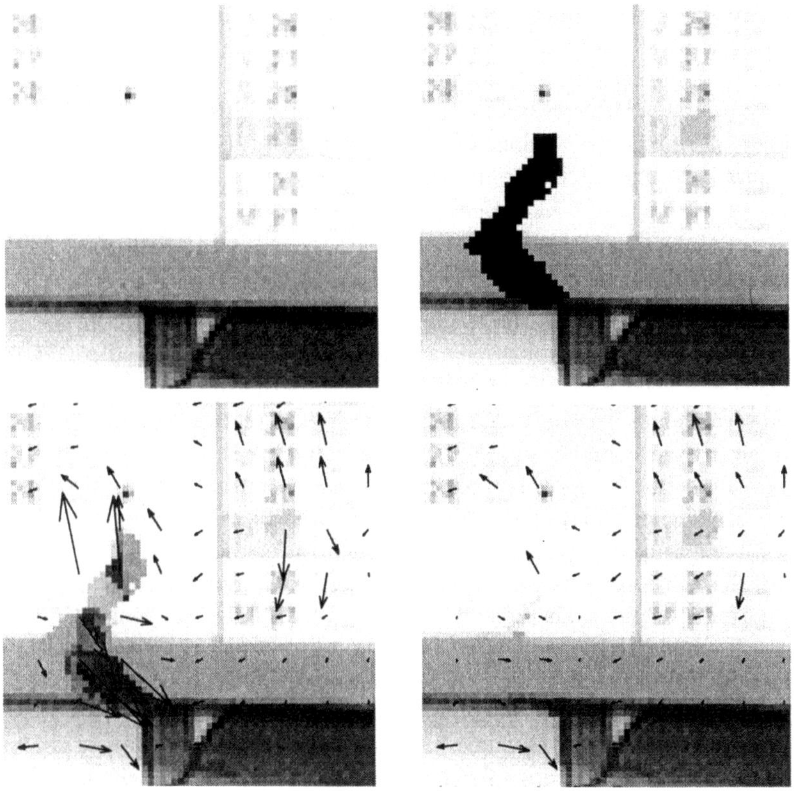

FIGURE 6.10. Top Row: Original (clean) frame 13 from MOBCAL (left), degraded frame 13 (right) showing large black blotch and smaller grey blotch to the upper right of the large blotch. Bottom Row: Motion vector field from MWBME superimposed on reconstruction using ML3Dex *without* motion correction (left) and *with* motion correction (right). The images have been brightened to improve contrast.

after motion correction) of the image shown in Figure 6.10 is illustrated in the bottom right hand image in that figure. A value of $\Lambda = 1.0$ was used and the initial motion estimates came from the MWBME with the same parameters as described previously. Figure 6.10 also shows the corrected motion field ($\mathbf{d}_{n,n-1}$ only) after just one pass of the algorithm presented above. The motion field was interpolated only where corruption occurred so blocks which did not contain corruption were left with their original motion estimate. This of course does not affect the subsequent interpolation since the uncorrupted regions were not processed. The interpolated motion field in the region of the large black blotch does not show 'true motion' everywhere. This can be improved with multiple iterations of the algorithm above. Even so, the interpolated vectors are very close to the 'true' motion for the smaller blotch, and at the extremities of the larger one. Furthermore the motion correction is good enough to dramatically improve the visual quality of the reconstruction.

It is true that in regions where there are motion discontinuities, this rough solution to the vector field correction problem is not sufficient. This can be surmounted by incorporating discontinuity states into the motion prior (See [118, 98]). There is then the additional problem of the required configuration of the image model for interpolation. Chapter 8 considers these issues in the case of more severe corruption of the image sequence.

6.6.2 Interpolation performance

Figure 6.11 shows quantitatively the performance of the median filters presented above with respect to RMS error of interpolation. The degraded MOBCAL sequence was used with the MWBME for generating initial motion estimates. The interpolation was only engaged at the sites of true corruption so as to decouple the performance of any particular detector that may have been used. Motion correction as described above was used to correct the motion information at the site of corruption *before* each median filter was applied. The RMS error was measured only over the sites which were reconstructed.

As far as RMS error is concerned the filters with more spatio-temporal support perform better than those without. Thus ML3Dex, LIN and CUBE perform well while ML3D and the BI filter do not remove much of the distortion. The CUBE filter performs slightly better than the others because of its heavy smoothing effect which causes low RMS errors. Visually, the LIN filter can actually perform quite well since it is best able to preserve image sharpness. However it has no tolerance for motion estimation errors, whereas the spatial extent of ML3Dex gives some robustness to these errors. The CUBE filter causes too much blurring of image detail in general. The MOBCAL sequence is not a good one for demonstrating the lack of sharpness in general using median type filters for interpolation. Much of the sequence consists of large flat areas for which the median filter is well tuned. However, Figure 6.12 shows visually the difference in sharpness in interpolation at one site where it is visible, the right hand bearing of the mobile. Here there is a bright reflection which is completely removed by the CUBE, and well preserved by LIN (because the motion correction was effective). The ML3Dex output is some kind of compromise between the two as expected. The ML3Dex is thus the safest to use in practice.

Figure 6.13 once again emphasizes the importance of motion correction. All the filters show improved performance when this is undertaken. The reader should appreciate that this is a vital step in designing systems for missing data reconstruction.

6.7 Real degradation

The processes discussed so far have been explored with respect to their performance on an artificially degraded sequence. In practice for real degradation,

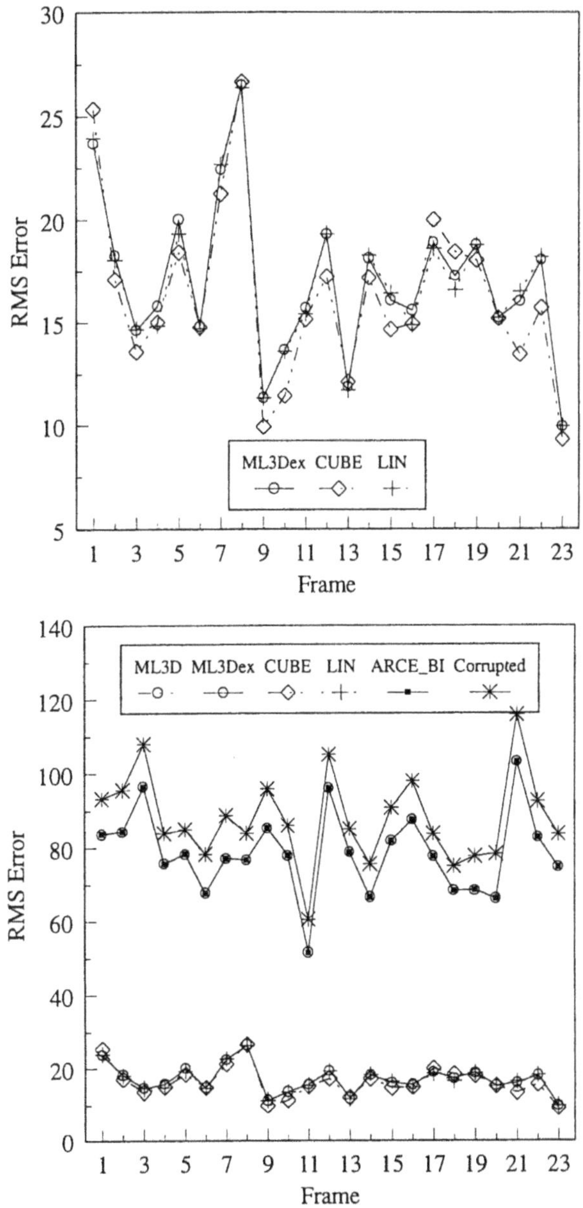

FIGURE 6.11. RMS Error for each frame of the degraded MOBCAL sequence reconstructed using several median based filters *with* motion correction. Top plot shows a comparison between the three more successful filters. Bottom plot compares all filters used as well as corrupted original.

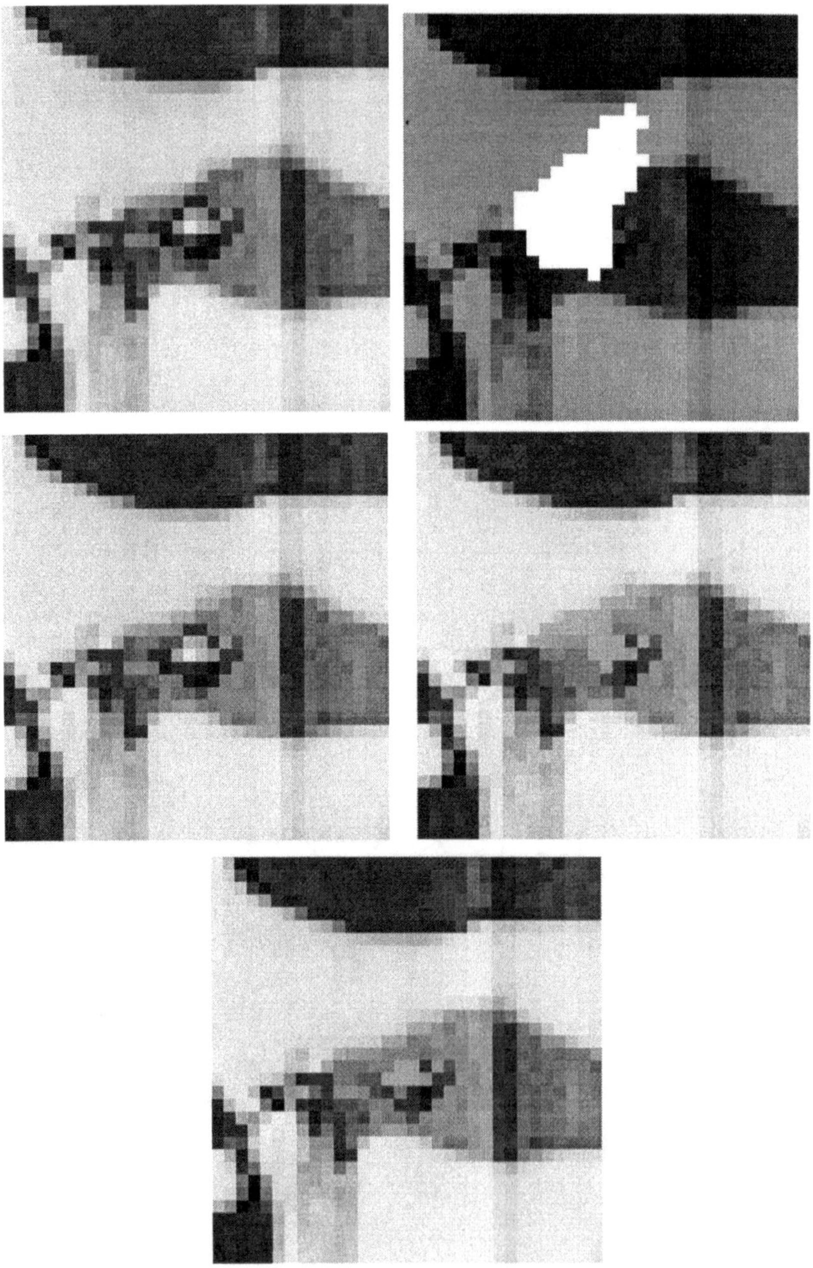

FIGURE 6.12. Top Row: Portion of frame 17 from clean MOBCAL sequence (left), corresponding location of degradation (white) of degraded MOBCAL (right). Middle Row: Interpolated region using LIN (left) and CUBE (right). Bottom: Interpolated region using ML3Dex.

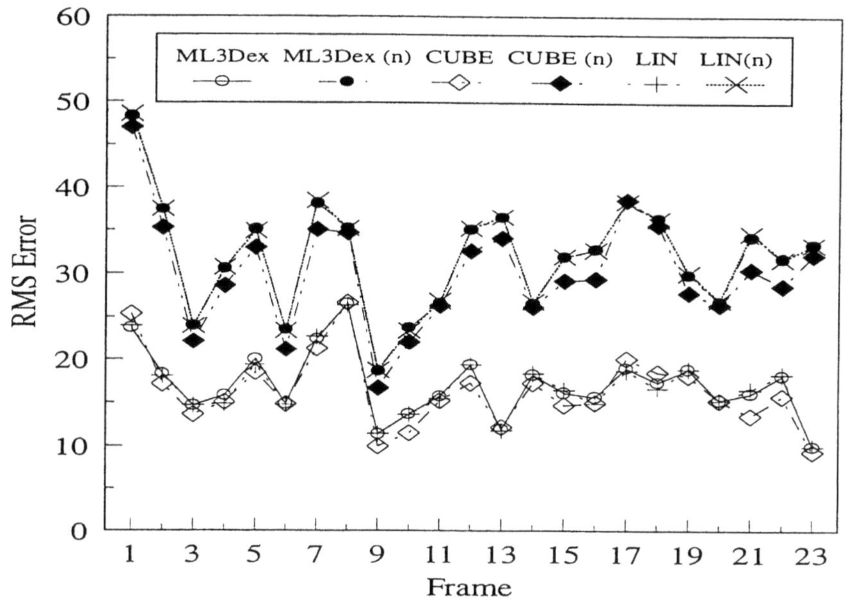

FIGURE 6.13. RMS Error for each frame of the degraded MOBCAL sequence reconstructed using several median based filters *with* and *without* motion correction. Filters implemented *without* motion correction are denoted '(n)' in the legend.

the level of corruption is on average not as severe as this, although the treatment of severe degradation is important. Figure 6.14 shows three frames from a sequence showing real degradation. The main blotches to be detected and removed are boxed in white. As can be seen, the edges of real blotches in film tend to be less sharp than in the artificial case. For video, however the dropouts tend to be as sharp as those shown in the artificial sequence. In a real system some attention must be paid to the fact that the actual extent of the blotch may be larger than that indicated by the temporal detectors presented here.

This problem can be solved simply by post processing the binary detection field with a *dilation* operator to give a pessimistic estimate of the blotch extremities. The dilation process can be achieved by one of two mechanisms. The simpler strategy is to treat the binary field as a continuous valued image and filter it with a 3×3 (for instance) FIR filter. The output is then thresholded to reset the field to 1 wherever the output value exceeds 0.5 for instance, or 0.0 for a very large dilation effect. A more controlled method is to apply a 3×3 structuring element and dilate the field morphologically. Thus each pixel in the input field is scanned and at each site that is set to 1 and also not surrounded by eight neighbours which are also set to 1, all the pixels in the 3×3 block are set to 1.

The effect of this dilation operation is shown in Figure 6.15. This figure shows the original degraded frame and the result of detection using ROD and SDIa using the MWBME. Also shown is the resulting field after dilation using the process described above. The dilated field is seen to cover the corrupted region

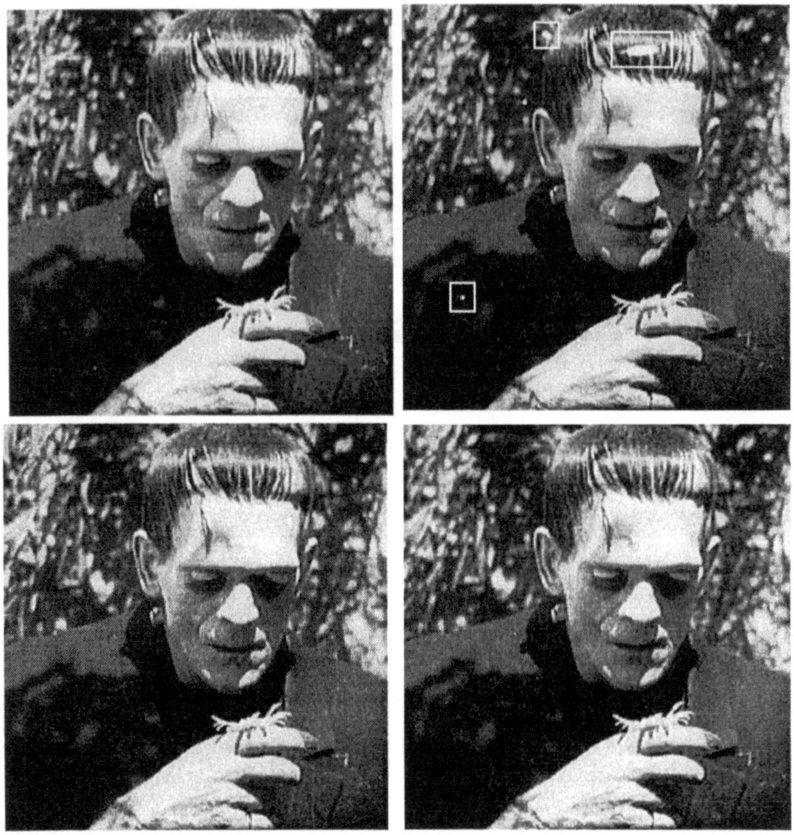

FIGURE 6.14. Clockwise from top left: Frames 1,2 of the FRANK sequence, Frame 2 restored using SDIa and ML3Dex with motion correction, Frame 3 of FRANK.

more completely. The advantage of this step is also that it is not necessary for the detector to detect the entirety of the blotch in the first place, as shown in the figure. As long as enough of the blotch is detected for the dilation operation to spread the detection field completely over the blotch, a successful interpolation is possible. This then implies that higher thresholds than that indicated by Figure 6.4 can be used and still generate a good detection field. The end result is fewer false alarms than would otherwise occur at low thresholds. The figure shows that it is possible for all the detectors discussed here to behave equally well in a real situation where the user has some control over the thresholds to be applied. In this instance the SDIp shows slightly fewer false alarms than SDIa. The ROD is more robust to the false alarms generated by the moving petals and also detects the boundary of the blotch more accurately than SDIa.

Figure 6.16 shows the motion field (a zoom on the largest corruption) before and after correction using the methods presented above. The effect of correction is to make the motion field in the corrupted region similar to that in the surrounding areas. Two interpolations are shown in Figure 6.17 using the ML3Dex

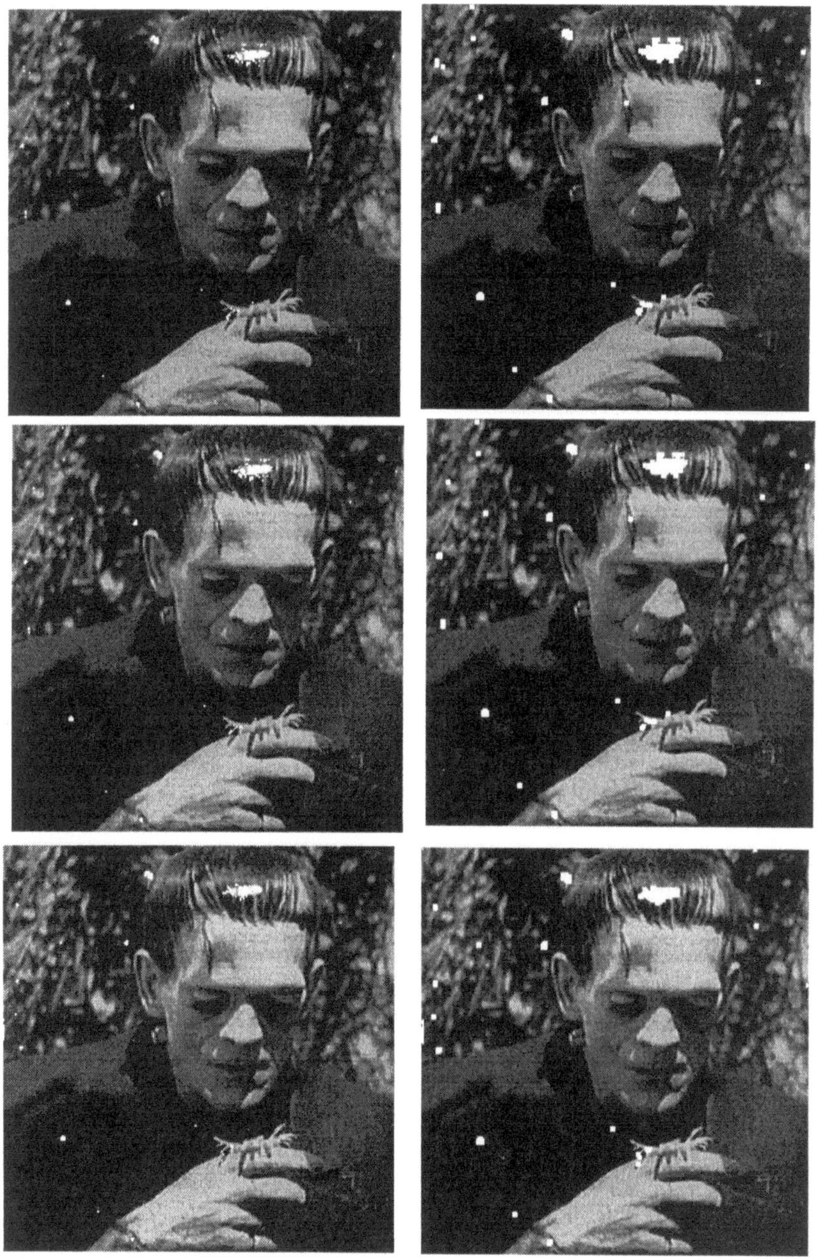

FIGURE 6.15. Detection on frame 2 of FRANK superimposed on original image. Left column: Raw detection field, Right Column: Dilated field. Top to bottom: Detection using SDIa ($E_t = 27$), SDIp ($E_t = 27$), ROD ($t_1 = 25$).

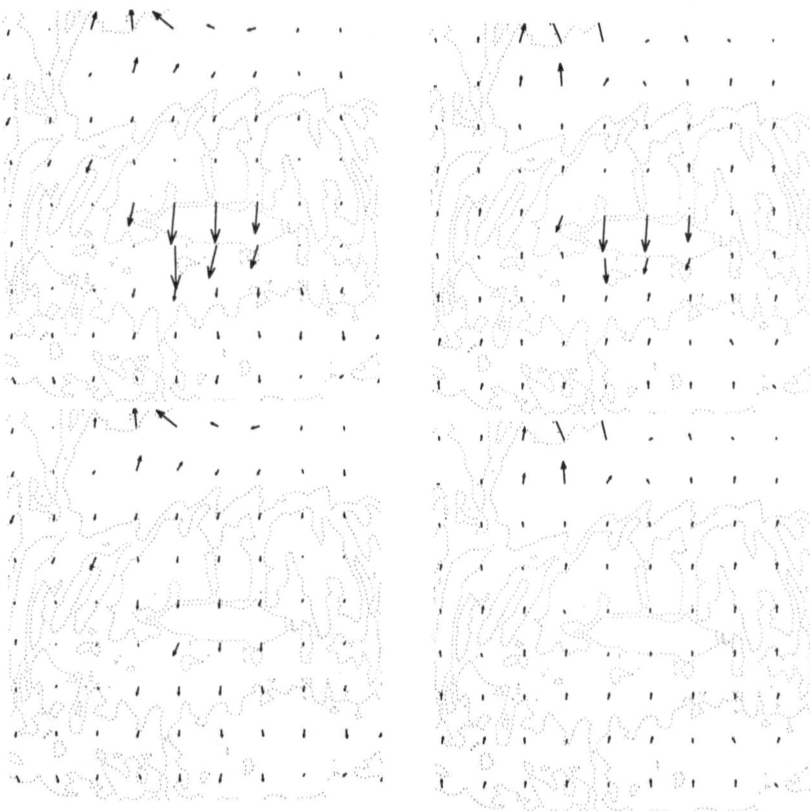

FIGURE 6.16. Zoom of estimated motion fields using MWBME on FRANK sequence. The vector fields are superimposed on a contour plot of the FRANK image which indicates the major distortion in the middle of the picture. The top row shows the fields $\mathbf{d}_{n,n-1}$ (left), $\mathbf{d}_{n,n+1}$ (right) estimated from the degraded data. The bottom row shows the corresponding fields after motion correction is applied.

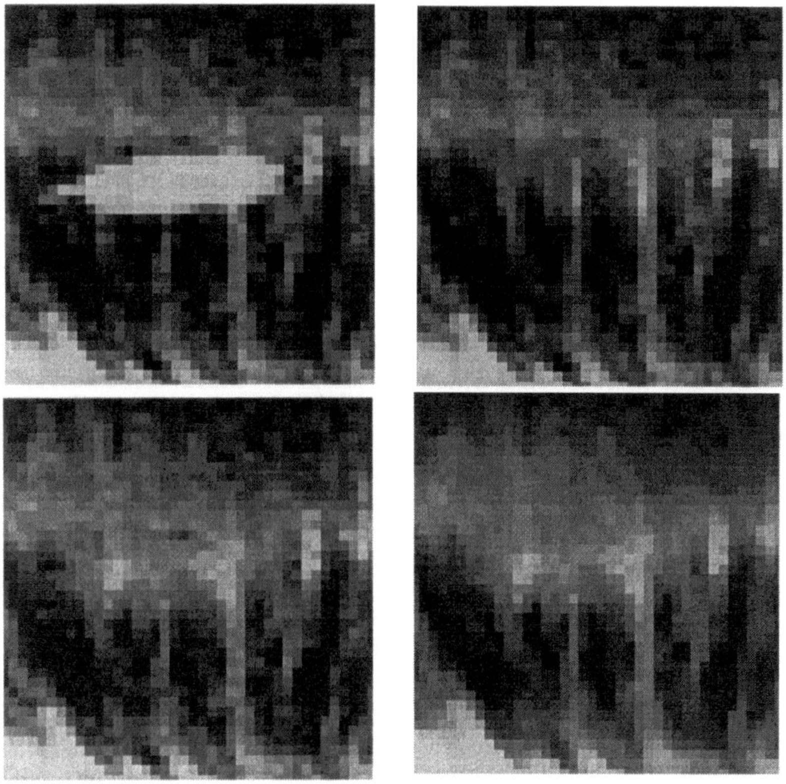

FIGURE 6.17. Zoom on reconstructed FRANK in region of major distortion. Clockwise from top left: Original degraded image, Restored using SDIa+ML3Dex with motion correction, Restored using ML3Dex globally *without* motion correction, Restored using SDIa+ML3Dex *without* motion correction.

with and without motion correction. This image shows much more texture than MOBCAL and the poor performance of ML3Dex *without* motion correction is clearly visible. Figure 6.17 also shows the result of using ML3Dex globally at every site in the image. The faded or blurred result should convince the reader that a global filtering process for removing missing data is undesirable.

6.8 Final comments

Figure 6.18 shows a block diagram of a controlled interpolation system that engages the median filter interpolation only when an impulse is detected. It is a summary of the three step system discussed in this chapter. The idea has been to detect missing data, then reconstruct the motion field in the missing regions and finally to reconstruct image data in these areas. The suitable positions for pre and post processors are also indicated. The system is a viable proposition for real-time implementation since each block is of relatively low complexity.

Note that the motion estimator remains the most complex part of this system. The system has been shown to be effective and has evolved from the simple two step 'detect then interpolate' processes discussed in [96, 91, 90] over the last six years.

An important point to be made is that in real sequences it is the false alarm rate which causes the most problems. They tend to occur in regions of difficult-to-estimate motion and subsequent interpolation stages can destroy the original motion. The effect is visually disturbing *if* the motion is large. This implies that it is sometimes better to err on the side of caution in setting the thresholds for the detectors in frames with fast moving objects in the motion picture.

There are still improvements which could be made to this framework. One clear desirable feature is that the motion correction stage should be able to assist in generating a better detection which then has a knock-on effect in generating better interpolation. This kind of joint estimation/feedback framework for all the unknowns at the same time is explored using a Bayesian methodology in the next chapter. That strategy reduces the need for user interaction to some extent by removing the need for thresholds on the DFD. It further introduces the use of specific models for the degradation concerned.

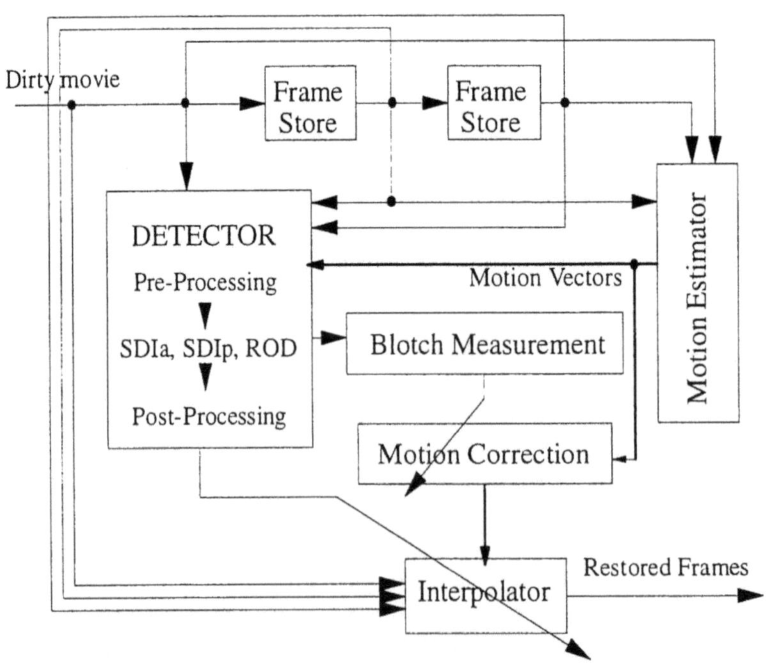

FIGURE 6.18. Components of the missing data reconstruction system.

7

Model Based Reconstruction for Missing Data

The previous chapter has introduced a number of simple but effective techniques for missing data removal. Those techniques were designed by exploiting the particular features of blotches, mainly temporal discontinuity. The methods employed were direct, in that they examined quantities which could be specifically related to the feature in question. The systems contained modules which addressed each problem as it arose. For example it was recognized that motion correction was vital for better image reconstruction and so a method was developed to correct motion, and kept separate from the subsequent image reconstruction. This chapter presents a framework for coherent decision making with respect to the missing data problem. It results in a process which identifies the missing regions at the same time as correcting the motion *and* image information. The framework employs models for the degradation process as well as the image sequence, in a quantitative fashion, in contrast to their use as 'mission statements' in the previous chapter. In some sense the model based technique presented here is a generalization of the methods introduced in the previous chapter, and the connections are explored after the technique is presented.

The chapter begins with the model statements and moves directly into the use of Bayesian inference to describe the blotch removal process in broad terms. The specific manifestation of the various probability distributions required are then described and some previous work is related. A scheme for solving the resulting optimization problem is then described. Results are presented which illustrate the difference between *typical* reconstruction and *maximal* reconstruction. The chapter also describes options for fast algorithms for implementing the model based interpolation process which is at the heart of the proposed system. These fast algorithms make the entire system much more practical than was previously possible.

7.1 The model for degradation

The observed degraded sequence may be modelled by a switching process such that the data that is the corruption is switched into the original clean data at specific sites. This switching process is monitored by the field (or image) $b(\mathbf{x})$. A corrupted site in $b(\mathbf{x})$ is set to 1, otherwise the site is 0. The model for the observed, corrupted image $G_n(\mathbf{x})$ is thus

$$G_n(\mathbf{x}) = (1 - b(\mathbf{x}))I_n(\mathbf{x}) + b(\mathbf{x})c(\mathbf{x}) \tag{7.1}$$

$c(\mathbf{x})$ is the corruption data and is switched in whenever $b(\mathbf{x}) = 1$.

This model was first introduced by Vaseghi [182] and employed in a Bayesian framework by O'Ruanaidh [151], Godsill and Rayner [54, 51]. They employed the model for click detection and removal in audio. In essence, the missing data detection problem is that of estimating $b(\mathbf{x})$ at each pixel site. The replacement model was employed deterministically in [96, 91] for image sequences. This idea of employing 'replacement noise' in a Bayesian framework for image sequences was first employed by Morris and Fitzgerald [119, 121, 120]. The connections between that work and the process presented here are illustrated later in this chapter.

7.2 The image data model

In chapter 2 the 3DAR model was introduced as a stochastic image sequence model. The stochastic nature of the model has interesting implications for textural interpolation even though it is unclear if it has any physical significance for image formation. The reader is warned against comparisons with the use of the AR model for modelling speech. In that instance the model has clear physical implications in terms of the resonant nature of the speech production process. For image sequences the model is perhaps better thought of in its prediction mode, as indicated in Chapter 2. As a predictor, it allows implicit and quantitative use of spatial and temporal information in the image sequence.

The model is defined on the original, uncorrupted image data. Its form is repeated here for convenience.

$$I_n(\mathbf{x}) = \sum_{k=1}^{P} a_k I_{n+\mathbf{q}_k^t}(\mathbf{x} + \mathbf{q}_k^s) + e(\mathbf{x}) \tag{7.2}$$

A particular pixel in frame n at site $\mathbf{x}$ is therefore predicted by a linear combination of pixels in the current and surrounding frames plus an added excitation or residual error $e(\mathbf{x})$. The P coefficients of the model, which define the linear combination, are denoted a_k for $k = 1 \dots P$. The pixels used in the prediction are called the support or neighbourhood of the model and are mapped by the P offset vectors $\mathbf{q}_k$, which have a spatial component $\mathbf{q}_k^s$ and a temporal component $\mathbf{q}_k^t$. An example model configuration is shown in Figure 2.5 in Chapter 2. In equation 7.2 it is assumed that the image data is already compensated for motion as indicated in Chapter 2, this keeps the notation simple. The prediction error or excitation, $e(\mathbf{x})$, is a Gaussian random variable having variance σ_e^2 and zero mean. This is described with the notation $e(\mathbf{x}) \sim \mathcal{N}(0, \sigma_e^2)$. It is assumed that these residuals are uncorrelated. (In Chapters 2 and 3 the symbol $\epsilon(\cdot)$ is used for the residual.)

7.3 A Bayesian framework for joint detection and reconstruction

It is required to solve for the unknowns $b(\mathbf{x}), c(\mathbf{x}), I_n(\mathbf{x})$ given the corrupted data $G_n(\mathbf{x})$. In addition, note that the motion information, $\mathbf{d}(\mathbf{x})$ is also an unknown since it is not available at the corrupted sites[1]. The other 'hidden' unknowns are the P AR model coefficients a_k, and σ_e^2. To effect a solution, it is required to manipulate the conditional probability $p(\theta | G_{n-1}(\mathbf{x}), G_n(\mathbf{x}), G_{n+1}(\mathbf{x}))$, which is the probability of all the unknowns (denoted by θ) given the observed data. Note that an implicit assumption is that the data is never corrupted in the same location in consecutive frames, therefore it is reasonable for the expression to be represented by $p(\theta | I_{n-1}(\mathbf{x}), G_n(\mathbf{x}), I_{n+1}(\mathbf{x}))$, since in the locations of interest, $G_{n-1}(\mathbf{x}) = I_{n-1}(\mathbf{x})$. The unknowns may then be identified as those which maximize that conditional expression resulting in what will be called *maximal* estimates (encompassing both Maximum Likelihood (ML) or Maximum-a-Posteriori (MAP) estimates). Alternatively random samples for the unknowns can be generated and used as *typical* estimates. Least Squares or MMSE estimates can be related to the ML or MAP solutions.

Proceeding in a Bayesian fashion, the conditional may be rewritten in terms of a product of a likelihood and a prior as follows:

$$p(\theta | I_{n-1}(\mathbf{x}), G_n(\mathbf{x}), I_{n+1}(\mathbf{x})) \propto p(G_n(\mathbf{x}) | \theta, I_{n-1}(\mathbf{x}), I_{n+1}(\mathbf{x})) p(\theta) \tag{7.3}$$

This posterior expression may be further expanded to yield

$$\begin{aligned} p(G_n(\mathbf{x}) | \theta, I_{n-1}(\mathbf{x}), I_{n+1}(\mathbf{x})) p(\theta) &\propto p(G_n(\mathbf{x}) | I_n(\mathbf{x}), c(\mathbf{x}), b(\mathbf{x})) \\ &\times p(I_n(\mathbf{x}) | \mathbf{a}(\mathbf{x}), \sigma_e^2, \mathbf{d}(\mathbf{x}), I_{n-1}(\mathbf{x}), I_{n+1}(\mathbf{x})) p(b(\mathbf{x}) | B) p(c(\mathbf{x}) | C) \\ &\times p(\mathbf{d}_{n,n-1}(\mathbf{x}) | D) p(\mathbf{d}_{n,n+1}(\mathbf{x}) | D) p(\mathbf{a}(\mathbf{x})) p(\sigma_e^2) \end{aligned} \tag{7.4}$$

---

[1]The subscripts have been dropped in the notation for the motion vector $\mathbf{d}(\mathbf{x})$ for simplicity and to emphasize that the arguments apply to both backward and forward motion information.

Note that the model coefficients, denoted by a vector $\mathbf{a}$, are allowed to vary across the image and so are a function of position. This helps account for spatial statistical non-stationarity. As usual, $\mathbf{d}_{h,k}(\mathbf{x})$ denotes the motion vector at site $\mathbf{x}$ mapping frame h into frame k. The symbols D, B, C denote local neighbourhoods of the respective variables. It is now necessary to assign the exact functional forms to the various terms.

7.3.1 The corruption likelihood

The first distribution on the right hand side of equation 7.4 is derived from the model for degradation stated in equation 7.1. Thus

$$p(G_n(\mathbf{x})|I_n(\mathbf{x}), c(\mathbf{x}), b(\mathbf{x})) \propto \delta(G_n(\mathbf{x}) - (1 - b(\mathbf{x}))I_n(\mathbf{x}) - b(\mathbf{x})c(\mathbf{x})) \tag{7.5}$$

The delta function is a direct consequence of the switching behaviour of the corruption process.

7.3.2 The original (clean) data likelihood

The second distribution in equation 7.4 is the likelihood of the original, clean image data at position $\mathbf{x}$, given its surrounding clean image data. It is derived directly from the model statement of equation 7.2. In later sections it will be seen that the model coefficients, $\mathbf{a}(\mathbf{x})$ are estimated in a blockwise fashion, thus the allowance for non-stationarity is on a block grid and not on a pixel grid. In other words, there is a single coefficient set to be estimated for each predefined block of pixels. Within a block, the residuals $e(\mathbf{x})$ will be a sample from a multivariate Gaussian distribution. Scanning the residuals in a raster fashion within a block allows the formation of an excitation vector $\mathbf{e}$. The joint distribution for these residuals in a block of size $N = B \times B$ pixels is

$$p(\mathbf{e}) = \frac{1}{\sqrt{2\pi\sigma_e^2}^N} \exp\left(-\frac{\mathbf{e}^T\mathbf{e}}{2\sigma_e^2}\right) \tag{7.6}$$

Note that to generate residuals at each of the N sites in a block, image data outside that block both in space and time would be necessary to provide the required support pixels. By scanning all the required data into a data vector $\mathbf{i}$, $\mathbf{e}$ may be expressed as $\mathbf{Ai}$, where $\mathbf{A}$ is a matrix of coefficients arranged to stack the residuals accordingly. This stacking is described in detail in section 7.8. It is necessary to express the likelihood in alternate forms depending on whether variables are manipulated on a pixel or block basis. On a block basis, the likelihood can be written as

$$p(\mathbf{i}|\mathbf{a}(\mathbf{x}), \sigma_e^2, \mathbf{d}(\mathbf{x}), I_{n-1}, I_{n+1}) \propto \frac{1}{\sqrt{2\pi\sigma_e^2}^N} \exp\left(-\frac{\mathbf{i}^T\mathbf{A}^T\mathbf{A}\mathbf{i}}{2\sigma_e^2}\right) \tag{7.7}$$

The manipulation of the alternate form is presented in Appendix D.

7.4 The priors

The remaining distributions encode the prior belief about the values of the various unknowns. For simplicity, a uniform prior is assigned to $\mathbf{a}$. This encodes a notion of no bias on the estimation of the model coefficients, and effectively removes $p(\mathbf{a}(\mathbf{x}))$ from equation 7.4. The variance σ_e^2 is assigned a non-informative prior $p(\sigma_e^2) \propto 1/\sigma_e^2$, following [151]. The remaining priors are more involved and deserve separate discussion.

7.4.1 The motion prior

The discussion in Chapter 2 has introduced the idea that the motion field should be smooth over most of the image. This smoothness constraint is violated at motion boundaries typically occurring at the edges of moving objects. In practice for this application, it is sufficient to encode the notion of smoothness in order to achieve implicit motion interpolation. The prior adopted for motion smoothness is a Gibbs Energy prior, for instance as introduced by Konrad and Dubois [98] and Stiller [162]. To reduce the complexity of the final solution the motion field is block based, with one motion vector being employed for each specified block in the image.

The prior for $\mathbf{d}_{n,n-1}(\mathbf{x})$, the motion vector mapping the pixel at $\mathbf{x}$ in frame n into frame $n-1$, is as follows:

$$p_d(\mathbf{d}_{n,n-1}(\mathbf{x})|\mathbf{S}_n(\mathbf{x})) \propto \exp - \left(\sum_{\mathbf{v} \in \mathbf{S}_n(\mathbf{x})} \lambda(\mathbf{v})[\mathbf{d}_{n,n-1}(\mathbf{x}) - \mathbf{v}]^2 \right) \tag{7.8}$$

where $\mathbf{v}$ is each vector in the neighbourhood represented by $\mathbf{S}_n(\mathbf{x})$, and $\lambda(\mathbf{v})$ is the weight associated with each clique. The situation is illustrated in Figure 7.1.

In order to discourage 'smoothness' over too large a range, $\lambda(\mathbf{v})$ is defined as $\lambda(\mathbf{v}) = \Lambda/|\mathbf{X}(\mathbf{v}) - \mathbf{x}|$ where $\mathbf{X}(\mathbf{v})$ is the location of the block providing the neighbourhood vector $\mathbf{v}$. This location is measured in terms of blocks. $\Lambda = 2.0$ in the experiments presented later.

7.4.2 The priors for corruption and detection

In practice, each region of missing data tends to have fairly constant intensity (see Figures 7.4, 7.13), therefore it is reasonable to place a similar energy prior on both the binary field $b(\mathbf{x})$ and the blotch value field $c(\mathbf{x})$. It is found that acknowledging discontinuities in these fields leads to much better behaviour. Edges in the c and b fields must correspond to edges in the image since the corrupted areas are generally well delineated from their surrounding by a marked grey scale transition. Thus a simple zero crossing edge detector employed on the degraded image will enable the rough configuration of an edge field that can be used subsequently to define the priors on the c and b fields.

The priors are therefore defined as follows:

$$p_c(c(\mathbf{x})|C) \quad \propto \quad \exp\left(-\sum_{z\in C}\lambda_z^c(1-u(z,\mathbf{x}))|c(\mathbf{x})-z|\right) \tag{7.9}$$

$$p_b(b(\mathbf{x})|B) \quad \propto \quad \exp\left(-\sum_{z\in B}\lambda_z^b(1-u(z,\mathbf{x}))|b(\mathbf{x})-z|\right) \tag{7.10}$$

where z is one of the values of the variables c or b in the 8 connected neighbourhood of the location $\mathbf{x}$, denoted C, B (as previously). $u(z,\mathbf{x})$ is set to 1 if there is a significant zero crossing between the location $\mathbf{x}$ in the image and location from which z is extracted. Thus the smoothness constraint is turned off across significant zero crossings in the image. Note that these priors are defined on the *pixel* resolution image grid, whereas the motion prior discussed previously is defined on a *block* basis. In the results shown later $u(\cdot)$ was configured using an edge detector employing Difference of Gaussians (see D.O.G. filters in [114]) with the gradient threshold set at 5.0, the variance of the Gaussian filters was $1.0, 1.6$ and the filter window sizes were 9×9.

There is an important alternative prior which may be used for the corruption field $c(\mathbf{x})$. It is the so called Gaussian Markov Random Field Prior (GMRF). It is defined as below

$$p_c(c(\mathbf{x})|C) \propto \exp\left(-\sum_{z\in C}\lambda_z^c(1-u(z,\mathbf{x}))(c(\mathbf{x})-z)^2\right) \tag{7.11}$$

The prior is in fact *almost* identical to an autoregressive model for $c(\mathbf{x})$ except that it defines the conditional distribution only. The use of this prior allows the samples for c to be generated from the well known Gaussian distribution, and therefore results in a very fast iterative scheme.

7.5 Solving for the unknowns

The solution is generated by manipulating $p(\theta|G_{n-1}, G_n, G_{n+1})$. For instance the MAP estimate is generated by maximizing the distribution with respect to the unknowns. Unfortunately, due to the nonlinear nature of the expression, a closed form solution to the optimization problem is not available. One could make approximations to the various terms involved in the expression in order to allow some kind of closed form solution, but fortunately, the Gibbs Sampler allows samples of the parameter vector to be drawn from this complicated distribution. These random samples can be manipulated numerically to yield the required estimate. For instance, the average of the samples yields the MMSE estimate.

7.5.1 The Gibbs sampler

Consider that $\mathbf{i}$ contains at least the missing pixels (indicated at $b(\mathbf{x}) = 1$) and their immediate AR support. The missing pixels are denoted $\mathbf{i}_u$ and the remaining pixels as $\mathbf{i}_k$. The Gibbs Sampler then operates iteratively, given some starting guess for the unknowns, by drawing random samples from the conditional posterior distribution for each unknown in turn:

$$\begin{aligned} &\mathbf{a} \sim p(\mathbf{a}|\mathbf{i}, \sigma_e^2, \mathbf{b}, \mathbf{d}, \mathbf{c}, \mathbf{g}); && \mathbf{b} \sim \mathbf{p}(\mathbf{b}|\mathbf{i}, \sigma_e^2, \mathbf{a}, \mathbf{d}, \mathbf{c}, \mathbf{g}); \\ &\mathbf{c} \sim \mathbf{p}(\mathbf{c}|\mathbf{i}, \sigma_e^2, \mathbf{a}, \mathbf{d}, \mathbf{b}, \mathbf{g}); && \sigma_e^2 \sim p(\sigma_e^2|\mathbf{a}, \mathbf{i}, \mathbf{b}, \mathbf{d}, \mathbf{c}, \mathbf{g}); \\ &\mathbf{d} \sim p(\mathbf{d}|\mathbf{a}, \mathbf{i}, \mathbf{b}, \sigma_e^2, \mathbf{c}, \mathbf{g}); && \mathbf{i}_u \sim p(\mathbf{i}_u|\mathbf{a}, \mathbf{i}_k, \sigma_e^2, \mathbf{b}, \mathbf{d}, \mathbf{c}, \mathbf{g}) \end{aligned} \qquad (7.12)$$

where $\mathbf{c}, \mathbf{b}$ etc are vectors containing the relevant parameters, and the location argument $\mathbf{x}$ has been dropped for simplicity. These conditionals can be derived by manipulation of the joint posterior given in equation 7.4. This sampling procedure is repeated until convergence is reached according to some suitable criterion. The framework allows straightforwardly for minimum mean-square (MMSE), maximum *a posteriori* (MAP) or sampled estimates of $\mathbf{i}_u$ by manipulation of the sampled values following convergence.

Given the 'non-informative' [22] prior distributions for $\mathbf{a}$, σ_e^2, introduced previously, the sampling operations for $\mathbf{a}, \sigma_e^2, \mathbf{i}_u$ involve simple random draws from well-known distributions. However some practical considerations encourage an altered sampling strategy.

7.5.2 Adaptations to the Gibbs sampler

The convergence of the Gibbs Sampler is generally improved if several unknowns are sampled jointly [109]. This is possible using the method of composition [151, 52]. In this scheme, a random draw from $p(\mathbf{a}, \sigma_e^2, \mathbf{d}|\theta_{-(\mathbf{a},\mathbf{d},\sigma_e)})$, for instance, is made possible by the decomposition

$$p(\mathbf{a}, \sigma_e^2, \mathbf{d}|\mathbf{b}, \mathbf{c}, \mathbf{i}) = p(\mathbf{a}|\mathbf{d}, \mathbf{b}, \mathbf{c}, \sigma_e^2, \mathbf{i})p(\sigma_e^2|\mathbf{d}, \mathbf{b}, \mathbf{c}, \mathbf{i})p(\mathbf{d}|\mathbf{b}, \mathbf{c}, \mathbf{i}) \qquad (7.13)$$

Note that $\mathbf{a}, \sigma_e^2, \mathbf{d}$ are conditionally independent of $\mathbf{g}$. Recall that $\theta_{-(\mathbf{a},\mathbf{d},\sigma_e)}$ is a vector containing all the parameters in θ excepting $\mathbf{a}, \mathbf{d}, \sigma_e$. The various composition terms can be derived by successively *integrating out* $\mathbf{a}$, then σ_e^2 from the posterior distribution.

Random draws from $p(\mathbf{a}, \sigma_e^2, \mathbf{d}|\mathbf{b}, \mathbf{c}, \mathbf{i})$ can therefore be implemented by drawing from $p(\mathbf{d}|\mathbf{b}, \mathbf{c}, \mathbf{i})$ followed by a draw from $p(\sigma_e^2|\mathbf{d}, \mathbf{b}, \mathbf{c}, \mathbf{i})$ using the value of $\mathbf{d}$ drawn previously, then similarly for $p(\mathbf{a}|\dots)$, using the samples just generated for $\sigma_e^2 and \mathbf{d}$. In this manner a joint draw for $\mathbf{b}$, $\mathbf{c}$, $\mathbf{i}_u$ is also achieved.

The expressions required to perform these joint draws can be derived [52] by employing the vector $\mathbf{e}$ as previously described, containing all the excitation terms from equation 7.2 within a block of data and expressing it as $\mathbf{e} = \mathbf{i} - \mathbf{I}\mathbf{a}$. This then leads to the following distributions required in the joint draw from

$p(\mathbf{a}, \sigma_e^2, \mathbf{d}|\mathbf{i})$:

$$\begin{aligned} p(\mathbf{a}|\mathbf{d}, \mathbf{i}, \sigma_e^2) &= \mathcal{N}_P(\hat{\mathbf{a}}, \sigma_e^2(\mathbf{I}^T\mathbf{I})^{-1}) \\ p(\sigma_e^2|\mathbf{d}, \mathbf{i}) &= \mathrm{IG}((N-P)/2, E(\hat{\mathbf{a}}, \mathbf{i}, \mathbf{d})/2) \\ p(\mathbf{d}|\mathbf{i}, D) &\propto \frac{E(\hat{\mathbf{a}}, \mathbf{i}, \mathbf{d})^{-(N-P)/2}}{|\mathbf{I}^T\mathbf{I}|^{1/2}} p(\mathbf{d}|D) \\ \text{where } \hat{\mathbf{a}} &= (\mathbf{I}^T\mathbf{I})^{-1}\mathbf{I}^T\mathbf{i} \\ \text{and } E(\hat{\mathbf{a}}, \mathbf{i}, \mathbf{d}) &= \mathbf{e}^T\mathbf{e} \end{aligned} \tag{7.14}$$

where N is the number of pixels in the image block and D (shorthand for $\mathbf{S}_n(\mathbf{x})$) represents a neighbourhood of vectors surrounding the sampled location. The derivation of these expressions can be found in Appendix D.

In practice the draws for $\mathbf{b}, \mathbf{c}, \mathbf{i}$ are performed jointly on a pixel by pixel basis, sampling from the expression

$$p(b(\mathbf{x}), c(\mathbf{x}), I_n(\mathbf{x})|\mathbf{d}, \mathbf{a}, \sigma_e^2, B, C, \mathbf{i}_{-\mathbf{x}}, \mathbf{g}) \tag{7.15}$$

where $\mathbf{i}_{-\mathbf{x}}$ denotes all clean image data *not* at site $\mathbf{x}$, but required for all prediction equations (equation 7.2) involving site $\mathbf{x}$. This joint draw is unusual because of the degenerate delta function in the likelihood, but since b is a binary field, c is at most an 8 bit field, and this draw is specified on the *pixel* grid, a feasible sampling scheme results. The distributions required for the composition sampling can be derived by integrating the posterior to yield

$$p(b(\mathbf{x})|B, \ldots) = \begin{cases} p_c(c = g_n|\ldots)p_b(b = 1|B) & \text{for } b(\mathbf{x}) = 1 \\ p_\iota(i = g_n|\ldots)p_b(b = 0|B) & \text{for } b(\mathbf{x}) = 0 \end{cases} \tag{7.16}$$

$$p(i_n(\mathbf{x})|b(\mathbf{x}), \ldots) = \begin{cases} \mathcal{N}(\hat{i}, \frac{\sigma_e^2}{\mathbf{a}_u^T\mathbf{a}_u}) & \text{for } b(\mathbf{x}) = 1 \\ \delta(g_n(\mathbf{x}) - i_n(\mathbf{x})) & \text{for } b(\mathbf{x}) = 0 \end{cases} \tag{7.17}$$

$$p(c(\mathbf{x})|b(\mathbf{x}), i_n(\mathbf{x}), \ldots) = \begin{cases} \delta(g_n(\mathbf{x}) - c(\mathbf{x})) & \text{for } b(\mathbf{x}) = 1 \\ p_c(c(\mathbf{x})|C) & \text{for } b(\mathbf{x}) = 0 \end{cases} \tag{7.18}$$

where B is the set of neighbourhood of detection indicators $\mathbf{b}(\cdot)$ surrounding the sampled location. See Appendix D for details.

The utility of the algorithm is increased if there already exist initial estimates for motion in particular. It then becomes possible to draw samples from the numerically evaluated p.d.f. for $\mathbf{d} \sim \mathbf{p}(\mathbf{d}|\mathbf{i}, \sigma_\mathbf{e}^2, \mathbf{a}, \mathbf{b}, \mathbf{c})$ in a local region around the current estimate. In essence, the actual procedure employed for sampling for $\mathbf{d}$ involves proposing 8 candidate vectors from the neighbourhood and from perturbing these current vector by ± 1 pixel. The motion sample is then drawn from this set by direct numerical evaluation of the probability distribution assuming that the probability of all other samples is zero. This is an application of the *Griddy Sampler* [168]. Initial motion estimates can be taken from any number of motion estimators currently available, the multiresolution gradient based technique [83] discussed in Chapter 2, is employed here.

7.6 The JOMBADI algorithm

The previous discussions have been somewhat abstract in that there was no specific indication about precisely what mathematical operations are required. This section presents a clearer step by step recipe for the procedure outlining in addition the chronology of the iterations. This process will be called the JOMBADI algorithm: **JO**int **M**odel **BA**sed **D**etection and **I**nterpolation. There are two major parts to the algorithm, the first part is the joint draw for $\mathbf{a}, \mathbf{d}, \sigma_e^2$ and the second is the draw for $\mathbf{b}, \mathbf{c}, \mathbf{i}_u$. The first joint draw is performed on a *block* basis since the motion and coefficient fields are expected to be generally smooth functions. The second draw is performed on a pixel basis in order to more carefully delineate the missing regions. Each draw is performed for all the variables across the whole image before moving on to the draw for the next set of variables. It is also possible to make the draws in turn at each site before moving onto the next, but sweeping through the image in the fashion described here leads to a simpler implementation. The description begins with a listing of the overall sequence of activities.

1. The first step is to generate a *kick start* of estimates for the sampler. This is achieved using a simple technique for detection and interpolation of the missing data. Any one of the techniques from the previous chapter can be employed, the combination of the WBME for estimating motion and the SDIa for detection is used here. Since the samples for the unknowns are generated using joint sampling strategies, it is found that using the degraded data as the start image for both the $c(\mathbf{x})$ and $I_n(\mathbf{x})$ fields is adequate. The main point of this step is actually to generate reasonable estimates of motion in the areas which contain *uncorrupted* data.

2. The image is divided into blocks and in each block one sample each for $\mathbf{d}_{n,n-1}$, $\mathbf{d}_{n,n+1}$, $\mathbf{a}$, σ_e^2 is drawn jointly, in that order. All blocks are visited in a 'checkerboard' fashion (see Besag [15]). This requires that the block motion neighbourhood employed for the next block processed must not overlap with the neighbourhood of the current processed block. Using the 8 nearest neighbourhood configuration presented above, the blocks are visited so that the next site is always at least 2 blocks away in one direction from the current site.

3. The image is now scanned *pixel* by *pixel* again in a checkerboard fashion. At each site a sample for $c(\mathbf{x}), I_n(\mathbf{x}), b(\mathbf{x})$ is drawn jointly. See Appendix D for the derivation of the exact sampling scheme. The samples are drawn based on the values of model coefficients, variance, and motion, that have been estimated in step 2 for the block in which the current site lies. Some overlap between the blocks would therefore lead to a more smooth variation in the sampled model coefficients, motion etc. The importance of the size of this overlap is not investigated here.

4. The last two steps are repeated until some convergence criterion is satisfied, or until enough samples of $b(\mathbf{x}), I_n(\mathbf{x})$ have been collected so that

it is deemed sufficient to determine some numerical estimate. Note that these samples must be collected after the chain has converged.

7.6.1 *The draw for* $\mathbf{d}_{n,n-1}$, $\mathbf{d}_{n,n+1}$, $\mathbf{a}$, σ_e^2

The joint draw for $\mathbf{d}_{n,n-1}$, $\mathbf{d}_{n,n+1}$, $\mathbf{a}$, σ_e^2, is achieved as follows. It is assumed that the 3DAR model is temporally causal, even if it is spatially non-causal. This allows the consideration of 3 frames of image data, I_{n-1}, I_n, I_{n+1}.

1. At the particular block site to be processed (denoted $\mathcal{B}$) a set of candidate vectors for $\mathbf{d}_{n,n-1}$ is assembled. The candidates are made up of the eight motion vectors in the eight nearest block neighbours and the eight vectors created by perturbing the current vector in $\mathcal{B}$ by ± 1.0 pixels horizontally and vertically. It is expected that perturbing by fractional displacements is not useful since the coefficients of the spatial support of the AR model in the previous frame will arrange themselves to yield some kind of optimal interpolation in the case of fractional motion.

2. For each candidate vector $\mathbf{v}_c$ the spatial smoothness energy is calculated as

$$s(\mathbf{v}_c) = \sum_{k=1}^{8} \lambda_k |\mathbf{v}_c - \mathbf{d}_{n,n-1}(\mathbf{x}, k)| \tag{7.19}$$

   where $\mathbf{d}_{n,n-1}(\mathbf{x}, k)$ is the kth vector in the eight connected neighbourhood of blocks.

3. For each $\mathbf{v}_c$, a set of coefficients $\mathbf{a}_c$ is calculated by solving the Normal equations using the data in the current block as well as in blocks in the next and previous frames. These blocks are located by compensating for motion using $\mathbf{v}_c$ for the backward vector, and the existing estimate for $\mathbf{d}_{n,n+1}$ as the forward vector. The various correlation terms are measured using prediction equations set up in the blocks in frame n and $n+1$. The data is normalized to zero mean before the coefficients are calculated. This mean is estimated from the block in the current frame and is kept the same for all the candidate vectors. The value of the mean can be stored for later use in sampling for $I_n(\mathbf{x})$.

4. As each set of coefficients is calculated, an error E_c, the sum squared prediction error is calculated as below.

$$E_c = \sum_{\mathbf{x} \in \text{volume}} \hat{e}^2(\mathbf{x}) \tag{7.20}$$

   where $\hat{e}(\mathbf{x})$ is the prediction error defined in equation 7.2, using $\hat{\mathbf{a}}$ calculated above. Note that this measurement takes place over a *volume* of image data, incorporating motion compensated blocks in frames $n-1, n, n+1$.

5. The probability of each candidate is generated using

$$p(\mathbf{v}_c|\mathbf{i}) \propto \frac{E_c^{\frac{N-P}{2}}}{|\mathbf{I}^T\mathbf{I}|^{1/2}} p_d(\mathbf{d}_{n,n-1} = \mathbf{v}_c) \tag{7.21}$$

yielding a list of unnormalized probability measurements for each $\mathbf{v}_c$. The values need not be normalized before proceeding. See Appendix D for details of the derivation of this term.

6. A sample for motion is numerically drawn from the distribution calculated above. This sample is denoted $\mathbf{v}^s_{n,n-1}$. The corresponding E_c from the list generated is denoted E_s.

7. This process is repeated for the forward motion vector sample $\mathbf{d}_{n,n+1}$, to yield $\mathbf{v}^s_{n,n+1}$.

8. A sample for σ_e^2, σ_s^2 is then drawn from an Inverted Gamma distribution.

$$p(\sigma_e^2|\mathbf{d},\mathbf{i}) = \mathrm{IG}((N-P)/2, E_s/2) \tag{7.22}$$

This is a more difficult draw and is achieved by the recipe reported in [151] as follows:

   (a) Generate a Gamma Deviate ($\mathcal{G}$) with $(N-P-1)/2$ degrees of freedom.

   (b) $\sigma_s^2 = E_s/(2\mathcal{G})$

9. A sample for the coefficients, $\mathbf{a}_s$ in block $\mathcal{B}$ is then drawn from

$$p(\mathbf{a}|\mathbf{d},\mathbf{i},\sigma_e^2) = \mathcal{N}_P(\hat{\mathbf{a}}, \sigma_e^2(\mathbf{I}^T\mathbf{I})^{-1}) \tag{7.23}$$

This is achieved by the following steps (see [151]).

   (a) Calculate the least-squares estimate of the coefficients $\hat{\mathbf{a}}$ using the Normal equations and the image sequence data compensated using $\mathbf{v}^s_{n,n-1}$, $\mathbf{v}^s_{n,n+1}$. This should have already been calculated when sampling for motion, so there is no need to recalculate here.

   (b) Calculate the *square root* of $\mathbf{I}^T\mathbf{I}$. This can be done using Cholesky decomposition [151, 142]. In fact, Cholesky decomposition can also be used for solving the Normal equations [142] since $\mathbf{I}^T\mathbf{I}$ is positive definite. That process would have already generated a square root as required. Denote this square root $\mathbf{S}$.

   (c) Solve $\mathbf{Sr} = \mathbf{n}$ for $\mathbf{r}$, where each element of $\mathbf{n}$ is drawn from $\mathcal{N}(0,1)$.

   (d) $\mathbf{a}_s = \hat{\mathbf{a}} + \mathbf{r}$

7.6.2 *Joint sampling for $c(\mathbf{x}), I_n(\mathbf{x}), b(\mathbf{x})$*

The joint sample for c, I, b is then achieved at each *pixel* site by the steps described as follows. Note that λ^b, λ^c are user defined hyperparameters. They control the connectivity of the fields $b(\cdot)$ and $c(\cdot)$. High values bias JOMBADI toward detection of flat regions only.

1. The first step is to draw a sample for $b(\mathbf{x})$. This is done by evaluating equation 7.16 for both cases $b(\mathbf{x}) = 0, 1$. The sample is then drawn numerically. Drawing the sample requires knowledge of Z_i, Z_c, the normalizing constants for the distributions $p_i(\cdot), p_c(\cdot)$, as well as the values of the edge field $u(\cdot)$ as discussed previously. Because $p_i(\cdot)$ is a univariate Gaussian distribution, Z_i is straightforward to determine (see equation 7.17). Depending on the choice for $p_c()$, evaluation of Z_c can be difficult. If $p_c(\cdot)$ is chosen to be a GMRF as in equation 7.11 then again, it is a univariate Gaussian and the normalizing constant is simple to determine. If it is chosen as in equation 7.9 then the simplest method to determine Z_c is to use direct numerical evaluation. It is possible to get a good estimate here since it is known that the image is 8 bit integer valued and so there are at most 256 allowable states for $c(\mathbf{x})$. Thus Z_c is determined by evaluating equation 7.9 at each of these states and summing to yield the normalizing constant.

2. Having drawn a sample for $b(\mathbf{x})$, samples for $i(\mathbf{x})$ and $c(\mathbf{x})$ are simple to generate from equations 7.17, 7.18. If $b(\mathbf{x}) = 1$ then a corrupted pixel has been detected and $c(\mathbf{x}) = G_n(\mathbf{x})$. $i(\mathbf{x})$ has to be interpolated, and a sample is drawn using expression 7.17 for $b = 1$. The mechanics of drawing the sample are simple. First generate the least squares estimate of the interpolant, $[\mathbf{a}_u^T \mathbf{A}_k \mathbf{i}_k / (\mathbf{a}_u^T \mathbf{a}_u)]$ and add to this Gaussian white noise of variance $\sigma_e^2 / (\mathbf{a}_u^T \mathbf{a}_u)$. This draw is the univariate equivalent of the multivariate Gaussian draw used for $\mathbf{a}$ above.

3. If $b(\mathbf{x}) = 0$ then the pixel is uncorrupted and $I_n(\mathbf{x}) = G_n(\mathbf{x})$, and a value for $c(\mathbf{x})$ must be drawn from its prior. Again this draw is straightforward if the prior for c is a GMRF, otherwise the draw can be performed using direct numerical evaluation of the distribution or a Metropolis step.

7.7 Relation to other techniques

The algorithm presented above can be seen as a generalization of almost all the processes for detection and removal of missing data that have been presented to date. This section treats in turn every known formulation of detection/reconstruction and shows how it relates to this general framework.

7.7.1 Early model based systems: 1989–1996

The use of stochastic models for interpolation of image sequences with missing data first appeared in [96, 92, 93]. The work employed the spatio-temporal AR model in a similar fashion to that presented by Vaseghi [182] and Veldhuis [184]. The model was used as part of a two stage system for detection and interpolation which can now be related to a Maximum Likelihood approach to these problems. Later work evolved to incorporate motion correction as an additional stage prior to interpolation of the missing data [83]. This algorithm can be derived as a suboptimal version of JOMBADI. It was the precursor to a joint approach to the missing data problem and introduced the first trappings of a Bayesian approach to missing data removal for image sequences. The work continued to evolve in [52] in which an algorithm for joint motion and image data interpolation was presented. The detection was later incorporated as part of the joint estimation process in [53, 84, 85]. The early processes are discussed next.

Maximum Likelihood detection/interpolation

The earliest work on model based image sequence reconstruction assumed that the motion estimation process was robust to the presence of missing data [92, 93, 86, 87, 96, 118]. Thus motion was not considered to be a parameter to be estimated. Indeed some authors continue to make this assumption [75, 28]. As illustrated in the last chapter, it is not a good assumption to make when dealing with missing regions of large size. The early approach separated detection and interpolation as two distinct processes.

Early model based Detection: MBD-1

Assume that the image is corrupted according to the following equation.

$$
\begin{aligned}
G(\mathbf{x}) &= I(\mathbf{x}) + b'(\mathbf{x}) && (7.24)\\
\text{where } b'(\mathbf{x}) &= \begin{cases} 0 & \text{With probability } (1-P_s) \\ s & \text{With probability } P_s \end{cases} && (7.25)
\end{aligned}
$$

Here, s is a randomly distributed grey level representing a blotch or impulse, and it occurs at random intervals in the frame with probability P_s. As in the previous chapter, it is required to detect the likely occurrences of $b'(\mathbf{x}) \neq 0$ in order to isolate the distortion for further examination. The key to the solution is to recognize that the undistorted image $I_n(\mathbf{x})$ obeys the AR model whereas $b'(\mathbf{x})$ does not.

Suppose that the model coefficients for a particular image sequence were known. The prediction error could then be considered to be noise with some correlation structure as defined in Appendix B. If $G(\mathbf{x})$ was filtered with the model prediction filter the output could be written as below

$$\begin{aligned} \epsilon'(\mathbf{x}) &= G(\mathbf{x}) - \sum_{k=1}^{P} a_k G(\mathbf{x} + \mathbf{q}_k) && (7.26) \\ &= I(\mathbf{x}) + b'(\mathbf{x}) && (7.27) \\ &\quad - \sum_{k=1}^{P} a_k I(\mathbf{x} + \mathbf{q}_k) - \sum_{k=1}^{P} a_k b'(\mathbf{x} + \mathbf{q}_k) \\ &= \epsilon(\mathbf{x}) + b'(\mathbf{x}) - \sum_{k=1}^{P} a_k b'(\mathbf{x} + \mathbf{q}_k) \end{aligned}$$

Equation 7.28 shows that the undistorted samples in the degraded signal are reduced to the scale of the error or residual sequence. The distorted samples can be reduced in amplitude if there are other distorted samples nearby but this does not occur often. Therefore thresholding the prediction error is a useful method of detecting local distortion.

[96, 92, 93] went on to illustrate that the support of the model affects the detectability of distortion. In the case of Dirt and Sparkle, because the distortion occupies a large spatial area, a model with spatial support in the current frame would only give large prediction errors at the edges of a blotch. Inside the blotch the residual would be reduced in magnitude due to the growing importance of the last term in equation 7.28. In practice, models with supports not in the current frame are more effective since the distortion is always local (even impulsive) in time but not necessarily as local in space. Furthermore, it was found beneficial to separate the thresholding operation to manipulate two prediction errors calculated from a backward and forward predictor engaged at the pixel to be tested. Thus a blotch is detected when *both* forward and backward prediction errors are greater than some threshold.

To see how this relates to the joint framework for detection and interpolation, delete the model parameters, $\mathbf{a}$, the motion information, $\mathbf{d}_{n,n-1}, \mathbf{d}_{n,n+1}$ and the unknown image data $\mathbf{i}_u$ from consideration as unknowns in equation 7.4. The values for these parameters are assumed to be generated sufficiently accurately from the 'kick start' motion estimator etc. Introduce two 'discontinuity' detection variables b_{n-1}, b_{n+1} at each site $\mathbf{x}$ in G_n. Missing data is detected at $\mathbf{x}$ when $b_{n-1}(\mathbf{x}) = b_{n+1}(\mathbf{x}) = 1$. Reformulate the likelihood expression as

$$\begin{aligned} &p(G_n | b_{n-1}(\mathbf{x}), b_{n+1}(\mathbf{x}), I_{n-1}) \propto \\ &\qquad \exp\left(- \frac{\epsilon_n^b(\mathbf{x})^2 (1 - b_{n-1}(\mathbf{x})) + \epsilon_n^f(\mathbf{x})^2 (1 - b_{n+1}(\mathbf{x}))}{2\sigma_e^2} \right) \end{aligned} \qquad (7.28)$$

where $\epsilon_n^b, \epsilon_n^f$ are the backward and forward prediction error of the particular AR predictor chosen, at the observed image site i.e.

$$\begin{aligned}\epsilon_n^b(\mathbf{x}) &= G_n(\mathbf{x}) - \sum_{k=1}^{P} a_k I_{n-1}(\mathbf{x} + \mathbf{q}_k) \\ \epsilon_n^f(\mathbf{x}) &= G_n(\mathbf{x}) - \sum_{k=1}^{P} a_k I_{n+1}(\mathbf{x} + \mathbf{q}_k)\end{aligned} \tag{7.29}$$

where all the data has been already compensated for motion and the 3DAR model has support only in the previous or next frame, e.g. $P = 1$ and $q_k = [0\ 0\ -1]$.

The problem is now to estimate the values of b_{n-1}, b_{n+1} at each pixel site. Obviously the ML estimate for b_{n-1}, b_{n+1} at a site will always be obtained when $b_{n-1} = b_{n+1} = 1$, an undesirable situation. To prevent this, a penalty for introducing $b = 1$ is employed as a prior on b, thus $p(b) \propto \exp(-\alpha b)$. The frame subscripts for b have been dropped to emphasize that this applies to both forward and backward directions. The penalty α will be introduced when $b = 1$, thus the ML estimate for b at a site is only 1 when $\frac{\epsilon_n}{2\sigma_e^2} \geq \alpha$. This is equivalent to thresholding the prediction error. In the early work, no attention was given to estimation of σ_e^2 since this is just a scaling factor and can be compensated for by altering the value of the threshold α. The use of these discontinuity variables is identical to their use as Occlusion Indicators employed by Konrad and Dubois [98]. In this situation, the SDIa detector results when $P = 1$, $\mathbf{q}_1 = [0\ 0\ -1]$ and $a_1 = 1.0$.

For a real sequence, the model coefficients are unknown. They were estimated from the degraded sequence using the Normal Equations (Least Squares estimate), see Appendix A. The choice of the spatial extent of the volume used is important. If the size of a block is comparable to the size of a particular blotch then the coefficients are heavily biased toward that distortion and the resulting detection efficiency is poor. This effect is enhanced when the model has spatial support in the current frame. In practice, models with spatial support in the current frame were avoided.

This formulation of the detection problem works well for audio signals. In that case the signal remains statistically stationary for a sufficiently large number of samples so that the Least Squares estimate for the model coefficients is reliable. For image sequences this is not so since the data employed for coefficient estimation is relatively heavily corrupted by missing data because small data volumes must be used. Thus this method for detection of missing data performed rather poorly, when the model coefficients had to be estimated from the degraded data. The reader should recognize that there exist more robust means of estimating the model coefficients using for example iteratively reweighted least squares or least median of squares [101, 100, 99]. These methods would yield substantial improvement in the performance of the deterministic algorithm described in this section. However these improvements are unlikely to cause performance to exceed that of the JOMBADI algorithm.

Early model based Interpolation: MBI-1

Both Vaseghi [182] and Veldhuis [184] introduced the use of AR processes for interpolation of missing data in audio signals. The idea was to estimate the values of the missing data as those which minimized the sum squared prediction error in the detected missing region. It was assumed that an estimate for the model coefficients was already found. The idea was extended to 3DAR models in [96, 93, 92] for image sequences.

It is assumed that the information in a block of $M \times M$ pixels is sufficiently statistically homogeneous for a single model to be used for that block. Within this block, assume that there are m pixel locations that are corrupted. These locations are arbitrary; they may be connected together as in a blotch, or they may be a set of dispersed impulses.

Further, it is assumed that the block can be modelled by a 3DAR model incorporating $M \times M$ blocks in the surrounding frames that are compensated for motion. Again, the intensity of the pixels in the block in the current frame may be written as

$$I(\mathbf{x}) = \sum_{k=1}^{N} a_k I(\mathbf{x} + \mathbf{q}_k) + e(\mathbf{x}) \tag{7.30}$$

Where again, $I(\cdot)$ represents pixel grey level, a_k are model coefficients, and $e(\mathbf{x})$ the model excitation (or prediction error). It is assumed that a set of N frames are considered. Allowing for a border of pixels at the edge of the $M \times M$ block in the current frame (so that $(\mathbf{x} + \mathbf{q}_k)$ will never result in a location outside the $M \times M$ block), an equation for the error at every pixel within a centred $B \times B$ block in that frame can be written as below.

$$\mathbf{e} = \mathbf{A}\mathbf{i} \tag{7.31}$$

where $\mathbf{i}$ represents an $NM^2 \times 1$ column vector of row ordered pixels from the N, $M \times M$ blocks, $\mathbf{e}$ is a $2B^2 \times 1$ column vector of errors (made up of prediction error equations in frames $n, n+1$), and $\mathbf{A}$ a matrix of coefficients satisfying the model equation at all the considered points. This coefficient matrix is of size $2B^2 \times NM^2$. The vector $\mathbf{i}$ contains intensities of both known and unknown pixels. If this vector is separated into two vectors $\mathbf{i}_u$ (u for unknown) and $\mathbf{i}_k$ (k for known), which represent the known and unknown pixel intensities, then equation 7.31 can be written as

$$\mathbf{e} = \mathbf{A}_k \mathbf{i}_k + \mathbf{A}_u \mathbf{i}_u \tag{7.32}$$

Here, $\mathbf{A}_k$, $\mathbf{A}_u$ are the coefficient matrices corresponding to the known and unknown data vectors. They are submatrices of the $\mathbf{A}$ matrix, made by extracting the relevant columns.

To derive an interpolation, $\mathbf{i}_u$ must be found. Following Vaseghi [182], this is done by minimizing the squared error $\mathbf{e}^T\mathbf{e}$ with respect to $\mathbf{i}_u$ as follows:

$$\begin{aligned}
\mathbf{e}^T\mathbf{e} &= [\mathbf{A}_k\mathbf{i}_k + \mathbf{A}_u\mathbf{i}_u]^T[\mathbf{A}_k\mathbf{i}_k + \mathbf{A}_u\mathbf{i}_u]^T \\
&= \mathbf{i}_k^T\mathbf{A}_k^T\mathbf{A}_k\mathbf{i}_k + \mathbf{i}_k^T\mathbf{A}_k^T\mathbf{A}_u\mathbf{i}_u + \mathbf{i}_u^T\mathbf{A}_u^T\mathbf{A}_k\mathbf{i}_k + \mathbf{i}_u^T\mathbf{A}_u^T\mathbf{A}_u\mathbf{i}_u \\
\frac{\partial\mathbf{e}^T\mathbf{e}}{\partial\mathbf{i}_u} &= 2\mathbf{A}_u^T\mathbf{A}_k\mathbf{i}_k + 2\mathbf{A}_u^T\mathbf{A}_u\mathbf{i}_u \\
\Rightarrow \mathbf{i}_u &= -[\mathbf{A}_u^T\mathbf{A}_u]^{-1}\mathbf{A}_u^T\mathbf{A}_k\mathbf{i}_k
\end{aligned} \tag{7.33}$$

Therefore, the solution for the interpolated pixels is given by equation 7.33.

This solution can be derived from equation 7.4 by deleting the model parameters, $\mathbf{a}$ and the motion information, $\mathbf{d}_{n,n-1}, \mathbf{d}_{n,n+1}$ from consideration as unknowns. The detection variables b have already been estimated from a separate process so they are 'given'. Employing the alternative form for the likelihood as

$$\frac{1}{(2\sigma_e^2)^{\frac{N}{2}}} \exp\left(\frac{-\mathbf{e}^T\mathbf{e}}{2\sigma_e^2}\right) = \frac{1}{(2\sigma_e^2)^{\frac{N}{2}}} \exp\left(-\frac{[\mathbf{A}_k\mathbf{i}_k + \mathbf{A}_u\mathbf{i}_u]^T[\mathbf{A}_k\mathbf{i}_k + \mathbf{A}_u\mathbf{i}_u]}{2\sigma_e^2}\right) \tag{7.34}$$

allows the derivation of the conditional $p(\mathbf{i}_u|\theta_{-\mathbf{i}_u})$ as follows:

$$p(\mathbf{i}_u|\theta_{-\mathbf{i}_u}) \propto \exp\left(-\frac{[\mathbf{i}_u - [\mathbf{A}_u^T\mathbf{A}_u]^{-1}\mathbf{A}_u^T\mathbf{A}_k\mathbf{i}_k]^T[\mathbf{A}_u^T\mathbf{A}_u][\mathbf{i}_u - [\mathbf{A}_u^T\mathbf{A}_u]^{-1}\mathbf{A}_u^T\mathbf{A}_k\mathbf{i}_k]}{2\sigma_e^2}\right) \tag{7.35}$$

Thus the ML estimate for the missing data which maximizes $p(\mathbf{i}_u|\theta_{-\mathbf{i}_u})$ is given by equation 7.33 which is also the least squares estimate.

Of course this solution implies knowledge of the model coefficients. These must be estimated from the corrupt data. In [96] it was recognized that the corrupted data in the block usually caused such a heavy bias in the LS estimation of AR coefficients that the estimates were unusable. The solution was to use the knowledge of the detected location of distortion to weight out the effect of the corruption at the missing sites detected. The Normal Equations were altered to solve for the AR parameters using weighted LS estimation. Recall that normally, the model coefficients are chosen to minimize the expected value of the squared prediction error at all points in the block considered. Because some of this data is now known to be missing, the prediction error at these points may be weighted to zero so that this data does not affect the estimation process. The general approach is to weight the prediction error by some function $w(\mathbf{x})$, prior to minimizing the squared weighted error.

For the purposes of coefficient estimation, the new prediction equation may be written as

$$e_w(\mathbf{x}) = w(\mathbf{x}) \sum_{k=0}^{P} a_k I(\mathbf{x} + \mathbf{q}_k) \tag{7.36}$$

where all the symbols have their usual meaning, $a_0 = 1.0$, and $e_w(\mathbf{x})$ is the weighted error at position $\mathbf{x}$. Recall that this chapter assumes that the data volume being used has already been compensated for motion and so the motion parameter has been omitted from the 3DAR model.

Minimizing the squared error $[e_w(\mathbf{x})]^2$ with respect to the coefficients, then yields the following set of N+1 equations:

$$\sum_{k=0}^{P} a_k E[(w(\mathbf{x}))^2 I(\mathbf{x} + \mathbf{q}_k) I(\mathbf{x} + \mathbf{q}_m)] = 0 \text{ for } m = 0 \ldots N. \tag{7.37}$$

Recall that $a_0 = 1.0$. Therefore, these equations may be written in matrix form as

$$\mathbf{R}_w \mathbf{a} = -\mathbf{r}_w \tag{7.38}$$

where $\mathbf{R}_w$ is a $P \times P$ matrix of correlation coefficients, and $\mathbf{r}_w$ a $P \times 1$ vector of correlation coefficients. Equation 7.38 is the weighted solution for the P model coefficients. The most obvious choice for the weighting function is a binary field set to 0 for all the blotch positions and 1 otherwise. This is found to be extremely effective in practice.

In some sense this approach of ML detection (MBD-1) followed by ML interpolation (MBI-1) is similar to an ICM solution to the problem of finding the MAP estimate of the unknowns in equation 7.4. The solution is suboptimal since only 1 iteration of ICM is allowed in each step and many of the unknowns such as the motion and AR coefficients are not included in the problem. Nevertheless the two step approach can work since it is possible to generate reasonable estimates for these 'hidden' unknowns.

Motion interpolation and sampled image interpolation: MBI-2

The first published work in which consideration was given to the problem of motion interpolation appeared in [83]. The idea that motion estimates were corrupted at sites of missing data was acknowledged in [96] but no solution was presented. In [83] the problem was set up as described by equation 7.4, assuming a previous detection process had already provided estimates for $b(\mathbf{x})$. Thus the joint posterior was as in equation 7.4 excepting the indicator variables b. The idea was to propose a practical solution and so rather than solve for all the parameters jointly using the Gibbs sampler for instance, each sub-problem was treated separately using an approximate ML approach. The system proposed was a three step one, first detect missing data, then correct motion at the sites of detected corruption, and finally employ the 3DAR model for interpolation of the image data using MBI-1. The detection was performed using the SDIa described previously, although any heuristic could be employed instead. Motion correction followed by weighting out the influence of a modified data likelihood at sites which were detected as corrupted by missing data. It was assumed that the blotch was delineated from its surround using a suitable tracking algorithm on the detection field $b(\mathbf{x})$. A further assumption was made that the motion was the same over the entire missing region and so only a single estimate for motion was required at the missing site. The modified expressions for motion interpolation were as follows (for the backward motion component):

$$p(\mathbf{d}_{n,n-1}|G_n, I_{n-1}, b(\mathbf{x}), D) \propto$$
$$\exp\left(-\frac{\sum_{\mathbf{x}\in B} w^2(\mathbf{x})(G_n(\mathbf{x}) - I_{n-1}(\mathbf{x}+\mathbf{d}_{n,n-1}))^2}{2\sigma_e^2}\right) p(\mathbf{d}_{n,n-1}|D) \quad (7.39)$$

where $w(\mathbf{x})$ is a weight attached to the site $\mathbf{x}$ which was set to 0 when $b(\mathbf{x}) = 1$ and 1 otherwise. σ_e^2 is the variance of the prediction error of the 1 tap temporal AR process ($a_1 = 1.0$) which results in the likelihood shown. An estimate for σ_e could be made by measurement from the data around the uncorrupted region. A block based motion estimator (MWBME) was used to generate motion estimates and because the missing data could span several blocks it was necessary to define a flexible clique arrangement for the motion prior $p(\mathbf{d}_{n,n-1}|D)$. In essence it is the same prior as that defined by equation 7.8, but modified so that only vectors which were in blocks which showed little corruption were included in the neighbourhood. This arrangement is shown in Figure 7.1. The steps for motion interpolation at a site of missing data were therefore

1. Delineate the contour of the missing region detected.
2. Centre a bounding box on the missing region containing at most 20% corrupted data.
3. Use as candidates for corrected motion in that box from the eight nearest block based motion vector neighbours which had *uncorrupted* (or almost so) blocks. Generate four more candidates by perturbing the current (erroneous) candidate by ± 1 pixel.
4. Choose the interpolated motion vector to be that candidate which maximizes the expression 7.39 above.

This approximate MAP estimate for the motion information can be seen to be a simplification of the draw for a motion sample in the JOMBADI algorithm. Rather than sampling, it is generating an estimate which maximizes the local conditional distribution for $p(\mathbf{d}_{n,n-1}|D, G_n, I_{n-1})$, given some modifications to the likelihood and prior to allow for the case that there exists no current estimate for the missing image data.

The image interpolation step that then followed is identical in purpose to the draw for $\mathbf{i}_u$ and $I_n(\mathbf{x})$ specified in equations 7.35 and 7.17. The idea was to illustrate that drawing a *sample* for the interpolated data enabled the better regeneration of texture than using the ML or Least Squares estimate for the missing data. Note that in the JOMBADI implementation enumerated previously, the image data is interpolated on a per pixel basis, whereas here the entire region of missing data is drawn using a multivariate draw from a Gaussian distribution. This fact has more implications for convergence of the algorithm since whether the samples are drawn on a pixel or block based basis makes no

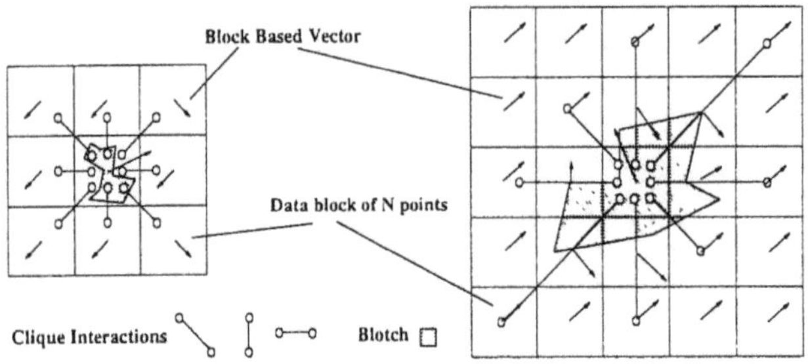

FIGURE 7.1. Left: Neighbourhood and cliques used for $p(\mathbf{d}_{n,n-1}(\mathbf{x}))$. Right: Altered neighbourhood used with a large blotch.

difference to the nature of the final interpolant. The draw for the interpolated data was made from the distribution specified in equation 7.35. Working on 1-D signals, O'Ruanaidh [151] was the first to point out that sampled interpolants for missing data were able to regenerate a typical excitation sequence, in contrast with LS or ML or MAP interpolants which cannot. He derived a recipe for drawing samples from the joint distribution for the interpolant using the AR model which is as shown in equation 7.35. The implications of this are explored when results are presented later in the chapter.

Joint image and motion interpolation: MBI-3

The work published in [52] investigated a half way point between the MBI-2 algorithm above and the JOMBADI algorithm. It presented a full implementation of the Gibbs sampler for estimating motion, missing image data, the 3DAR parameters and the model variance σ_e^2, assuming that the detection was performed separately possibly using one of the heuristics introduced in the previous chapter. Using composition sampling, joint draws for $\mathbf{a}, \sigma_e^2, \mathbf{d}$ were interspersed with draws for the missing data $\mathbf{i}_u$. It was not necessary to delineate regions of missing data for processing and the assumption that the motion was the same over the entire missing region was therefore relaxed. This process is therefore identical to the JOMBADI algorithm except that the detection variables $b(\mathbf{x})$ are assumed to already be set and so are removed from consideration together with the hidden corruption layer $c(\mathbf{x})$.

7.7.2 The Morris system: 1994

Robin Morris [118] was the first to consider the use of a Bayesian framework for missing data detection and removal in image sequences. His approach was a two process strategy: detect first, then reconstruct in the missing regions. His strategy differed from that presented here in that he did not consider a hidden corruption process, nor did he consider that the motion reconstruction problem was also an issue. At the time (1994) these were typical assumptions being made by Kokaram [89, 93, 81, 89, 86, 87], and represent an implicit acknowledgment that the main problem being treated was relatively small size corruption, up to 20 $\times$ 20 pixels. This was followed by the assumption that a multiresolution motion estimator would generally be robust to missing data of this size. It is now accepted that the accuracy of this assumption depends greatly on the brightness of the corruption and the surrounding image texture, as well as the motion estimation algorithm being used.

The detection process evolved from the notion that the requirement for a detector of missing data is one which could identify *areas* of high temporal image discontinuity both forward and backward in time. Morris introduced two indicator variables $b_{n-1}(\mathbf{x})$ and $b_{n+1}(\mathbf{x})$. $b(\mathbf{x}) = 1$ indicated the presence of a temporal discontinuity at $\mathbf{x}$ between frame n and $n-1$, it was 0 otherwise. Therefore $b_{n-1}(\mathbf{x}) = b_{n+1}(\mathbf{x}) = 1$ indicates that the pixel at $\mathbf{x}$ was corrupted. The detection problem was that of estimating these indicator variables at each pixel site, assuming that the motion information was reliable. An Ising prior density was used to organize the indicators spatially. This is identical in purpose to equation 7.10. The image data likelihood was influenced by the indicator variables as follows:

$$p(G_n(\mathbf{x})|I_{n-1}, \mathbf{d}) \propto \exp\left(-\lambda_b[G_n(\mathbf{x}) - I_{n-1}(\mathbf{x} + \mathbf{d}_{n,n-1}(\mathbf{x}))]^2(1 - b_{n-1}(\mathbf{x}))\right) \tag{7.40}$$

where all the terms have their usual meaning and a similar expression exists for $b_{n+1}(\mathbf{x})$. This likelihood expression may be recognized as a specific manifestation of equation 7.6 by setting $\lambda_b = 1/(2\sigma_e^2)$, selecting a 1 tap, purely temporal AR model, and evaluating expression 7.6 at one site only.

The equation 7.40 is identical to the usual form for configuration of occlusion proposed by Konrad and Dubois [98], in that b is the occlusion variable. Morris employed a penalty for setting the occlusion site to 1 since setting this value would always maximize the likelihood in any MAP estimation process. Thus the prior for b was made up of both the spatial term and a penalty, α as follows:

$$p(b_{n-1}(\mathbf{x})|B) \propto p_b(b_{n-1}(\mathbf{x}))\exp\left(-\alpha b_{n-1}(\mathbf{x})\right) \tag{7.41}$$

α can be related to a threshold on the displaced pixel difference, $(G_n(\mathbf{x}) - I_{n-1}(\mathbf{x} + \mathbf{d}_{n,n-1}))$ for detecting discontinuity. When $\lambda_b(G_n(\mathbf{x}) - I_{n-1}(\mathbf{x} + \mathbf{d}_{n,n-1})) \geq \alpha$ it is more likely that b_{n-1} is set to 1. This is similar to a soft decision (soft threshold) operation. Using this prior and the likelihood as shown, the posterior distribution can be written

$$p(b_{n-1}(\mathbf{x})|B, I_{n-1}) \propto p(G_n(\mathbf{x})|I_{n-1}, \mathbf{d})p_b(b_{n-1}(\mathbf{x}))\exp(-\alpha b_{n-1}(\mathbf{x})) \tag{7.42}$$

This is the same as equation 7.4 except ignoring motion, the hidden corruption layer and the unknown data as parameters. Adopting a 1-tap purely temporal 3DAR model with coefficient $a_1 = 1.0$ in equation 7.4 makes the transformation complete. The same expressions can be written for b_{n+1}. In effect the detection problem has been restated as finding occlusion in both the forward and backward directions, normally an invalid state but important here because missing data causes this 'invalid' state. To perform interpolation, Morris employed a GMRF as the image model. Data was interpolated at locations detected as missing in the first stage. He went on to propose Simulated and Mean Field Annealing processes for solution of the resulting optimization problem for both the detection and interpolation stages. In the results to follow, the ICM algorithm [15] is employed (for speed).

7.7.3 The N.T.U. systems: 1994–1997

The Nanyang Technological University (N.T.U.) in Singapore was introduced to the problem of Motion Picture Restoration in a collaborative project with the Signal Processing Laboratory at Cambridge. Some of the algorithms in [96] were implemented on massively parallel DSP arrays in an effort to improve the speed of the model based process presented there. The team of C. Man Nang, D. Krishnan, S. Kalra and G. Wooi Boon (at N.T.U.) worked on adjustments to the model based routines in [96] to improve the performance in occluded/uncovered areas, as well as in statistically non-stationary image areas. Recently they have developed modifications to the priors employed by Morris [118] to increase the robustness of the Morris system to errors in motion estimation. In what follows their contributions to the missing data removal problem in image sequences are considered in turn.

Bi-directional prediction

Occlusion and uncovering of objects has not been directly incorporated into the expressions derived so far. In these regions, if no suitable motion estimate can be found, then the tendency of the 3DAR interpolation process is to give a blurred or 'compromise' interpolant, provided of course that missing data has been detected there. In [19] Wooi Boon et al employ bi-directional prediction to achieve missing data reconstruction and so attempt to overcome this problem. The idea is that in this situation a prediction using the chosen 3DAR model, will yield reconstructed data with different prediction error depending on whether the prediction is made based on the previous or the next frame. Thus if two reconstructions for the missing patch are generated using data in the previous or the next frame only, then the reconstruction with the lower sum squared prediction error is used as the interpolant. Of course, doing this everywhere in the image has jettisoned the knowledge that at each missing data location occurring in areas which have not been occluded/uncovered; both initial (past) *and* final (future) conditions exist simultaneously to constrain the interpolant. Nevertheless, in practice this approach to image reconstruction is found to give visually pleasing results which are better than at least one of the interpolation processes above (MBI-1) in occluded regions.

The alternative derivation of this approach is as follows. Consider using a 3DAR model with taps only in the previous frame, and then using prediction equations only in the current frame to generate an interpolant at the site of missing data. Then the matrix $[\mathbf{A}_u^T \mathbf{A}_u]$ in MBI-1 (for example) becomes the identity matrix and the estimate of the interpolated data $\mathbf{i}_u$ becomes the predicted values of the image sequence in the missing locations.

This process of bi-directional prediction can be related to the configuration of hidden occlusion parameters as employed by Morris [122]. The details are not important here.

Block based non-stationary prediction

MBI-1 and MBI-2 are limited in the size of distortion that they can treat. It is assumed in those algorithms that the missing region is surrounded spatially by known data within some defined bounding box. In that box it is assumed that the motion is the same throughout. In large missing areas this assumption is violated and it is likely that motion may be different from one area to another. Furthermore the assumption that a single AR model must be used across the box is no longer valid and this may lead to distortion in the interpolated data. To get around this problem Kalra et al define a block structure over the image in the same way as in MBI-3 and the JOMBADI algorithm, in which a single set of AR coefficients is used in each block [75]. Given that their interpolation process is actually a prediction in the missing region, it is then a simple matter to reconstruct missing data in each model block since there is no knowledge of spatial consistency of reconstruction. For instance, in the case of a block containing only missing data which is itself surrounded by missing data, the interpolation equations for MBI-1 and MBI-2 generally become ill-conditioned whenever the models have some spatial support in the current frame. This does not arise with models without this support, such as the 5-tap 3DAR model containing 5 points of support in a '+' shape in the previous frame. This ill-conditioning is as a direct result of the lack of spatial initial conditions in the current frame in this situation. However, if the prediction equations giving the interpolation are restricted to the current frame only *and* the model is chosen to have support in the previous frame only, then as stated above the interpolation becomes a prediction from data in the previous or next frame only and then the situation is better conditioned.

Of course for large missing regions it is likely that the motion estimate is poor and so some motion correction is important. Thus for large missing regions using bi-directional prediction or not is likely to be immaterial when faced with this more fundamental problem. The N. T. U. team have not yet employed the benefits of motion correction.

An improved Morris system

In [28] Man Nang et al recognized that perhaps the biggest reason for false alarms when using purely temporal detectors of missing data, such as the SDIa or the Morris system, was erroneous motion estimation at moving edges. The JOMBADI algorithm is immune to this problem since it refines the motion estimate as part of the detection/interpolation exercise, but in two step sys-

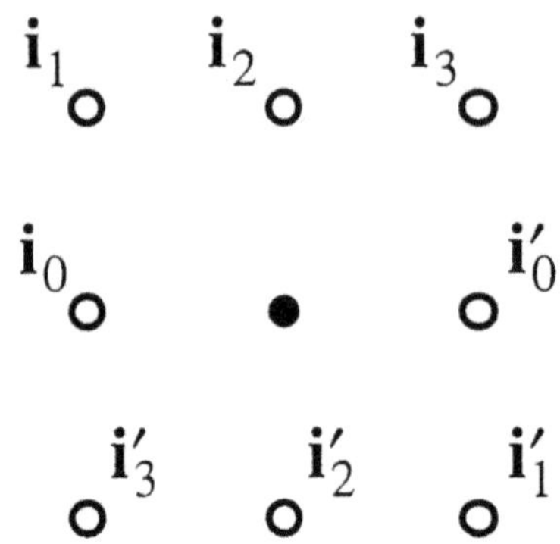

FIGURE 7.2. Eight connected neighbourhood (∘) used for detecting a moving edge at the central site (•).

tems like those introduced in the previous chapter, false alarms due to motion remains an issue. The problem manifests simply because at a moving edge the spatial resolution of the edge as well as the possibly block based nature of the motion estimation process chosen can cause motion estimates to be wrong by slight amounts e.g one pixel. This may lead to erroneously high measurements for displaced frame differences at the edge resulting in false detection of missing data. To get around this problem in the Morris system a data dependent prior was introduced which increased the penalty for setting a 1 in the discontinuity field whenever a moving edge was detected. The moving edge detector was a simple deterministic process which compared spatial and temporal gradients in motion compensated data. The Bayesian reader may question this solution since the prior is data dependent which perhaps violates the understanding that a prior encodes the expected nature of unknown values *before* the data is observed. Nevertheless, the solution does yield lower false alarms and is an improvement over the Morris system.

The prior is defined as

$$p(b_{n-1}(\mathbf{x})|B) \propto p_b(b_{n-1}(\mathbf{v},\mathbf{x}))\exp\left(-\alpha(\mathbf{x})b_{n-1}(\mathbf{x})\right) \tag{7.43}$$

$$\propto \exp\left(-\sum_{\mathbf{v}\in\mathcal{V}}\lambda(\mathbf{x})|b(\mathbf{x})-b(\mathbf{v})|\right)\times\exp\left(-\alpha(\mathbf{x})b_{n-1}(\mathbf{x})\right) \tag{7.44}$$

where $\mathcal{V}$ defines the eight connected neighbourhood around $\mathbf{x}$, and $\lambda(\mathbf{x}),\alpha(\mathbf{x})$ are modified weights which are determined by the output of a moving edge detector $\phi(\mathbf{x})$ as follows:

$$\begin{aligned}\lambda(\mathbf{x}) &= \lambda(\mathbf{v})+\phi(\mathbf{x})\\ \alpha(\mathbf{x}) &= \alpha+\phi(\mathbf{x})\end{aligned}$$

This prior increases the penalty term and spatial 'glue' in the detection field whenever a moving edge is detected. It therefore suppresses detection at moving edges. It is no substitute for actually solving for motion as part of the detection process.

The moving edge detector is described in the following steps (see [28] for further information).

1. Locate significant edges in the image. In [28] this was done by calculating the 4 directional spatial gradients across the eight connected neighbourhood at each site $\mathbf{x}$. Refer to Figure 7.2 to see the situation. If any pair of differences $|i_n - i'_n|$ is greater than a threshold t_d (25 was used), then $\mathbf{x}$ is flagged as a possible edge. Define the pair selected as i_e, i'_e. If not more than one other pixel in the neighbourhood was also flagged as a possible edge then the 3 pixel pair i_e, i, i'_e is flagged as a true edge.

2. To distinguish moving edges from edges that are stationary within an object, another thresholding step follows. For each site $\mathbf{x}$ that was flagged as an edge pixel, each pixel in its neighbourhood which was not similarly flagged is tested. Denote these neighbourhood sites as $\mathbf{z}$. Site $\mathbf{x}$ is flagged as a *moving* edge when

$$\begin{aligned} &|I_n(\mathbf{z}) - I_{n-1}(\mathbf{z} + d_{n,n-1}(\mathbf{z}))| < t'_d \\ \text{OR } &|I_n(\mathbf{z}) - I_{n+1}(\mathbf{z} + d_{n,n+1}(\mathbf{z}))| < t'_d \end{aligned} \tag{7.45}$$

is satisfied for all the pixel sites $\mathbf{z}$ in the neighbourhood of $\mathbf{x}$. t'_d is another threshold to be set. Thus a moving edge is detected when the DPD (Displaced Pixel Difference) is large on one or the other side of the edge.

3. The output of the moving edge detector is defined as

$$\phi(\mathbf{x}) = 90(i_e - i'_e)^2/t_d^2 \tag{7.46}$$

The choice of the factor 90 in the expression for $\phi(\cdot)$ seems somewhat arbitrary but is no doubt justified from experiment.

7.7.4 Related work in audio

A number of extremely high quality audio reconstruction techniques have been developed at the Signal Processing Laboratory at Cambridge since the mid eighties. The work began with Sayeed Vaseghi [182] who developed essentially deterministic techniques for restoration, but laid down the fundamental approach to using AR models for reconstruction of missing data. Around the same time similar techniques were developed by Veldhuis [184]. Simon Godsill has been responsible for the development of Bayesian approaches to the problem [51, 54]. He has designed a number of tools for joint detection and reconstruction of missing data.

7.7.5 The Rank Order Detector: 1996

In [123] Nadenau and Mitra presented a technique for detection of missing data which relied on tests performed on a ranked array of displaced pixel differences called the ROD (Rank Order Detector). It was outlined in the previous chapter. Its improved performance over other heuristics is due mainly to its increased robustness to motion estimation errors at moving edges. The first step involves ranking the pixel values in the ROD neighbourhood. Displaced pixel differences are calculated at the current site using each pixel in the ROD neighbourhood (after some rejection of outliers) in the previous and next frames. In a sense it is evaluating the likelihood defined by equation 7.6 or 7.40 at several candidate motion locations. This gives some information about the spread of the likelihood distribution for the detection field $b(\mathbf{x})$ at a particular site. It is difficult to justify the actual steps used in the ROD detector on Bayesian grounds, but the posterior distribution used for evaluation of the detection field could be of the form

$$p(b(\mathbf{x})|G_n, I, \mathbf{d}) \propto \exp\left(-\frac{(G_n(\mathbf{x}) - I(\mathbf{d}))^2(1 - b(\mathbf{x})}{2\sigma_e^2}\right) \exp\left(-\alpha b(\mathbf{x})\right) \tag{7.47}$$

where $I(\mathbf{d})$ represents the value of a pixel in the spatio-temporal neighbourhood defined by the ROD (suitably compensated for motion), it may be in the previous or next frame. $\mathbf{d}$ is a candidate motion vector and α is a penalty which could be related to a threshold as described in the previous section on the Morris system. Thus for each candidate motion or each pixel in the ROD neighbourhood the value of α determines whether the MAP estimate of $b(\mathbf{x})$ (given that particular motion candidate) is 1 or not as discussed previously. If *all* candidates set $b(\mathbf{x}) = 1$ then there is evidence for a true missing pixel at $\mathbf{x}$. This procedure does not result in the ROD detector as it is described in [123] since that involves three thresholds which further examine the spread of the likelihood distribution. However, it is hoped that this discussion has shed some light on the relation between the ROD detector and the JOMBADI algorithm.

7.7.6 Early work at the BBC: 1982

As was stated in the previous chapter, Richard Storey was perhaps the first to design and implement an automatic detection system for dirt and sparkle in archived motion pictures. In his work [164, 163] he describes a non-motion compensated version of the SDIp detector introduced in the previous chapter. The detector relied on the useful observation that the sign of the displaced frame difference at a site of missing data would usually be the same both forward and backward in time. It is difficult to relate this use of sign information to the models employed for the image sequence in this chapter. It is especially difficult to see how this sign information is being employed (if at all) by the JOMBADI algorithm or indeed by any of the inferential Bayesian procedures described previously. However, it is possible to derive the Storey SDIp detector from a Bayesian perspective by noting that it actually detects missing data with less accuracy (albeit with much less false alarm) than any of the detectors discussed previously. In practice it can perform better than the SDIa simply because its assumption about the sign of the displaced pixel difference is correct most of the time, resulting in perhaps a better tradeoff between correct detection and false alarm rate. This suggests that it is using more prior knowledge about blotches and the behaviour of motion estimators in the presence of blotches. If a prior for the detection variable is introduced as

$$p(b(\mathbf{x})) \propto \exp\left(-\lambda[1 - b(\mathbf{x})\mathrm{sgn}(f)\mathrm{sgn}(b)]\right) \tag{7.48}$$

where $\mathrm{sgn}(f)$ refers to the sign of the forward displaced pixel difference at $\mathbf{x}$ and $\mathrm{sgn}(b)$ the same for the backward difference, then the detection of blotches is suppressed when the signs are not the same, which is the desired action. The λ would give a soft threshold behaviour. Using this prior instead of the spatial smoothness prior used for the Morris detection system above (or in the JOMBADI process), would result in the SDIp as required.

7.8 Fast algorithms

The crucial step in generating an interpolant using the 3DAR model is generating the Least Squares Estimate for $\mathbf{i}_u$. This section presents fast algorithms for solving for this estimate using the expression

$$\mathbf{i}_u = [\mathbf{A}_u^T \mathbf{A}_u]^{-1} \mathbf{A}_u^T \mathbf{A}_k \mathbf{i}_k \tag{7.49}$$

Or equivalently, solving for $\mathbf{i}_u$ from

$$[\mathbf{A}_u^T \mathbf{A}_u] \mathbf{i}_u = \mathbf{A}_u^T \mathbf{A}_k \mathbf{i}_k \tag{7.50}$$

The matrices have already been defined. Only the 2-D interpolation case is considered here for simplicity, the arguments apply directly to more dimensions without any complications. Figure 7.3 shows a grid of 4×4 pixels in which the centre block of 2×2 pixels is missing, indicated by a dark square. The discussion begins with the use of a 3 tap spatial AR model employed for reconstruction. The model topology is as shown in Figure 7.3 and it is assumed that the three coefficients a_1, a_2, a_3 have already been estimated. At the locations in the grid marked by $\bullet$, 9 prediction equations can be written as follows, for each image intensity value $i_{h\ k}$ where h, k is the (row, column) position of that pixel. These

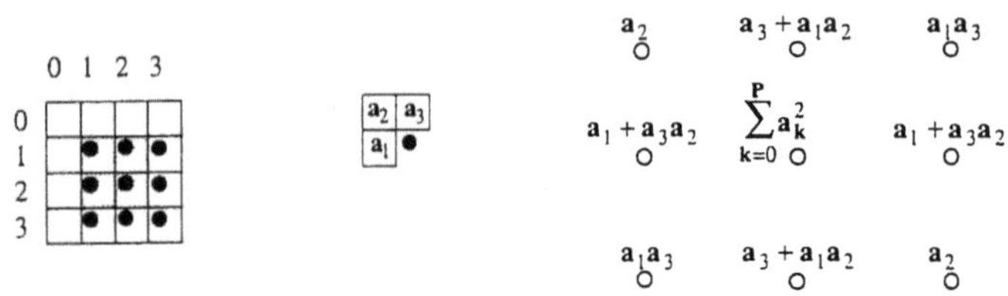

FIGURE 7.3. Left to right: Block of 4×4 pixels showing centre missing 2×2 block, prediction error mask showing filter taps a_1, a_2, a_3 with $a_0 = 1.0$ at centre labelled with •, and filter resulting from convolving the prediction mask with itself.

coordinates are shown on the figure.

$$\begin{bmatrix} e_{11} \\ e_{12} \\ e_{13} \\ e_{21} \\ e_{22} \\ e_{23} \\ e_{31} \\ e_{32} \\ e_{33} \end{bmatrix} = \begin{bmatrix} & & & & & * & * & & & * & * & & & & & \\ a_2 & a_3 & 0 & 0 & a_1 & 1 & 0 & 0 & 0 & 0 & 0 & 0 & 0 & 0 & 0 & 0 \\ 0 & a_2 & a_3 & 0 & 0 & a_1 & 1 & 0 & 0 & 0 & 0 & 0 & 0 & 0 & 0 & 0 \\ 0 & 0 & a_2 & a_3 & 0 & 0 & a_1 & 1 & 0 & 0 & 0 & 0 & 0 & 0 & 0 & 0 \\ 0 & 0 & 0 & 0 & a_2 & a_3 & 0 & 0 & a_1 & 1 & 0 & 0 & 0 & 0 & 0 & 0 \\ 0 & 0 & 0 & 0 & 0 & a_2 & a_3 & 0 & 0 & a_1 & 1 & 0 & 0 & 0 & 0 & 0 \\ 0 & 0 & 0 & 0 & 0 & 0 & a_2 & a_3 & 0 & 0 & a_1 & 1 & 0 & 0 & 0 & 0 \\ 0 & 0 & 0 & 0 & 0 & 0 & 0 & 0 & a_2 & a_3 & 0 & 0 & a_1 & 1 & 0 & 0 \\ 0 & 0 & 0 & 0 & 0 & 0 & 0 & 0 & 0 & a_2 & a_3 & 0 & 0 & a_1 & 1 & 0 \\ 0 & 0 & 0 & 0 & 0 & 0 & 0 & 0 & 0 & 0 & a_2 & a_3 & 0 & 0 & a_1 & 1 \end{bmatrix} \begin{bmatrix} i_{00} \\ i_{01} \\ i_{02} \\ i_{03} \\ i_{10} \\ i_{11}^* \\ i_{12}^* \\ i_{13} \\ i_{20} \\ i_{21}^* \\ i_{22}^* \\ i_{23} \\ i_{30} \\ i_{31} \\ i_{32} \\ i_{33} \end{bmatrix} \quad (7.51)$$

In the above, the unknown data is marked with a '*' and these elements constitute the unknown data vector $\mathbf{i}_u$. The other elements, which are known, are assembled into the column vector $\mathbf{i}_k$. This makes it easier to recognize the composition of the matrices $\mathbf{A}_k$ and $\mathbf{A}_u$ in the large coefficient matrix on the right hand side. The columns of coefficients which make up the matrix $\mathbf{A}_u$ are similarly marked with a '*' and the other unmarked columns constitute $\mathbf{A}_k$. The particular topology of the sites chosen for prediction error equations is selected since these are *all* the sites at which the prediction filter employs at least one tap of *unknown* data. After reading this section in its entirety, the reader may verify that increasing the extent of the prediction error sites beyond this limit does not change the interpolated values at all. However, decreasing the number of sites used would affect the interpolant. In that case there would not necessarily be enough boundary conditions to constrain the interpolant sufficiently. This would mean that it blended less smoothly with the known data points around it.

The first step in calculating $\mathbf{i}_u$ is to compute $\mathbf{A}_k\mathbf{i}_k$. A direct matrix multiply would require of the order of $9 \times 12 \times 2$ operations. Generally, given $M \times M$ missing pixels in an $N \times N$ pixel data block with this model, requires $(N-1)^2 \times (N^2 - M^2) \times 2$ operations to perform this matrix multiply. However, writing out the multiply in full allows a much faster operation.

$$\mathbf{A}_k\mathbf{i}_k = \begin{bmatrix} a_2 & a_3 & 0 & 0 & a_1 & 0 & 0 & 0 & 0 & 0 & 0 & 0 \\ 0 & a_2 & a_3 & 0 & 0 & 0 & 0 & 0 & 0 & 0 & 0 & 0 \\ 0 & 0 & a_2 & a_3 & 0 & 1 & 0 & 0 & 0 & 0 & 0 & 0 \\ 0 & 0 & 0 & 0 & a_2 & 0 & a_1 & 0 & 0 & 0 & 0 & 0 \\ 0 & 0 & 0 & 0 & 0 & 0 & 0 & 0 & 0 & 0 & 0 & 0 \\ 0 & 0 & 0 & 0 & 0 & a_3 & 0 & 1 & 0 & 0 & 0 & 0 \\ 0 & 0 & 0 & 0 & 0 & 0 & a_2 & 0 & a_1 & 1 & 0 & 0 \\ 0 & 0 & 0 & 0 & 0 & 0 & 0 & 0 & 0 & a_1 & 1 & 0 \\ 0 & 0 & 0 & 0 & 0 & 0 & 0 & a_3 & 0 & 0 & a_1 & 1 \end{bmatrix} \begin{bmatrix} i_{00} \\ i_{01} \\ i_{02} \\ i_{03} \\ i_{10} \\ i_{13} \\ i_{20} \\ i_{23} \\ i_{30} \\ i_{31} \\ i_{32} \\ i_{33} \end{bmatrix} \tag{7.52}$$

As can be seen, $\mathbf{A}_k$ is very sparse. Examining each row reveals an equivalent filtering operation. Each row is in fact made up of the coefficients of the prediction error filter also shown in Figure 7.3, but with the important modification that filter taps which occur at sites of missing data are set to 0. In other words the operation $\mathbf{A}_k\mathbf{i}_k$ yields the same result as filtering the block of data with the prediction error filter, except leaving out taps which occur at missing locations. Furthermore the filtering operation is carried out only at the sites at which the prediction error equations have been set up, indicated by • in Figure 7.3. For this 3-tap filter (4 taps including the current one), the number of operations required is now roughly $4 \times (N-1)^2 \times 2$ in general, and $4 \times 9 \times 2$ here. This represents a saving of a factor of 3 for this small example, but $\frac{N^2-M^2}{4}$ in general, i.e. from order N^2 to order $P+1$ where P is the order of the AR model chosen, in this case 3. Furthermore the large matrix $\mathbf{A}_k$ need not be stored at all, thus reducing memory requirements to the storage of the filter coefficients and the filtered data only.

The next step is to calculate $\mathbf{A}_u^T\mathbf{i}_f$, where $\mathbf{i}_f = \mathbf{A}_k\mathbf{i}_k$ has just been calculated. Again a simple filtering operation gives the same result and reduces the computation time as well as the memory requirements enormously.

$$\mathbf{A}_u^T\mathbf{i}_f = \begin{bmatrix} 1 & a_1 & 0 & a_3 & a_2 & 0 & 0 & 0 & 0 \\ 0 & 1 & a_1 & 0 & a_3 & a_2 & 0 & 0 & 0 \\ 0 & 0 & 0 & 1 & a_1 & 0 & a_3 & a_2 & 0 \\ 0 & 0 & 0 & 0 & 1 & a_1 & 0 & a_3 & a_2 \end{bmatrix} \begin{bmatrix} i^f_{11} \\ i^f_{12} \\ i^f_{13} \\ i^f_{21} \\ i^f_{22} \\ i^f_{23} \\ i^f_{31} \\ i^f_{32} \\ i^f_{33} \end{bmatrix} \tag{7.53}$$

The elements of $\mathbf{i}_f$ have been placed onto the same grid as defined in figure 7.3 at the locations of the prediction error sequence, hence their notation $i^f_{h\ k}$. The reader can now notice that the rows of $\mathbf{A}_u^T$ are actually the coefficients of the *mirrored* prediction filter, also shown in Figure 7.3. Furthermore, the output of this operation (in this case) is just 4 values corresponding to the 4 missing pixel sites. The mirror prediction filter is therefore applied to the missing sites only, using the filtered data $\mathbf{i}_f$ placed there, with all other sites on the grid (which were not prediction error sites) set to 0. This again implies that the matrix $\mathbf{A}_u$ need not be stored and also the number of operations has dropped from order $(N-1)^2$ to $P+1$, the order of the AR model chosen, in this case 3. The saving as far as memory requirements goes (for both $\mathbf{A}_u$ and $\mathbf{A}_k$) is from $(N-1)^2N^2$ elements down to $P+1$, a massive reduction!

The final steps involve calculating $\mathbf{A}_u^T\mathbf{A}_u$ (see equation below) and then taking the inverse, or solving the set of simultaneous equations. Writing out this operation shows that in fact every element of $\mathbf{A}_u^T\mathbf{A}_u$ is actually one coefficient from a filter which results from convolving the prediction error filter with itself. Therefore there need never be any multiplication involved in calculating $\mathbf{A}_u^T\mathbf{A}_u$ aside from calculating the initial convolved prediction filter. $\mathbf{A}_u^T\mathbf{A}_u$ is found by simply distributing the coefficients of the convolved prediction filter inside an $M \times M$ matrix. The process of distribution is defined by the spatial arrangement of the missing pixels on the original grid.

$$\mathbf{A}_u^T\mathbf{A}_u = \begin{bmatrix} \sum_{k=0}^{P} a_k^2 & a_1 + a_2a_3 & a_3 + a_1a_2 & a_2 \\ a_1 + a_2a_3 & \sum_{k=0}^{P} a_k^2 & a_1a_3 & a_3 + a_1a_2 \\ a_3 + a_1a_2 & a_1a_3 & \sum_{k=0}^{P} a_k^2 & a_1 + a_2a_3 \\ a_2 & a_3 + a_1a_2 & a_1 + a_2a_3 & \sum_{k=0}^{P} a_k^2 \end{bmatrix} \tag{7.54}$$

Recalling that $a_0 = 1.0$, the rows of $[\mathbf{A}_u^T\mathbf{A}_u]$ above can be seen to be made up of the coefficients of the prediction filter convolved with itself as shown in Figure 7.3. The filter is applied only at the locations known to be missing, corresponding to each row of $\mathbf{A}_u^T\mathbf{A}_u$, and taps which use data at sites which were not sites of prediction equations (indicated by • in Figure 7.3) are set to zero.

At this point there are two mechanisms which could be used for finding the least squares estimate of $\mathbf{i}_u$. The simple approach is to calculate the matrix $[\mathbf{A}_u^T\mathbf{A}_u]$ as above and then solve the set of equations 7.50 using Cholesky decomposition. This would also give as a by-product the square root $\sqrt{\mathbf{A}_u^T\mathbf{A}_u}$ as required for generating a *sample* of the interpolated data from the multivariate posterior distribution. However, as it is known that the action of multiplying by $[\mathbf{A}_u^T\mathbf{A}_u]$ is the same as filtering at the unknown data sites by the convolved prediction filter, the conjugate gradient method could be used to solve the system of equations instead. This would not yield the required square root as a by-product for sampling, but could be much quicker than Cholesky decomposition for generating the least squares interpolant, particularly for large amounts of missing data.

7.8.1 Single site interpolation

The JOMBADI algorithm requires a single site update for the missing data, which requires the estimation of the least squares interpolant for a single missing pixel given a completely known data neighbourhood. For a single missing site, the matrix $\mathbf{A}_u$ is actually a column vector $\mathbf{a}_u$. The reader may verify this by repeating the arguments in the previous section for a single missing site. Using all the possible prediction equations which involve the single missing pixel, $\mathbf{a}_u^T\mathbf{a}_u$ is simply $\sum_{k=0}^{P} a_k^2$. Furthermore the arguments above which show that the various matrix multiplications are simple filtering operations also apply here and result in a trivial algorithm for interpolating a single pixel. Finally, the required square root for sampling from the posterior distribution for a single pixel is just $\sqrt{\sum_{k=0}^{P} a_k^2}$.

7.9 Performance

As in the previous chapters, the behaviour of the algorithms described so far are illustrated here using both artificial and real degraded movies. The MOBCAL sequence was degraded with blotches according to equation 7.10 with an additional penalty term to suppress the presence of extremely large degradation. The degraded sequence (BCALA) as well as the original sequence (GC256) can be found on the CD-ROM accompanying this book[2]. The sequences are created from 256×256 portions of the MOBCAL sequence. The blotches look realistic[3] and an interesting set of three frames (frames 7,8,9) are shown in Figure 7.4. The bottom right image shows in white the areas that were corrupted by blotches. In the experiments which follow, the first 12 frames of the sequence were used. The initial motion estimate was obtained using the MWBME with 3 multiresolution levels, blocksize of 9×9 pixels (2 pixel overlap) for the first 2 levels and 5×5 for the final one, ill-conditioning threshold 100.0, motion threshold 5.0 grey levels per pixel, and the variance of the Gaussian kernel used for pyramid generation was set to 1.0.

The performance of the algorithms can be examined from the standpoint of detection accuracy and interpolation fidelity. The JOMBADI algorithm performs both these tasks while the other methods require two step systems i.e. detection followed by interpolation. The detection accuracy of the various detectors proposed in this and the previous chapter is summarized in figure 7.5. It shows a receiver operating characteristic (ROC) for each of the detector systems including the JOMBADI system. As in the previous chapter, the correct detection rate was measured as the ratio between the number of pixels correctly detected as corrupted and the total number of corrupted pixels. The false alarm rate is the ratio between the number of pixels detected as corrupted which were in fact not corrupted, and the number of uncorrupted pixels in the

[2]The reader is invited to use these sequences in testing or comparing algorithms.

[3]The careful observer will note that they are not exactly the same as real blotches.

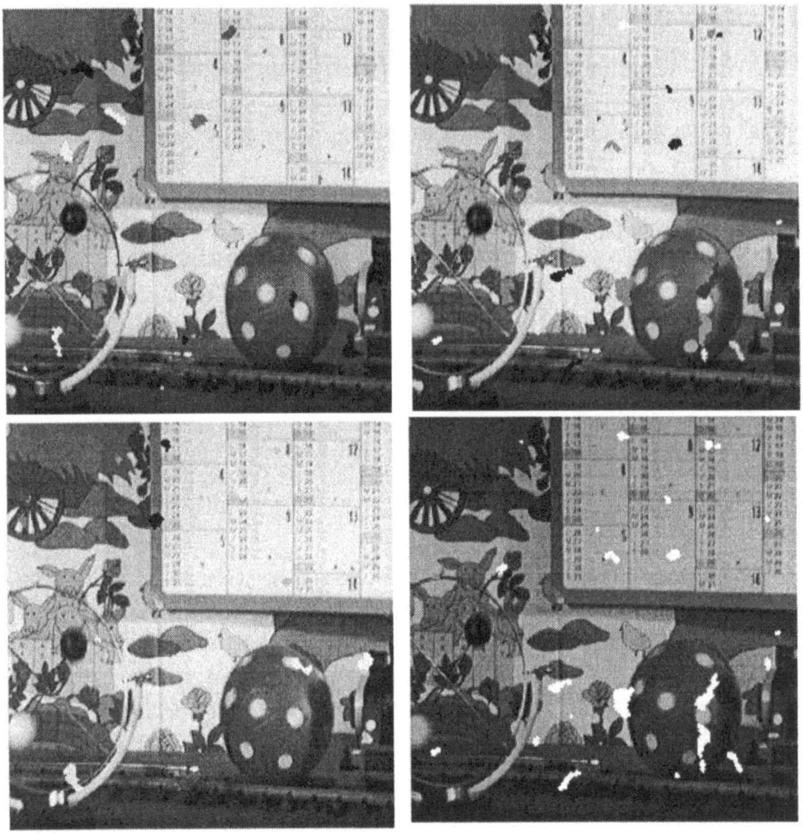

FIGURE 7.4. Clockwise from top left: Frames 7,8 of the degraded MOBCAL sequence, Frame 8 with location of distortion overlayed in white (image darkened to enhance contrast with indicated blotches), Degraded frame 9.

image. The rates for each frame were then averaged over all the frames used in the experiment. Each point on the ROC represents measurements made with a particular setting for the user defined parameter(s) required by the various detectors. The curves shown for the ROD, SDIa and SDIp are the same as those in the previous chapter. The figure shows that JOMBADI outperforms the other detectors by a large margin, whether 1 tap or 5 tap (in a + shape in the previous frame) 3DAR models are used. None of the other detectors can operate at 86% accuracy with less than 10^{-4} false alarm, for this false alarm rate their accuracies drop to less than 50%.

To generate the ROC for the Morris and JOMBADI detectors it is necessary to perform several different runs with several sets of hyperparameter values. In both algorithms there are 2 hyperparameters that need setting, for JOMBADI they are $\lambda_{\underline{z}}^{c}, \lambda_{\underline{z}}^{b}$, for Morris it is λ_b, α. It is sensible to presume that with judicious choice of values for these parameters it is possible to obtain performance which would cover much of the ROC space. The important factor is to determ-

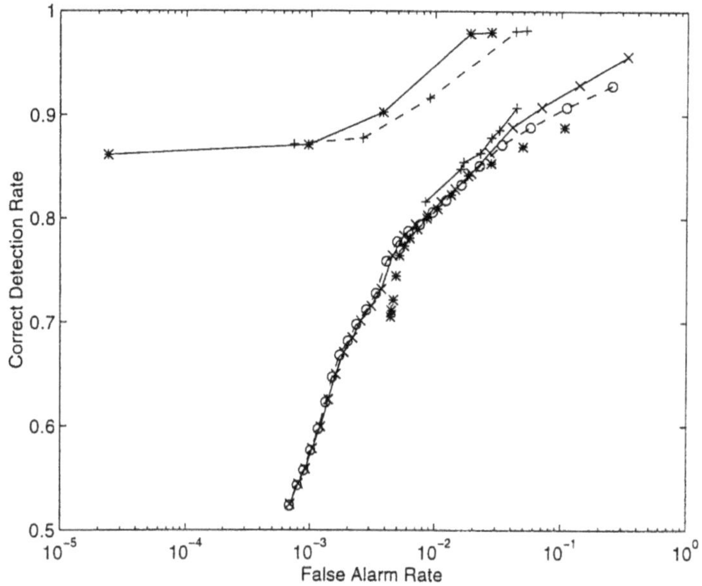

FIGURE 7.5. Receiver Operating Characteristics for various detectors. Morris (+–), SDIa (×–), SDIp (o- -), ROD (∗ · · ·), JOMBADI (∗–) 1 tap 3DAR process, (+- -) 5 tap 3DAR process.

ine whether many of the values for these parameters actually give reasonable performance. Five pairs of values were tried for JOMBADI, yielding the observed characteristic, and 7 pairs for the Morris system. In JOMBADI λ^c_z, λ^b_z are set by scaling a circularly symmetric weighting function (values inversely proportional to the radius from the centre of the eight connected neighbourhood) with a single value λ^c, λ^b. The values used for each point (from top right of the graph to bottom left) were the same for both models used as follows: $(\lambda^c, \lambda^b) = (0.1,1.0), (0.15,1.0), (0.15,4.0), (0.4,8.0), (1.0,8.0)$. The reader can observe therefore, that as the strength of the spatial 'glue' increases, the detection performance of JOMBADI suffers in correct detection but gains much with respect to false alarm. This is because more small corruption is being missed, but the larger blotches are being better delineated. Since much of the corruption is greater than 5×5 pixels in area the performance is overall better. The detection result from JOMBADI was created by counting the number of times each site in $b(\mathbf{x})$ was set to 1 as samples were generated. Sites that have been set to 1 more for more than 75% of the iterations after a 10 sample burn in, are taken as being set to 1. 50 iterations were used.

The hyperparameter pairs used for the Morris ROC (again for points from from top right to bottom left) were: $(\lambda_b, \alpha) = (4.0,1.0), (1.0,5.0), (4.0,5.0), (4.0,8.0), (4.0,10.0), (8.0,10.0), (8.0,20.0)$. The behaviour is not quite the same as for JOMBADI. Increasing the spatial 'glue' (λ_b) does help reject false alarms at the expense of detecting less small blotches but increasing the temporal penalty α causes more performance loss.

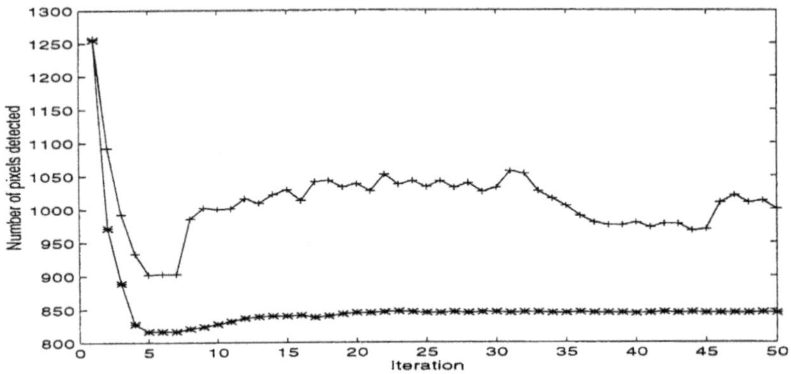

FIGURE 7.6. Convergence of JOMBADI using a 5 tap (+) and 1 tap (*) 3DAR model.

The form of the ROC's in Figure 7.5 is overall very satisfying. The Morris system performs better than the deterministic systems primarily because it introduces the notion of spatial connectivity into the indicator field. It therefore suppresses the detection of single pixel blotches which abound in the primarily temporal detectors like SDIa, SDIp and ROD (although the ROD does use some spatial information). The JOMBADI system introduces mainly two other features into detection, the spatial connectivity of the blotches themselves *and* the correction of motion. These two features account for most of the improvement in detection performance. The reader would expect that as the order of the 3DAR model increases, the JOMBADI performance should improve. It is not clear that this is occurring from the ROC's shown. Recall that the detector is only one aspect of JOMBADI, however, and the sampler was terminated after 50 iterations. Thus with increased model order it is feasible that the 'burn in' time for the sampler is more. This fact is illustrated in Figure 7.6 which shows the convergence of JOMBADI on frame 8, with hyperparameters corresponding to the leftmost point on the ROC trace in figure 7.5. Convergence is measured only by the number of pixels flagged as corrupted as the algorithm precedes, the plot cannot say anything about accuracy. The convergence of the system using both the 1 tap 3DAR model and the 5 tap is shown. It is clear that the 1 tap system has already converged after just 20 iterations, but it is uncertain whether even after 50 iterations the 5 tap system has converged. The reason for this has not yet been fully investigated, but it is suspected that the lack of priors on the allowed 3DAR models implies that it may be possible for unstable models to be proposed through random sampling for the coefficients. An unstable model may cause poor interpolation which in turn encourages pixels to be detected as corrupted when they are not. The stability of multidimensional systems is a complex problem since there exists no straightforward multidimensional factorization theorem [107]. The introduction of priors which encourage stable multidimensional models is an area for future investigation. Nevertheless, it is encouraging in figure 7.5 that the 5 tap process still performs well and even appears to bottom out at the same level as the 1 tap process. The reader

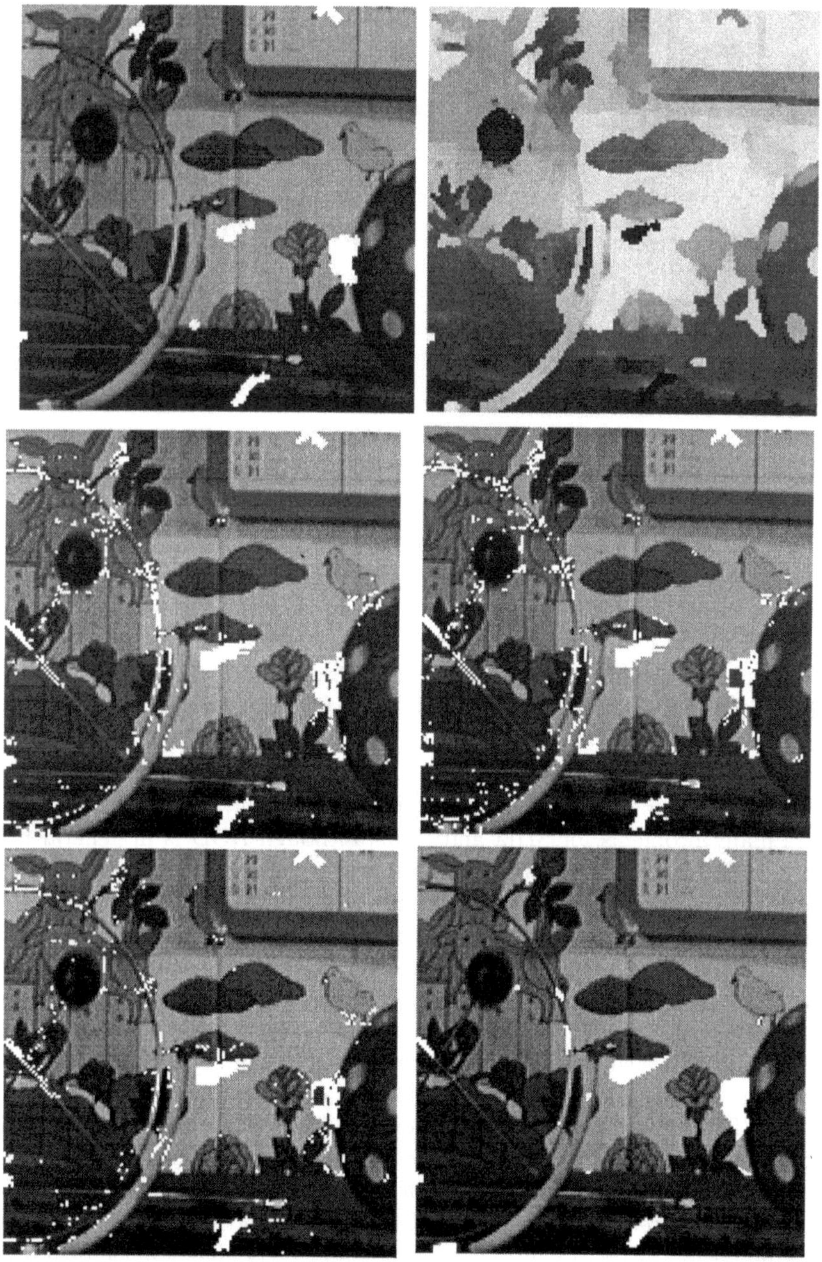

FIGURE 7.7. Detected blotches using various detectors superimposed (on slightly darkened image) in white. Top row: JOMBADI with 1 tap AR process (left), estimated $c(\mathbf{x})$ not darkened (right). Middle row: SDIa (left) SDIp (right). Bottom row ROD (left) Morris (right).

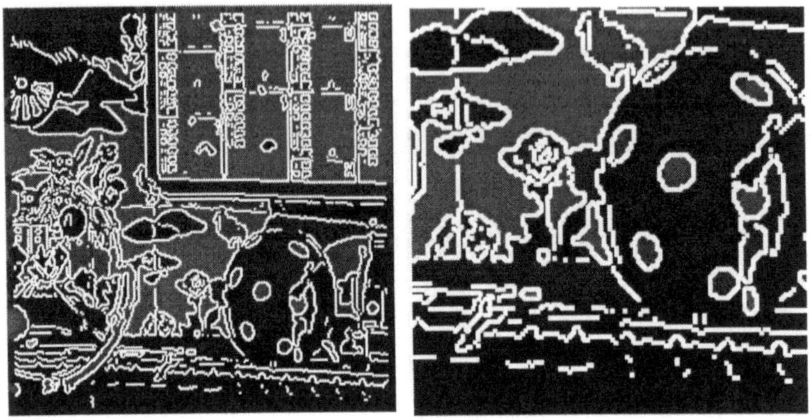

FIGURE 7.8. Edge fields (white) used for JOMBADI on frame 8 of degraded MOBCAL, Left: entire frame, Right: zoom on portion used for previous figures.

is reminded at this stage that JOMBADI is both a detector and an interpolator, therefore the behaviour of a particular model is to be assessed on both detection and interpolation, it turns out that in general increasing the order of the 3DAR model can improve the interpolation.

To give some visual feel for the areas detected, Figure 7.7 shows the results of detection on frame 8 (zoom) using JOMBADI ($\lambda^c, \lambda^b = 1.0, 8.0$), SDIa and ROD. The thresholds for ROD and SDIa were chosen so that their operating point on the ROC was almost the same at $P_c = 0.8$, $P_f = 0.015$ for this particular frame. The thresholds were 9 and 14 respectively. The spinning mobile causes many problems for SDIa and ROD, although ROD is in general more robust to the motion estimation errors in these regions. The Morris detector performs very well, achieving a good compromise between JOMBADI and the other detectors. JOMBADI is almost completely immune to the problems of motion estimation errors because of the strength of the spatial priors, and its ability to correct motion simultaneously. The figure also shows a sample (after 50 iterations) of $c(\mathbf{x})$. It is as expected. Where a blotch is detected the field is equal to the observed image, and otherwise it is a sample from the Laplacian type prior, which tends to contain large flat areas separated by gross edges. The edge field used for this frame is shown in figure 7.8, superimposed on the image itself and dilated to improve the contrast against the background image. It shows that the rough configuration of the edge field using a simple application of a DOG filter is good enough to be used for JOMBADI. Only edge sites indicating vertical and horizontal edges are shown, the diagonal elements were derived from these in order to keep computation low in the sampling steps.

The early model based detection systems are not compared here since they suffer badly from bias in the estimation of AR coefficients. In any case, the JOMBADI algorithm is in some sense the correct way of estimating these coefficients. No results are shown for the N.T.U. improvement to the Morris detection system.

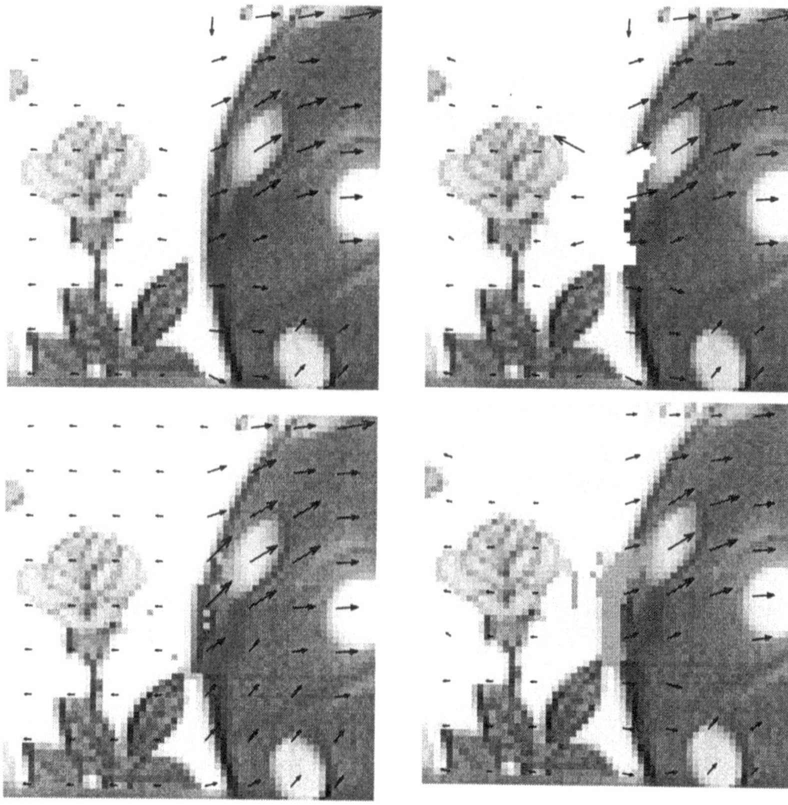

FIGURE 7.9. Clockwise from top left: Motion field using MWBME on clean original data, Motion field from MWBME with corrupted data, Motion field after motion interpolation using MBI-2 with corrupted data, Motion field after 50 iterations of JOMBADI using 1 tap AR process with corrupted data. *The vector fields in the bottom row are superimposed on the corresponding restored image.*

7.9.1 Interpolation

There are two signals being interpolated in the missing data problem, the image data *and* the motion field. Figure 7.9 illustrates the 50th sample of the motion field resulting after 50 iterations of JOMBADI on the degraded MOBCAL image and compares this with the motion field resulting after using the motion interpolator described as part of MBI-2 [83]. Also shown is the 'kick start' motion field (top right) and the motion field which results if the MWBME was used on the original, clean MOBCAL sequence (top left). The frame 8 is shown as before, and the zoom is on this particular region because it shows a moving edge at the boundary of the moving ball. The backward motion (from frame 8 into frame 7) is shown and the correct motion estimate is approximately 1.0 pixels to the left for the background and some clockwise rotation of the ball. The JOMBADI motion field is superimposed on the restored image and the MBI-2 is superimposed on the image resulting from using SDIa (threshold set to 14) followed by motion correction and the ML3Dex spatio-temporal median filter as in the previous chapter. Recall that there is one motion vector per 9×9 pixel block, the vector for each block is drawn to originate from the centre of the block. The images used have been brightened a little to improve the contrast of the vector graphic.

The motion field from MBI-2 is a definite improvement over the 'kick start' motion field, and it is even an improvement over the top left motion field. But it only corrects the motion field where the detector (SDIa) has detected a problem (see Figure 7.7). This is a sensible approach since the motion field is only corrupted where there is missing data. However it may occur that the erroneous motion causes no detection where there is a blotch. The JOMBADI algorithm in some sense corrects for motion everywhere and so is more tolerant to this problem. The motion field that results from JOMBADI is much improved in the flat background areas, since the algorithm has managed to propagate the background motion everywhere in the background despite the lack of texture in some areas. The moving edge of the ball is not exactly delineated in either method. This is due in part to the block based resolution of the vector field and also to the absence of knowledge about occlusion and uncovering. Nevertheless the edge is delineated sufficiently to result in a good interpolation as shown.

Image interpolation behaviour is compared quantitatively in Figure 7.10. The plot measures the RMS error between the original clean image and the reconstructed image for several interpolating processes. To compare the JOMBADI process fairly, and decouple its behaviour as a detector, the detection field was fixed to the known areas of corruption in the artificially degraded MOBCAL sequence. The other interpolators similarly acted only in known areas of degradation. The RMS error was measured only over the corrupted regions. The interpolation used from the JOMBADI algorithm was the average of 40 samples from the Gibbs Sampler after a 10 sample burn in. It would not be sensible to compare a single sample from the Gibbs Sampler with respect to RMS of reconstruction since a random sample from the posterior for the missing data is likely to have a higher RMS error than the LS interpolant despite the fact that it may look better. In other words, MBI-1, MBI-2 etc. attempt to generate some approximation to the true image data which optimizes some criterion resulting in a Least Squares or MAP estimate. There is absolutely no guarantee that a random sample from the distribution for the missing data (generated by JOMBADI) would be close to the mode. However the average of random samples from the distributions would give something close to the MMSE (Minimum Mean Squared Error) estimate for the missing data, and this can be fairly compared with the results from MBI-1, MBI-2 etc.

As can be seen, the model based interpolators generally performed better than the median structure. The median system was implemented by using the ML3Dex filter in place of the model based reconstruction step in MBI-2. The bi-directional predictor sometimes performed better than either MBI-2 or JOMBADI (e.g. frames 2,12). Overall the early model based interpolators performed similarly on this sequence. The sometimes better performance of the bi-directional predictor is due to its ability to overcome occlusion/uncovering

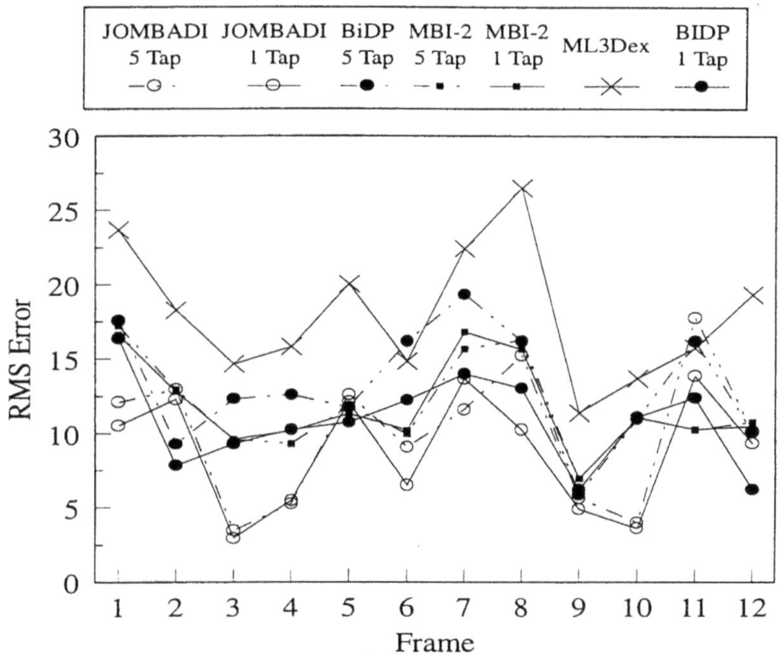

FIGURE 7.10. RMS Error for each frame of the degraded MOBCAL sequence reconstructed using several systems, JOMBADI interpolation is the average of 40 samples after 10 sample burn in, BiDP refers to the Bi-Directional interpolator as discussed in section 7.7.3.

problems near the boundary of moving objects. Neither JOMBADI nor MBI-2 as proposed have any provision for temporal image discontinuities like these, even though it is a relatively simple adjustment to the algorithms proposed. Since the blotches occur on moving edges only some of the time, the improvement shown by the bi-directional interpolator is intermittent.

The importance of correcting for motion is highlighted quantitatively in figure 7.11 using the MBI-2 system with and without (MBI-1) the motion correction step. The Bi-Directional predictor system is also shown with and without motion compensation. All systems show improvement with motion correction. The reader should be convinced at this stage that motion correction is a vital part of missing data removal in image sequences, regardless of the exact nature of the interpolator employed.

Figure 7.12 gives some visual feel for the interpolations possible using various schemes and also highlights the differences. The figure shows a zoom including the same portion of MOBCAL used to display motion estimation results previously. The bi-directional predictor shows an extremely good result at the moving edge as compared to the 'compromise' interpolation achieved by the 5 tap interpolator in JOMBADI and MBI-2. However, JOMBADI performs better elsewhere because it is designed to cope (to some extent) with the inhomogeneous nature of images (due to the block based model parameter sets)

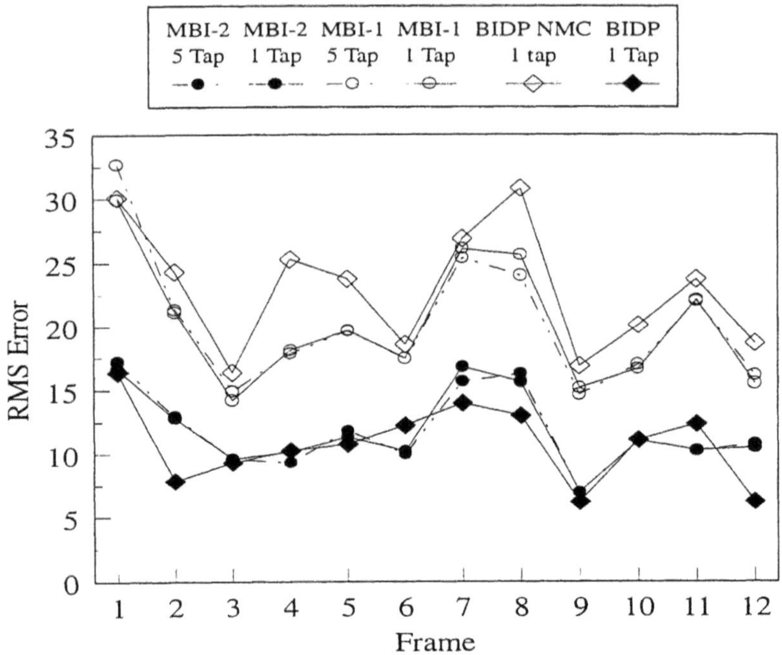

FIGURE 7.11. RMS Error for each frame of the degraded MOBCAL sequence reconstructed using several model based systems with and without motion correction. BiDP NMC refers to Bi-Directional Prediction *without* motion correction.

and is able to derive better parameter values. The figure also illustrates an important point with respect to the artificial degradation used here. The sequence was not processed to prevent blotches from occurring in the same or nearly the same location in consecutive frames. Therefore it is quite possible that the simple interpolation schemes would match consecutive blotches in the motion correction step. This detrimentally affects the subsequent interpolation, since it is assumed that the motion compensated data in the surrounding frames is uncorrupted. JOMBADI is more robust to this problem. This phenomenon occurs on the right hand side of the moving ball where MBI-2 and Bi-directional Prediction interpolate erroneously. The reader can verify that blotches in Figure 7.4 have occurred close together in consecutive frames by observing the frame previous to the one shown here. See the interpolated regions on the right hand side of the ball. Note that the median interpolation is not that bad although it does have problems across moving edges. As a final observation, the image in the bottom right of the figure shows an interpolation using MBI-1 i.e. MBI-2 *without* motion correction. The visibly poor result again emphasizes that without motion correction the image model, regardless of its form, is hard pressed to generate a good reconstruction. The reader is invited to view the sequences on the CD-ROM where more of these phenomena can be observed.

MOBCAL is a poor sequence for illustrating the visual difference between a *sample* from the Gibbs sampler used in the JOMBADI algorithm and the MAP

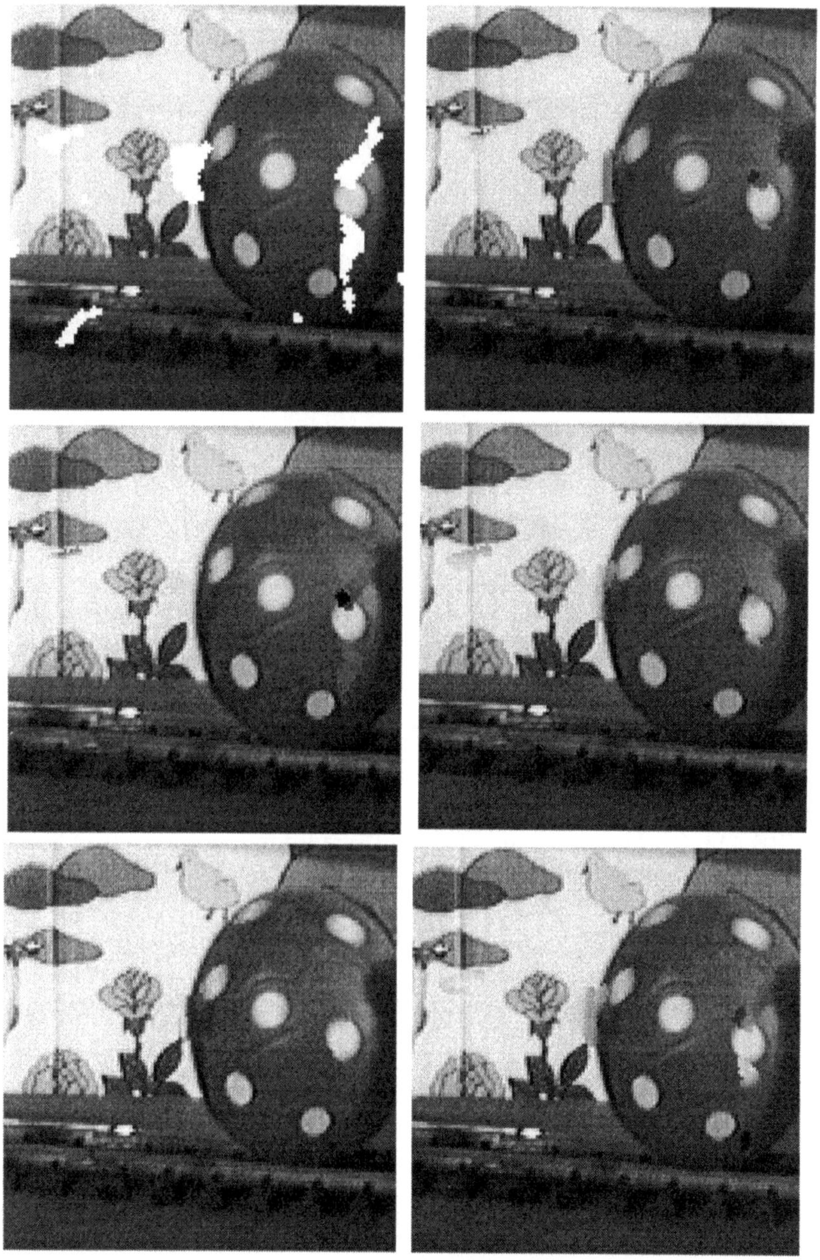

FIGURE 7.12. Top Row: Locations interpolated, ML3Dex interpolation. Middle Row: Bi-directional Prediction, MBI-2 both using 5 tap 3DAR model. Bottom Row : JOMBADI using 5 tap 3DAR model, MBI-1 with 5 tap 3DAR model.

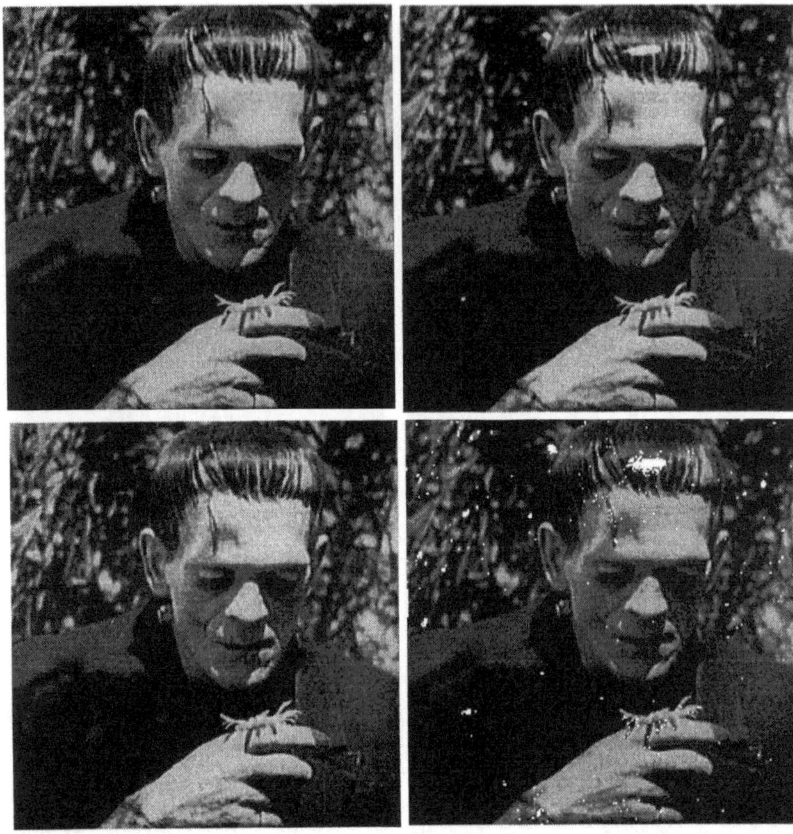

FIGURE 7.13. Clockwise from top left: Real Degraded frames 1,2 of FRANK, Detection on Frame 2 using ROD with thresholds 12, 39, 55, frame 3 of FRANK.

or LS interpolation. This is because MOBCAL has little visible texture or noise i.e. little 'stochastic' image content. This phenomenon is better illustrated when real degradation in archived sequences are discussed in the next section.

7.9.2 Real degradation

It is acknowledged that the blotches used in the degraded MOBCAL sequence are somewhat contrived, in that they are completely flat, have very sharp edges and occur with a very high probability. In real situations the blotches are more soft at their edges, presumably due to diffusion/wear of film material, they contain more noise texture within them and occur less frequently although with much varying sizes. Thus the actual behaviour of the various systems described in this chapter is different in a real situation. As usual, when dealing with real dirty movies, the only mechanism for evaluation of the performance of any restoration system is a visual assessment. Figure 7.13 shows three frames from a corrupted film, sequence FRANK (size 256 $\times$ 256 pixels). It also shows the

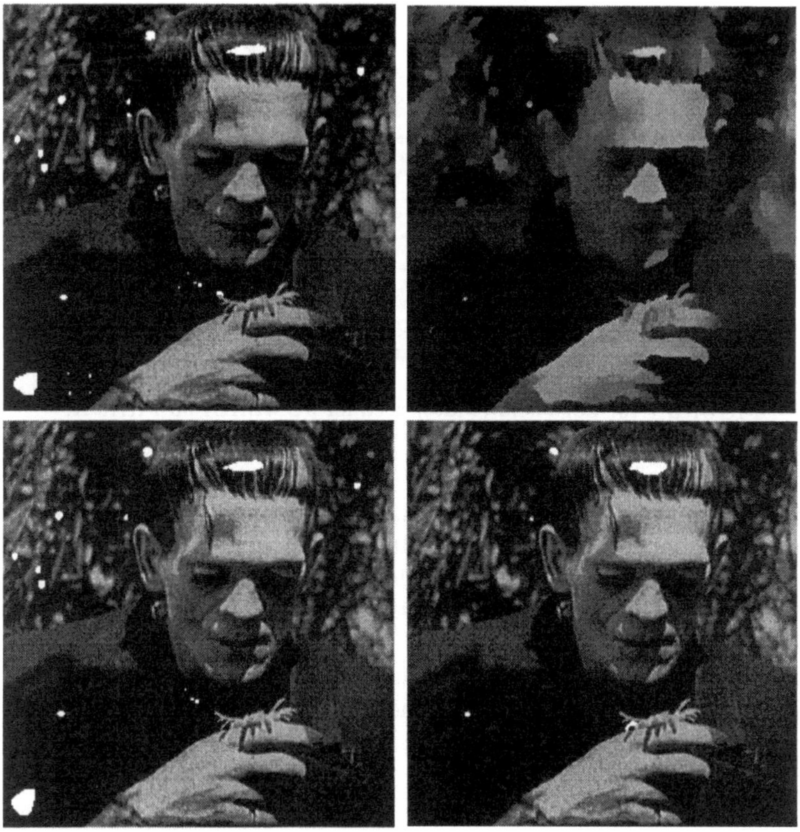

FIGURE 7.14. Clockwise from top left: Detection (white) using JOMBADI with 5 tap 3DAR ($\lambda^b = 4.0, \lambda^c = 0.15$), Estimated $c(\mathbf{x})$, Detection using Morris System ($\lambda_b = 4.0, \alpha = 8.0$), Detection using JOMBADI with 1 tap 3DAR ($\lambda^b = 4.0, \lambda^c = 0.15$).

result of detection using the ROD detector on frame 2 of the sequence (using the MWBME for initial motion estimation). The main corruptions (large one in the head) are detected, but the difficult motion of the petals of the flower and the trees in the background both cause false alarms.

Figure 7.14 shows the detection result using JOMBADI (5-tap 3DAR model with '+' shape in previous frame and 1-tap) and the Morris system again using the MWBME for initial motion estimation. Both processes were run for 50 iterations. The detection field for JOMBADI was set to 1 wherever the sites were set to 1 more than 75% of the time after a 10 sample burn in. The JOMBADI result is extremely accurate (in both cases) while the Morris result shows quite good performance even though it overestimates the extent of the blotch in the head, underestimates the extent of the other smaller distortions, and erroneously detects parts of the flower as corrupted. The JOMBADI algorithm is able to draw upon its knowledge of the 'flatness' of blotches to stop the

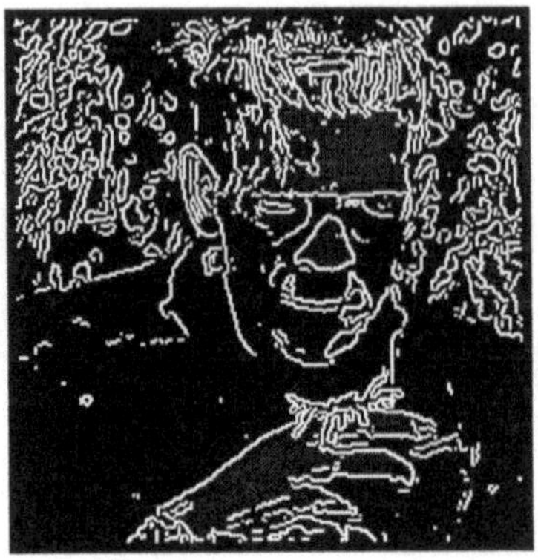

FIGURE 7.15. Edge field (white) used for JOMBADI on FRANK.

propagation of the detection field outside the boundaries of the blotch. The edge field used is shown in Figure 7.15. It has been dilated to improve the contrast to allow the reader more comfortable viewing. The edge field is clearly able to delineate the large corruptions and assists in stopping the 'flow' of indicator variables outside the corrupted region. In the background region, however, the relatively flat profile of the leaves, and their unusual motion, causes a number of them to be erroneously flagged as blotches by JOMBADI. The Morris detector however draws no information from the actual blotch intensity profile, and as they are of small area, it does not flag them. This also accounts for the fact that JOMBADI is able to detect a very poorly contrasted patch of corrupted data (in the lower left hand corner of the image) whereas this is missed by the Morris detector. However as the patch is difficult to see, it is arguable whether it is important to remove it or not.

The FRANK sequence is a good one for illustrating the difference between a *sample* posterior for the missing data and the Least Squares or MMSE estimates that could be generated. The differences are best illustrated using MBI-2 since the ability of JOMBADI to allow for statistical non-homogeneity of the image causes only a small visible difference between the two types of interpolants. The top row of Figure 7.16 shows the result of using MBI-2 and a 5 tap 3DAR model ('+' shape in the previous frame as usual), combined with the detection output of JOMBADI shown in Figure 7.14. The detection result from JOMBADI was dilated twice to give a larger area to interpolate in order to further emphasize the visual result. The boxed area delineates the interpolated region. The left interpolation was generated by using the Least Squares interpolant after motion correction, and the right hand image is a sample from the posterior for the missing data, i.e. a *sample* from the distribution shown in equation 7.35. The sample was drawn with the constraint that it must be within ± 2 standard deviations from the Least Squares interpolant. The reader should be able to see that the left hand side interpolant is *oversmooth* whereas the right hand side better represents the surrounding 'texture', and appears to be more *active*. Whether this texture is primarily noise or not is a separate issue.

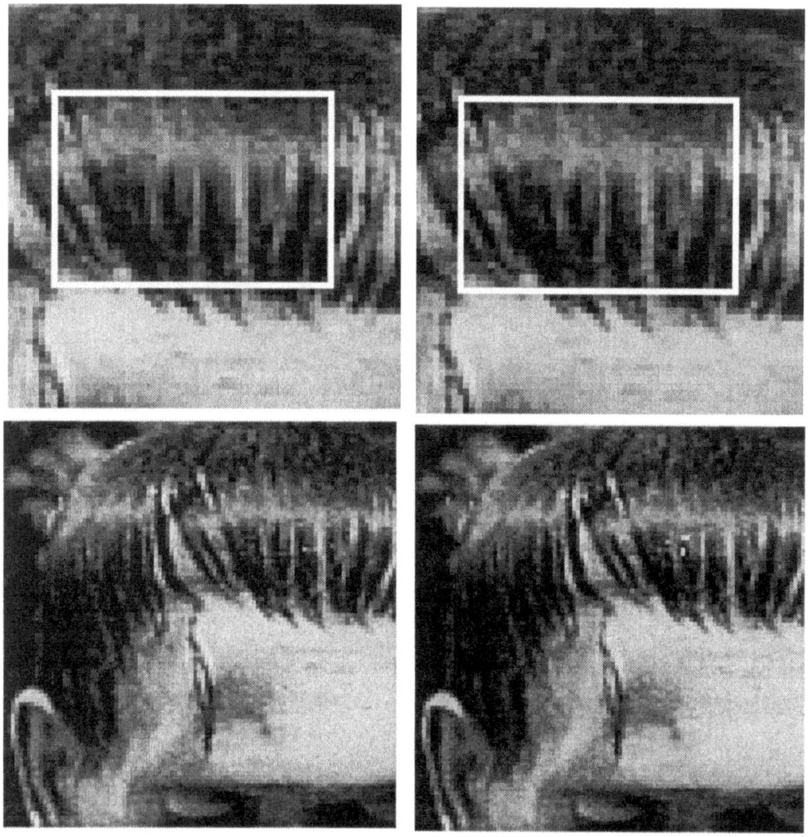

FIGURE 7.16. Top row: Zoom on reconstruction of major corruption on frame 2 of FRANK using MBI-2, least squares interpolation (left), sample from posterior (right). Bottom Row: JOMBADI with 5 tap 3DAR model (left), Restoration using MBI-2 and ML3Dex for interpolation (right).

This is the essential difference between 'maximal' interpolation and 'sampled' interpolation. A random sample from the distribution for the missing data is more 'typical' in some sense than the optimal or mean estimate. Note that it is possible for the sample to be visually poor since it is a random quantity and there is a finite but low probability of poor interpolants. Therefore in practice the samples are drawn within $\pm s$ standard deviations from the mean where s is typically between 0.5 and 2.5. For all the results shown in this chapter $s = 2.0$.

The bottom row of Figure 7.16 compares the JOMBADI result of interpolation with the result of using MBI-2 with the ROD detector and ML3Dex after motion correction. The median result shows poor textural reproduction when compared with any of the other model based interpolators. Note that the right hand restoration has not quite completely removed the blotch. This is because the ROD detection did not quite detect all of the blotch. A practical solution to this problem (which usually occurs) is to dilate the detection field before re-

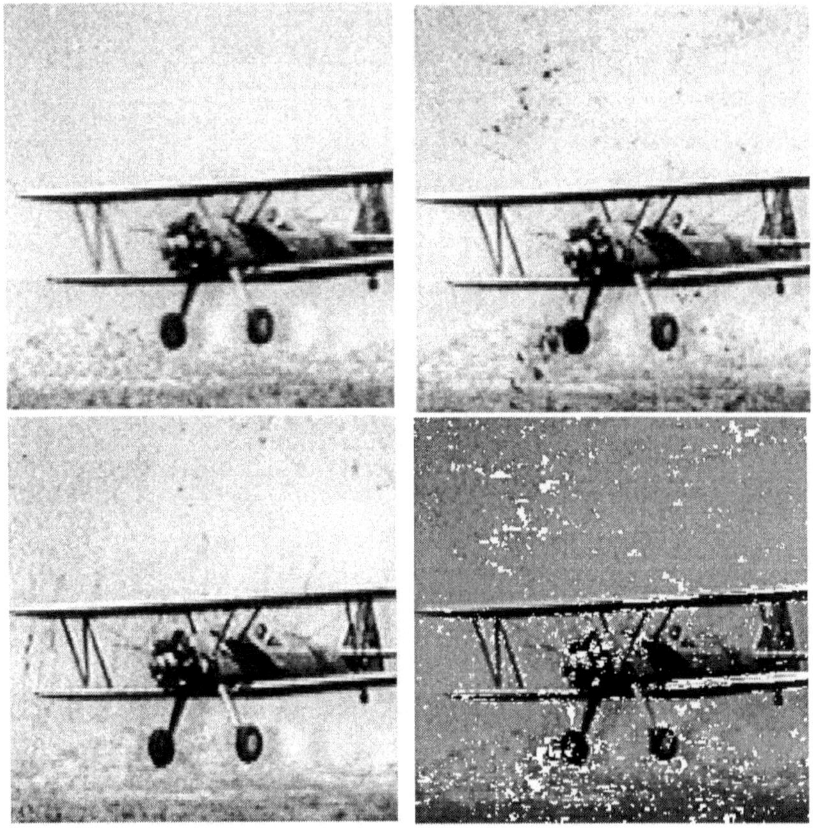

FIGURE 7.17. Top row: 256 × 256 portion of frames 1,2 from real degraded sequence CHAN4. Bottom Row: frame 3 from sequence (left), Detection result using SDIp (threshold = 15.0).

constructing the image area [96, 91]. This gives a pessimistic estimate of blotch location and helps to account for 'blurry' edges as well. This was discussed in the previous chapter.

The FRANK image shows fairly well defined corruption, even though the reader will observe that the blotches are not as sharp as in the artificially generated case. Figure 7.17 (a 256 × 256 portion of a 512 × 512 image) shows a more difficult situation. The corruption appears to be caused by a fingerprint smudge. Furthermore the image contains substantial film grain noise. The corrupted areas are therefore certainly not well delineated. Some detection results are shown in Figures 7.17 and 7.18 along with the edge field used for JOMBADI. The edge field was dilated to improve the visibility of the edge locations to the reader. The deterministic detectors e.g. SDIa, ROD, SDIp suffer from the problems of erroneous motion estimation, grain noise, and the movement of fine features such as the struts connecting the biplane wings. ROD detects more of the wing struts as false alarms than the SDIp. JOMBADI shows quite

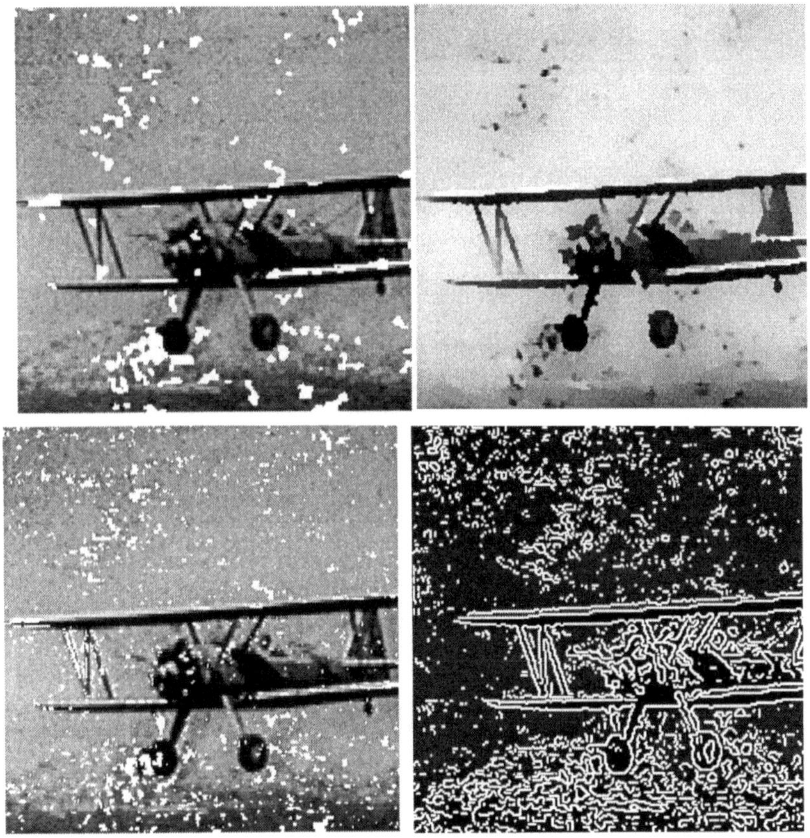

FIGURE 7.18. Top row: JOMBADI detection (left), 100th sample of $c(\mathbf{x})$. Bottom Row: ROD detection with thresholds=12, 39, 55 (left), Edge field used for JOMBADI (right).

good performance even in this case and the estimated $c(\mathbf{x})$ is again as expected. The lower wing in the original dirty image shows heavy aliasing. This causes all the detectors to flag parts of the wing as false alarms since the aliasing pattern changes between the frames. In a sense the spatial connectivity prior for JOMBADI causes the detection to backfire here since it connects the detected wing elements together. However, the reconstructed image shows no ill effects.

Figure 7.19 shows a zoom on two interpolation results again indicating the superior quality of the model based interpolator, JOMBADI with a 5 tap (+ shape as usual) 3DAR model. The difference pictures are shown to indicate the areas interpolated. The engine is sharper after JOMBADI because it has not been tampered with as much as when the ROD was used for detection. The zoom is quite large in order to allow the reader to appreciate the differences between the two reconstructions. The CD-ROM contains the image data for more close inspection.

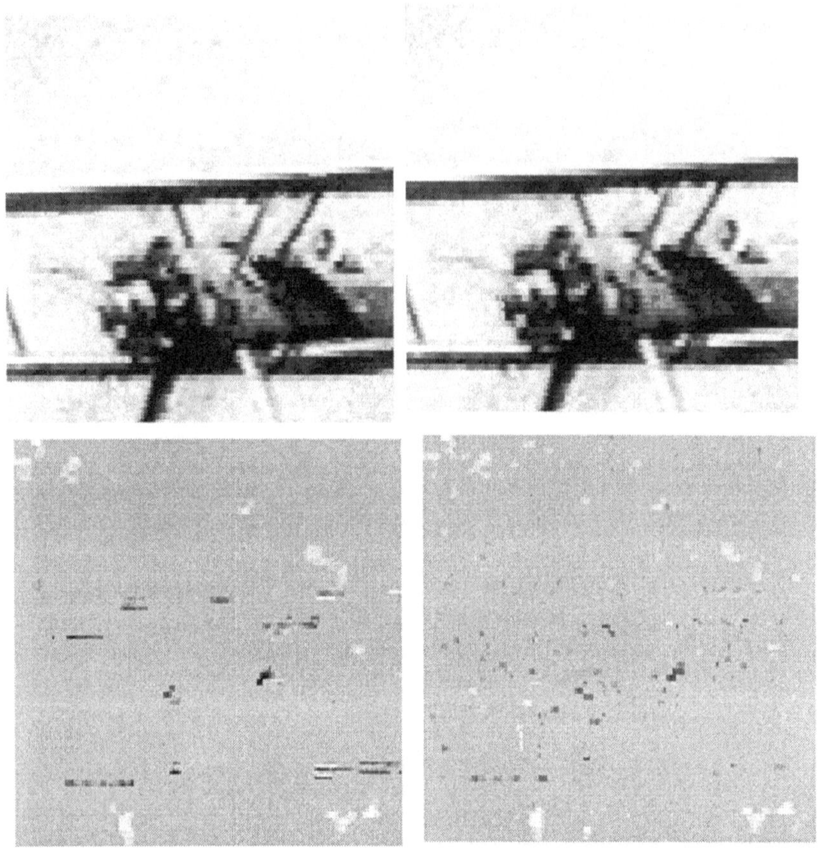

FIGURE 7.19. Restorations. Top row: 100th JOMBADI sample of $i(\mathbf{x})$(left), MBI-2 with motion correction using ROD as detector followed by ML3Dex as interpolator (right). Bottom Row: Difference between restoration and original degraded frame 2 portion.

7.10 Summary comments

Several model based reconstruction algorithms for video have been presented. The overall best performer is the JOMBADI algorithm particularly for its robustness in detection of blotches. The MBI-2 algorithm can perform well when the corrupted region is relatively small and the motion is not so complicated as to affect the deterministic detector used. MBI-2 systems which use median filters for interpolation instead of the model based interpolator, cannot guarantee a good reconstruction of image material. The median interpolant tends to look too 'faded' or blurred. The reader should be convinced at this stage that motion correction, whether as a separate step or an intrinsic part of the algorithm, is a crucial part of good reconstruction algorithms for image sequences. Note that the algorithms presented here can also be used for digital dropout errors in video as well as Dirt and Sparkle in film.

Considering computation however, the JOMBADI algorithm is undoubtedly the highest in computational load followed by MBI-2 with full model based reconstruction, MBI-2 with Bi-Directional prediction and finally MBI-2 with median filtering. However, there are a number of adjustments which could be made to JOMBADI to improve speed, for example relying solely on the simultaneous motion estimation/blotch detection aspect and using a 1-tap purely temporal 3DAR model with the coefficient set to -1.0. The reconstruction fidelity would suffer, but for most cases of relatively small distortion the quality would be good enough. The three step systems: detection, motion correction, interpolation, are viable real-time system proposals which could be implemented in current technology for broadcast purposes. Of course these processes are more user dependent and would require some user intervention when the image sequence becomes complicated due to warping motion or shadow effects etc.

Computational cost aside, it is the spirit of joint estimation in algorithms like JOMBADI which show the way toward completely automatic detection and treatment of missing data. The JOMBADI algorithm is powerful because of its ability to allow coherent communication between the various sub-processes which perform image and motion interpolation. The remaining problem of dealing with more complicated motion can be treated by changing the underlying image motion model and incorporating warping techniques, for example [24]. Dealing with a wider range of distortion, particularly film degradation such as smudges and transparent blemishes remains an interesting area for future work. Smudges may be dealt with by incorporating different priors for the blotch intensity profile, perhaps to incorporate the low pass effect of some blurring function. This would simultaneously allow dealing with 'real' blotches which do not necessarily have sharp edges, as was shown at the end of the chapter. Transparent blemishes are more difficult and may require more in-depth knowledge of their formation.

There is an argument that the spatial support for the 3DAR model may not be useful most of the time in reconstructing missing data in image sequences, although it does act as an optimal interpolator for fractional motion. This is simply because each frame is mostly similar to the ones around it in time. However in areas where the motion model breaks down, it is the spatial component which allows some kind of 'compromise' interpolant which is not visually poor, such as in areas of occlusion and uncovering. Purely temporal interpolation tends to yield very poor reconstructions when this occurs. Future developments to the techniques presented here would include knowledge of occlusion and uncovering in order to restrict the need for compromise interpolation further. The usually good performance of the bi-directional prediction process in areas of occlusion/uncovering illustrates that there is much to be gained by allowing for temporal discontinuities.

The reader is reminded at this stage, that despite the attempt to include non-homogeneous interpolation into JOMBADI (for instance) the procedures for interpolation described in this and the previous chapter are really only suitable for 'small' size distortion (say up to 20×20 pixels for PAL resolution TV images). When entire frames or large portions of frames are missing, it is difficult to rely only on spatial motion smoothness to interpolate the missing motion information. Also useful spatial image information is then sometimes too sparse to use. The next chapter presents techniques for interpolating these huge missing areas by relying almost solely on motion smoothness and discontinuity constraints to interpolate the missing image data.

8

Large Area Reconstruction

The previous chapters have considered the missing data problem in terms of relatively small sizes of distortion. The idea that motion correction plays an important role even with moderate sizes of corruption was also presented. This chapter focuses attention on the situations in which severe corruption occurs. In such situations a large proportion of the image can be lost. One example is the case of extreme physical abrasion, as in Figure 8.1. The problem also manifests due to loss of video signal in transmission. This is illustrated in the second frame of Figure 8.2, where loss of video signal has caused the video recording device (Sony Betacam Broadcast Standard VTR) to duplicate the last successfully recorded line until the signal resumed. The result is a swathe of duplicated lines (up to 50% of the vertical height) in every other field of video. The solutions presented in previous chapters are not well equipped to deal with this situation. In the case of the 'local' processes such as the two-step median and early 3DAR reconstruction (MBI-1, MBI-2) techniques, the missing area is now too big to be successfully assigned one motion vector. The joint motion estimation, detection and reconstruction technique can fare much better since it acknowledges the non-stationarity of the image and motion contents at the outset and so can implicitly estimate the separate moving objects that may be involved. However, that solution is still block based, and this limits its ability to define the edges of moving regions. Furthermore, it relies heavily on spatial motion information to achieve motion reconstruction. In this case of large corruption, it will take some time before the reliable motion information (at the extremities of the blotch) can propagate into the affected region. However, it is possible to deploy the joint solution presented in the previous chapter in this situation by making the necessary adjustments. These include allowing for occlusion as well as using a

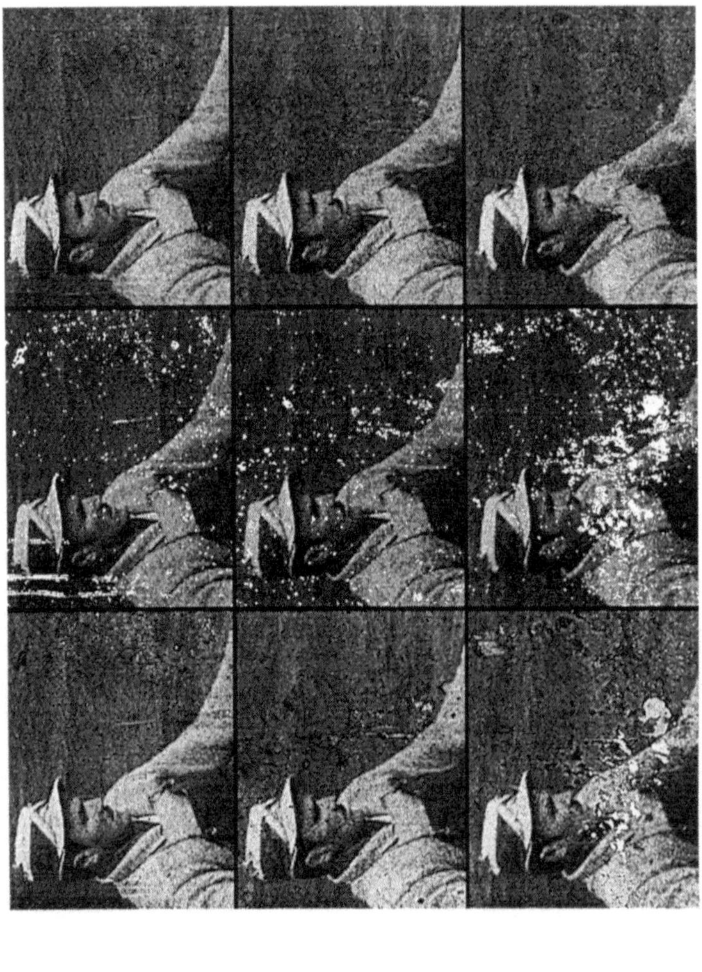

FIGURE 8.1. An example of extreme physical corruption of the film. Image is 360×287. Columns left to right: Original Degraded set of three frames showing large corruption particularly in the third frame, Detection using ROD detector superimposed on the original, Reconstruction using the algorithm discussed in this chapter.

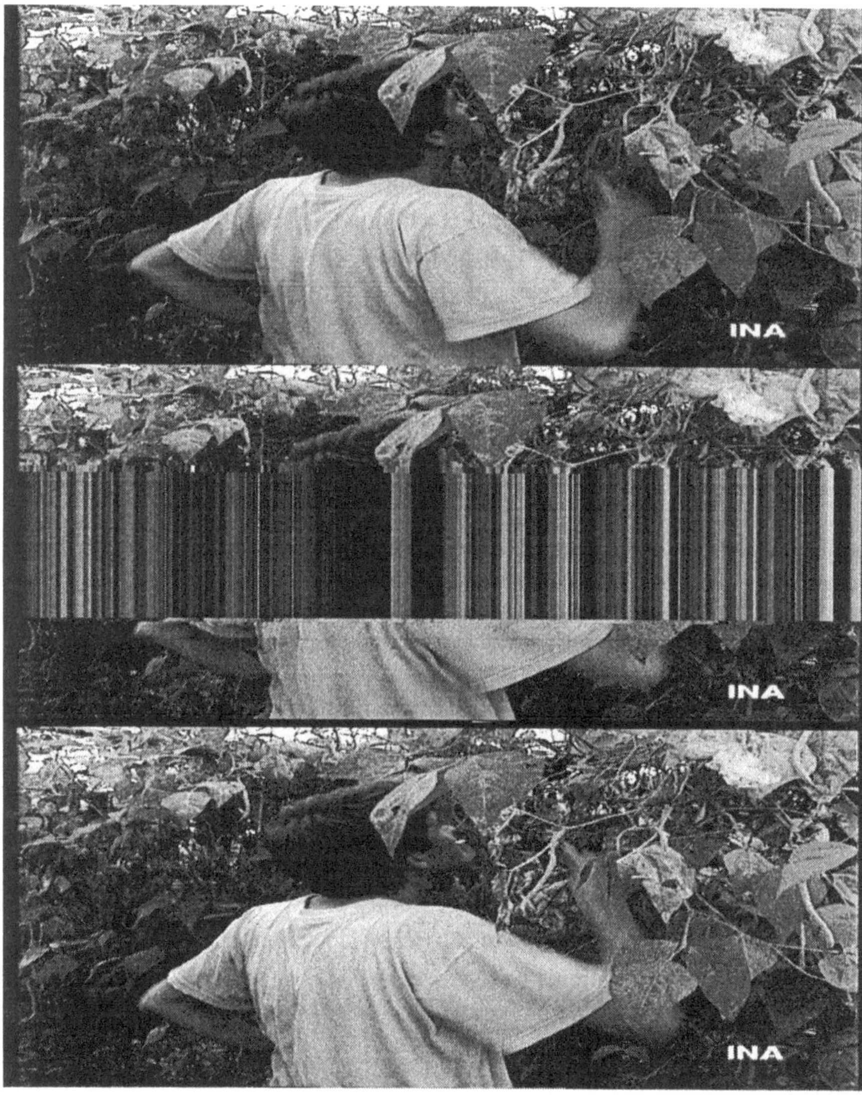

FIGURE 8.2. An example of severe dropout. Images are 720×288.

pixel based motion reconstruction strategy instead of a block based one. This would necessarily incur some computational overheads.

In an effort to limit the computational complexity of an algorithm for large area reconstruction, this chapter begins by assuming the basic image sequence model as follows:

$$I_n(\mathbf{x}) = I_{n-1}(\mathbf{x} + \mathbf{d}_{n,n-1}(\mathbf{x})) + e_n(\mathbf{x}) \tag{8.1}$$

where the variables have their usual meaning and $e_n(\mathbf{x}) \sim \mathcal{N}(0, \sigma_e^2)$. This model implies that the image contents in any missing portion of a frame can be deduced by reconstructing the motion field in that area. Once the motion is reconstructed, the missing data can be estimated in effect by 'looking up' the corresponding known image data in the surrounding frames. Of course, a large missing region is probably going to obliterate some moving object in an interesting sequence and so a good reconstruction technique must also attempt to reconstruct motion discontinuities as well. Since the resulting problem contains many interacting variables, a Bayesian approach is used to organize the algorithm design.

The chapter continues with some consideration of the nature of motion fields and introduces a new prior for occlusion. The data likelihood and various informative priors are combined into the posterior distribution for the unknown data. The final algorithm is then proposed as an iterative refinement process using a *griddy* Gibbs Sampler [168]. The modifications necessary to treat the case of repeated temporal data loss (as in Figure 8.2) are then presented. The results describe both visual and quantitative performance. The algorithms presented here assume that the missing region has already been detected. The extensions required to perform joint detection and reconstruction under this framework are discussed in the section 'Final Comments'. Some of the work discussed here has already been introduced in earlier chapters. It is repeated here in order to avoid breaking the thread of the discussion.

8.1 An overview of the motion interpolation problem

It is educational at this point to reflect on the problem statement. Consider Figure 8.3. The figure shows the true motion field between 5 frames of a sequence centred on frame n, as well as the original, clean image data in all frames. The problem to be solved is similar to that addressed in the previous chapters. *Given known missing areas in frame n, and known, clean original data in frames $n-1$, $n-2$, $n+1$, $n+2$, what is the original data at the required locations in frame n?* The missing areas are so large that a joint image/motion reconstruction algorithm may take a long time to converge since there are many ambiguities to be resolved. Thus the proposed idea is to use the vector fields for frames $n+1$, $n-1$ to reconstruct the missing vector field for frame n. Having done this, some unspecified image data interpolation technique can be employed for image data reconstruction. Since the missing area

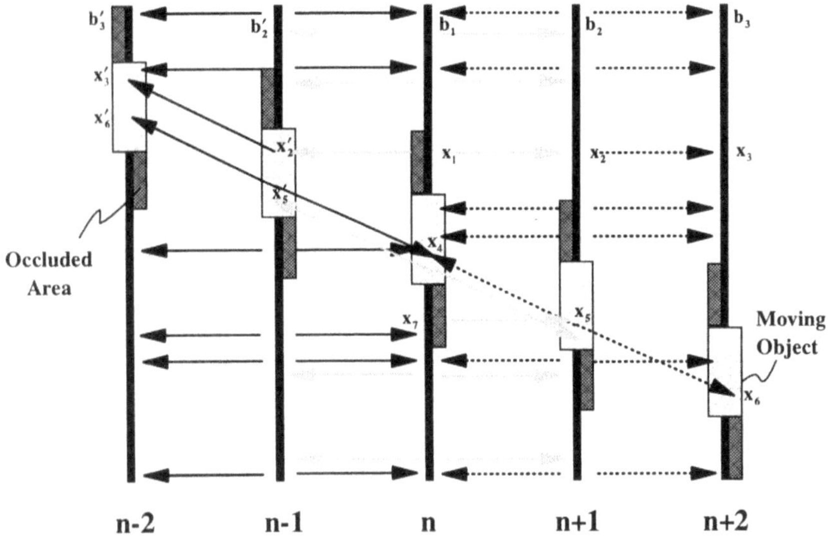

FIGURE 8.3. Five frame view of an object undergoing translational motion.

may be very large, the occlusion of objects may also need to be identified. This is so because in order to interpolate image data at sites of occlusion, it must be made clear which image frames do not contain the information needed. If occlusion is omitted from the image interpolation step, the resulting reconstructed image is distorted in those areas. While this is not a serious problem when the occluded region to be reconstructed is small (say 10×10 pixels), the human observer is more sensitive to the poor reconstruction when the region is large. Therefore, two occlusion variables, $O_{n,n-1}(\mathbf{x}_u)$, $O_{n,n+1}(\mathbf{x}_u)$ are associated with each site of unknown data $\mathbf{x}_u$, in frame n. $O_{n,n-1}(\mathbf{x}_u)$ takes on a value of 1 (or ON) when the data at $\mathbf{x}_u$ in frame n is occluded in frame $n-1$ and it is 0 (or OFF) otherwise. $O_{n,n+1}(\mathbf{x})$ is defined similarly. For instance, $O_{n,n-1}(\mathbf{x}_1) = 1$, $O_{n,n+1}(\mathbf{x}_1) = 0$ in Figure 8.3.

Because frame n is corrupted, the vector fields involving frame n have also been corrupted. Hence the problem is further constrained to disallow the use of any vectors $\mathbf{d}_{n,n-1}(\mathbf{x})$ and $\mathbf{d}_{n,n+1}(\mathbf{x})$. Assuming low acceleration, $\mathbf{d}_{n-1,n-2}(\mathbf{x}) = -\mathbf{d}_{n-1,n}(\mathbf{x})$. Thus in figure 8.3, the motion mapping $\mathbf{b}_2 \to \mathbf{b}_3$ ($\mathbf{x}_5 \to \mathbf{x}_6$) is identical to $\mathbf{b}_2 \to \mathbf{b}_1$ ($\mathbf{x}_5 \to \mathbf{x}_4$). Applying this together with the mirror constraint described in section 8.2.1 implies that motion vectors mapping frame $n+1$ to $n+2$ (for instance) can be employed (properly motion compensated) as an estimate of the missing vector field from n to $n+1$. This rule could be used in Figure 8.3, for instance, to allow the vectors $\mathbf{x}_5 \to \mathbf{x}_6$ and $\mathbf{x}'_5 \to \mathbf{x}'_6$ to be used as the interpolated motion at location $\mathbf{x}_4$, mapping frame n into $n+1$ and $n-1$ respectively. A similar situation exists for $\mathbf{b}_1, \mathbf{b}_2, \mathbf{b}_3, \mathbf{b}'_2, \mathbf{b}'_3$, at the top of the figure. However, there are ambiguities even in this perfect motion case.

The problems which arise by employing this simple interpolation scheme are illustrated by considering Figure 8.4. This figure illustrates the result of back-

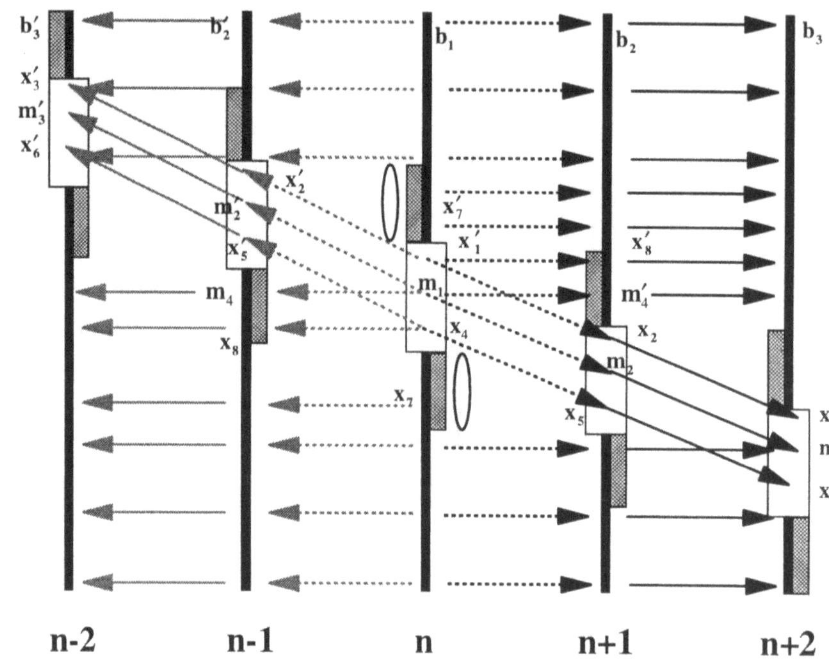

FIGURE 8.4. Simple interpolation of the motion field with best case motion.

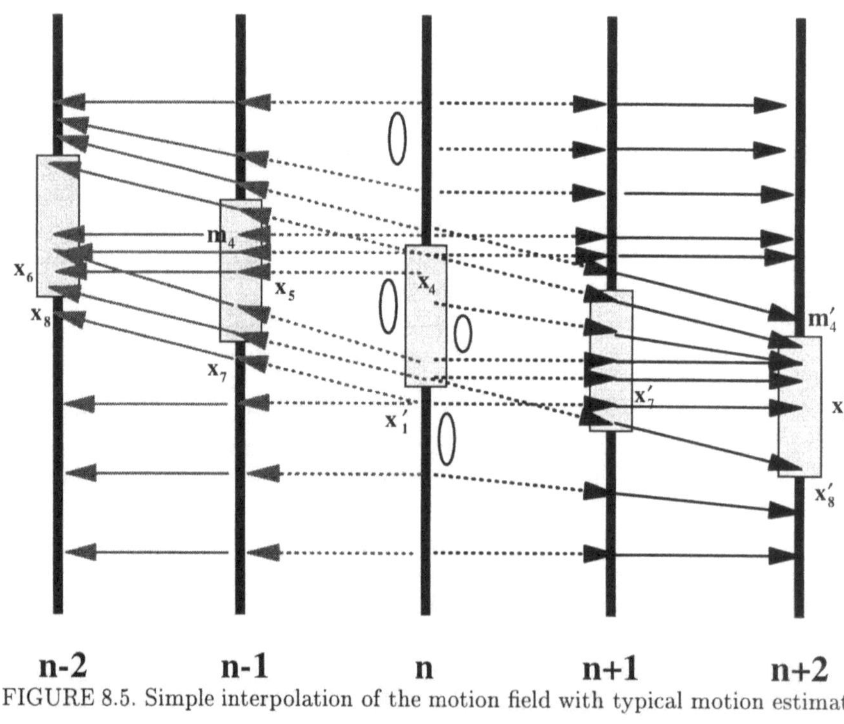

FIGURE 8.5. Simple interpolation of the motion field with typical motion estimates.

projecting all the solid vectors along their motion trajectories to the frame n, there to be used as an estimate for the missing motion field. The dashed vectors show the motion vectors which result from manipulating the solid vectors. For the most part, the technique is effective e.g. at $\mathbf{b}'_2$, $\mathbf{b}_1$, $\mathbf{b}_2, \ldots$. This is particularly true away from the moving object. However, in areas influenced by occluding regions there are problems resulting in multiple motion vectors being assigned to the same location. For instance, the vector $\mathbf{x}_2 \rightarrow \mathbf{x}_3$ ($\mathbf{x}'_5 \rightarrow \mathbf{x}_{6'}$) and the 'no motion' vector at $\mathbf{x}'_8$ ($\mathbf{x}_8$), are both assigned to position $\mathbf{x}'_1$ ($\mathbf{x}_4$). There can be only one vector at a particular location and this implies choosing the correct candidate. Fortunately on the opposite side, there is just one hit due to the trajectory $\mathbf{x}'_2 \rightarrow \mathbf{x}'_3$ ($\mathbf{x}_5 \rightarrow \mathbf{x}_6$). Applying the assumption of no acceleration for selecting the best of the two candidates on the right hand side does select the correct triplet of mappings $\mathbf{x}'_2 \rightarrow \mathbf{x}'_1 \rightarrow \mathbf{x}_2$ ($\mathbf{x}'_5 \rightarrow \mathbf{x}_4 \rightarrow \mathbf{x}_5$) as the interpolated motion. In stark contrast, assuming no acceleration does not help at $\mathbf{m}_1$, where there are 2 hits on both sides, due to vectors at $\mathbf{m}'_4$, $\mathbf{m}_2$, $\mathbf{m}_4$, $\mathbf{m}'_2$. The most that can be done with this constraint alone is to select the best two candidates for possible trajectories : $\mathbf{m}'_2 \rightarrow \mathbf{m}_1 \rightarrow \mathbf{m}_2$ and $\mathbf{m}_4 \rightarrow \mathbf{m}_1 \rightarrow \mathbf{m}'_4$. Only the first is correct, the second selection of vectors will probably cause background to be interpolated where an object would exist. It is here that spatial smoothness of the motion field helps to choose the correct candidate set, since only the former selection chooses candidates which are similar to their local neighbourhood (provided that has been estimated correctly at some point). Note that this circumstance of multiple hits on both sides of the central frame will only occur in areas undergoing very large motion (relative to their size) and might be regarded as an extreme case. Note that 'holes' in the interpolated vector field form in the occluded regions. This happens at the opposite side of $\mathbf{x}_7$, $\mathbf{x}'_7$, and is denoted by an ellipse. In this ideal case, they can be easily identified as occluded regions since the opposite side has a single hit.

It may be thought that a simple set of rules for arbitrating between choices of interpolated vectors can be used in a real situation when some available motion estimation process has provided the vector fields between frame pairs $n-1, n-2$ and $n+1, n+2$. Figure 8.5 shows this situation. The motion estimated using a block based motion estimator e.g. the MWBME is shown in the outer frame pairs. These simple motion estimates show the expected problems of motion haloes just at the edges of moving objects e.g. $\mathbf{m}'_4, \mathbf{x}_7$. In addition the motion is contrived to illustrate that a lack of texture in the interior within the object could lead to no motion being detected there. The net effect is that of diverging or converging vector fields in areas which should show purely translational motion. This occurs at $\mathbf{x}_5, \mathbf{x}_8, \mathbf{x}'_5, \mathbf{x}_{8'}$. When used to form an interpolation, this results in no interpolated motion at those regions indicated by ellipses, for example below $\mathbf{x}_4$. This can yield apparent occlusion in areas when there is none, e.g. at the top of frame n. Alternatively there could arise no interpolated vectors on either side of the frame, e.g. below $\mathbf{x}_4$.

It is therefore understandable that the problem of successfully arbitrating between the correct and incorrect interpolation becomes very difficult in this real case. Essentially, in a real situation, the problem of multiple 'hits' becomes much more likely, as does the occurrence of 'holes' in the interpolated vector fields on both sides of the interpolated frame. The conflict rapidly becomes difficult to resolve. To avoid this, a more coherent approach to the problem is employed. Before launching into a presentation of a Bayesian framework for motion field interpolation, a number of constraints on motion and image data are presented in probabilistic terms.

8.2 Constraints on motion

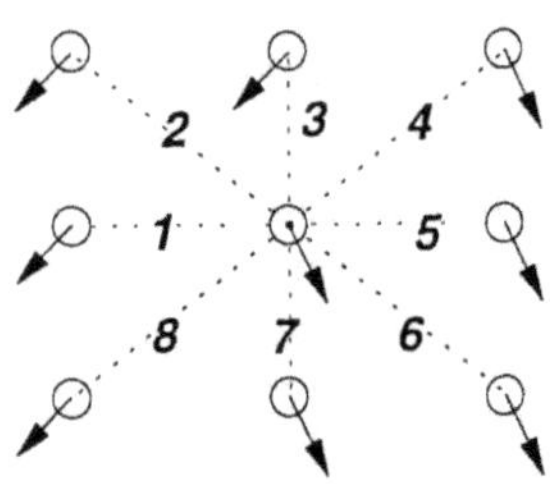

FIGURE 8.6. Cliques used for spatial motion smoothness. $\mathbf{q}_{1,\ldots,8} = [-1\ 0], [-1\ -1]$ etc.

Consider the situation illustrated in Figure 8.7. It shows an 'edge on' view of a rectangular object undergoing translational motion as it traverses two frames. On the left frame pair is superimposed the *true* motion field of the two frame sequence. The vectors map the current frame n into frame $n-1$ and vice-versa. Assuming rigid objects, this diagram illustrates that over the entire region occupied by an object the motion does not vary (particularly here with translational motion). In general, it is a good assumption that over small local regions the motion field is spatially smooth. This constraint has been used many times in the literature to overcome the aperture effect [65, 129, 44, 131]. The spatial smoothness constraint may be introduced as a penalty on motion vectors which cause an increase in the local spread of motion. Proceeding in a Bayesian fashion [104, 98, 13], the spatial smoothness constraint may be expressed in terms of a prior probability distribution for a single motion vector given its neighbourhood of vectors. A typical, simple Gibbs Energy prior [98, 13] will suffice here. Therefore, the prior probability of a particular motion vector $p(\mathbf{d}_{n,n-1}|\mathbf{D}(\mathbf{x}))$ at some location $\mathbf{x}$ given its eight nearest neighbours, denoted by $\mathbf{D}(\mathbf{x})$, is defined as

$$\begin{aligned} p_s(\mathbf{d}_{n,n-1}(\mathbf{x})|\mathbf{D}(\mathbf{x})) &= \frac{1}{Z_s}\exp-\left[\sum_{k=1}^{k=8} w_k \mathcal{V}(\mathbf{d}_{n,n-1}(\mathbf{x}) - \mathbf{d}_{n,n-1}(\mathbf{x}+\mathbf{q}_k))\right] \\ \text{where } \mathcal{V}(\mathbf{x}) &= \begin{cases} |\mathbf{x}| & \text{for } |\mathbf{x}| \le v_t \\ v_t & \text{for } |\mathbf{x}| > v_t \end{cases} \end{aligned} \tag{8.2}$$

where $\mathbf{q}_k$ is the vector offset to the kth neighbourhood vector, w_k is a weight assigned to the kth pairwise interaction, Z_s is the normalizing constant and the function $\mathcal{V}(x)$ is as defined. The situation is illustrated in Figure 8.6, where the horizontal and vertical components of $\mathbf{q}_k$ are also defined.

Ideally, there ought to be a facility for turning off the spatial motion smoothness at motion boundaries. Configuration of an explicit edge process to allow this would be assisted by knowledge of the underlying *unknown* image data. To allow practical convergence of the final algorithm, discontinuities at edges are

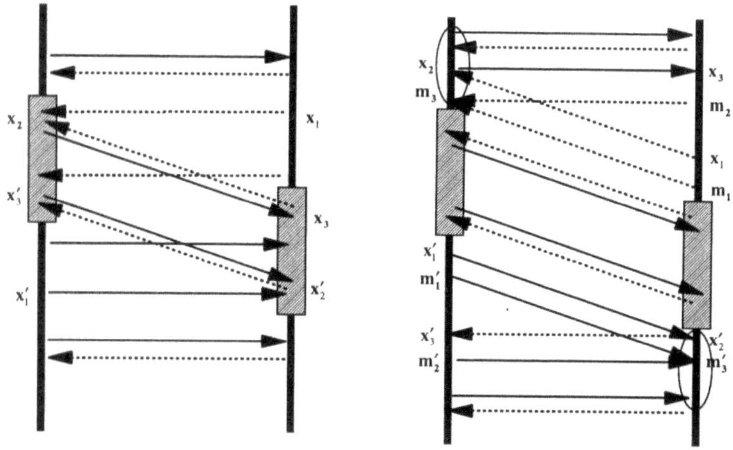

FIGURE 8.7. Edge-on view of a rectangular object undergoing translational motion.

sacrificed in favour of implicit allowance for edges. This takes the form of an adjusted potential function for the clique interaction $\mathcal{V}(x)$ as defined. This is a robust 'operator' in some sense as it places a ceiling on the effect of motion vectors in the neighbourhood which cause a discontinuity greater than v_t in this case. Thus it is tolerant to discontinuities, and so allows some kind of implicit edge identification in the motion field. The use of discontinuity adaptive priors such as these are discussed in some detail in [104, 103, 105].

8.2.1 Local temporal smoothness

Consider Figure 8.7 once more. Rather than define temporal motion smoothness across several frames assuming little acceleration, it is possible to define a kind of *correspondence* smoothness using the forward and backward vector pairs. Where there is motion without occlusion (temporal discontinuity), the vectors in each pair are very similar (for the ideal case, the same). However, at a site which is occluded in the next frame, this *mirror* effect is violated. This effect was noticed by Elias et al [40]. Therefore in the figure, $\mathbf{x}_3$ maps to $\mathbf{x}_2$ via $\mathbf{d}_{n,n-1}(\mathbf{x}_3)$ and conversely, $\mathbf{x}_2$ maps onto $\mathbf{x}_3$ via $\mathbf{d}_{n-1,n}(\mathbf{x}_2)$. Thus there is no occlusion i.e. $\mathbf{d}_{n,n-1}(\mathbf{x}_3) = -\mathbf{d}_{n-1,n}(\mathbf{x}_2)$. In contrast, although $\mathbf{x}_1$ maps to $\mathbf{x}_2$, $\mathbf{x}_2$ does not map to $\mathbf{x}_1$. Thus the image data at $\mathbf{x}_1$ has been occluded in frame $n-1$. The use of this mirror effect in a prior distribution for the motion vector given its *temporal* neighbourhood allows the detection of occlusion using motion alone. This is of importance here since the image data is unknown at sites $\mathbf{x}_u$ of the required frame n.

$$
\begin{aligned}
p_t(\mathbf{d}_{n,n-1}(\mathbf{x})|\mathbf{d}_{n-1,n}, \sigma_v^2, O_{n,n-1}(\mathbf{x})) &= \\
\frac{1}{Z_t}\exp -\left[\frac{1}{2\sigma_v^2}|\mathbf{d}_{n,n-1}(\mathbf{x}) + \mathbf{d}_{n-1,n}(\mathbf{x}+\mathbf{d}_{n,n-1}(\mathbf{x}))|^2(1-O_{n,n-1}(\mathbf{x}))\right] \quad (8.3)
\end{aligned}
$$

A simple proposal for this prior is defined as in equation 8.3, where $O_{n,n-1}(\mathbf{x})$ has been introduced to indicate that the image data at $\mathbf{x}$ in frame n is occluded in frame $n-1$ when $O_{n,n-1}(\mathbf{x}) = 1$. It is otherwise zero. Z_t is the normalizing constant. Overlooking this switching variable, the prior is a Gaussian distribution with variance σ_v^2 and places a bias toward a vector which is similar to the motion compensated mirror vector $\mathbf{d}_{n-1,n}(\mathbf{x}+\mathbf{d}_{n,n-1}(\mathbf{x}))$ when the data is not occluded. In contrast, when $O_{n,n-1} = 0$ the entire argument of the exponent is forced to zero, hence the probability distribution is then uniform, implying that all vectors at $\mathbf{x}$ are equally likely, as far as temporal smoothness is concerned. This behaviour effectively *turns off* the temporal interaction between mirror pairs when data is occluded.

The right hand pair of frames in Figure 8.7 shows motion estimates from a typical motion estimation algorithm (such as the WBME introduced earlier) which does not incorporate occlusion estimation or any motion priors. It illustrates that the use of the mirror argument still holds even if there is a vector halo effect near the borders of moving objects. Consider the mirror violation of the trajectories $\mathbf{x}_1 \rightarrow \mathbf{x}_2 \rightarrow \mathbf{x}_3$ $(\mathbf{x}'_1 \rightarrow \mathbf{x}'_2 \rightarrow \mathbf{x}'_3)$ which successfully denote occlusion despite the erroneous motion indicated by $\mathbf{x}_1 \rightarrow \mathbf{x}_2$ $(\mathbf{x}'_1 \rightarrow \mathbf{x}'_2)$. In addition, note that 'many to one' mappings may also be used to indicate occlusion as is illustrated by the triplet $\mathbf{m}_1, \mathbf{m}_2 \rightarrow \mathbf{m}_3$ $(\mathbf{m}'_1, \mathbf{m}'_2 \rightarrow \mathbf{m}'_3)$. However, this prospect for another Occlusion related prior is not pursued here.

8.3 The occlusion prior

The delineation of occluded regions will contribute to sharp interpolation of missing regions which occur in the region around a moving object. To this end it is important to identify useful configurations for a prior or a rule for how occlusion fields are usually observed. It is clear that occluded regions will not occur as isolated pixels separated by large space. There is some inherent spatial smoothness to the occlusion field. To capture this observation it is simple to adopt a similar form to the spatial motion smoothness prior.

$$p_o(O_{n,n-1}(\mathbf{x})|\mathbf{O}(\mathbf{x})) = \frac{1}{Z_o} \exp - \sum_{k=1}^{8} \lambda_k |O_{n,n-1}(\mathbf{x}) - O_{n,n-1}(\mathbf{x}+\mathbf{q}_k)| \tag{8.4}$$

where $\mathbf{O}(\mathbf{x})$ denotes the Occlusion values at the eight nearest neighbours of the considered position, the weights λ_k have been chosen to be different from w_k, and Z_o is again a normalizing constant. There is one remaining issue about the occlusion parameters affecting the probability of image data. This is considered more appropriately in section 8.5.2.

8.4 The posterior distribution

As is evidenced by the discussions above, designing a set of rules for choosing the correct vector at each site is unlikely to suceed in severe cases. The need to design any list of rules can be avoided by resorting to choosing the best vector based on local probability densities. The unknowns to be estimated are $\mathbf{d}_{n,n-1}(\mathbf{x}_u)$, $\mathbf{d}_{n,n+1}(\mathbf{x}_u)$, $O_{n,n-1}(\mathbf{x}_u)$, $O_{n,n+1}(\mathbf{x}_u)$ at the sites of known corruption $\mathbf{x}_u$. Ideally, estimates for these variables can be made by choosing those which maximize the joint probability distribution for all these variables across all sites $\mathbf{x}_u$ simultaneously. One may choose to do this deterministically, starting from an initial estimate one hopes to be close to the global maximum. Alternatively, as is the case here, one may choose to solve the problem numerically by generating samples from the joint distribution and then manipulating these samples accordingly to yield the MMSE estimate or to locate the mode of the distribution.

In the standard manner (see e.g. [47]) this joint problem can be decomposed into problems of smaller dimensionaltiy by employing the Gibbs Sampler. The problem of drawing samples from the joint distribution is then made possible by drawing samples from the local conditional distributions for the unknown data at each site $\mathbf{x}_u$ given previously drawn samples in the local neighbourhood. At each site the draws may be further decimated so that just one variable at a time is sampled i.e. first $d_{n,n-1}(\mathbf{x}_u)$ then $O_{n,n-1}(\mathbf{x})$ followed by $d_{n,n+1}(\mathbf{x}_u)$ and $O_{n,n+1}(\mathbf{x})$. At each site $\mathbf{x}_u$ it is required to draw a sample of $\mathbf{d}_{n,n-1}(\mathbf{x}_u)$ and $O_{n,n-1}(\mathbf{x}_u)$ from

$$p(\mathbf{d}_{n,n-1}(\mathbf{x}_u), O_{n,n-1}(\mathbf{x}_u)|\mathbf{D}(\mathbf{x}_u), \mathbf{O}(\mathbf{x}_u), \sigma_v^2, \sigma_e^2, I_{n-1}, I_{n+1}) \tag{8.5}$$

This is the conditional distribution of two unknowns at site $\mathbf{x}_u$, given the current state of all other variables in the local neighbourhood both temporally and spatially. $\mathbf{D}(\mathbf{x})$, $\mathbf{O}(\mathbf{x})$ are the current spatio-temporal and spatial neighbourhoods of the motion and occlusion respectively around site $\mathbf{x}$. A similar expression can be written for $\mathbf{d}_{n,n+1}(\mathbf{x}_u)$ and $O_{n,n+1}(\mathbf{x}_u)$. To simplify the notation, define a vector of parameters $\mathbf{K}$

$$\mathbf{K} = [\mathbf{D}(\mathbf{x}_u), \mathbf{O}(\mathbf{x}_u), \mathbf{d}_{n-1,n}, \sigma_v^2, \sigma_e^2, I_{n-1}, I_{n+1}]$$

describing the state of the local neighbourhood. These parameters are not 'unknown', they are the current values of the various variables at sites not including the current site being processed. Note again that the problem posed is motion interpolation in the presence of missing image data at $\mathbf{x}_u$ i.e. $I_n(\mathbf{x}_u)$. Hence $I_n(\mathbf{x}_u)$ is not available.

The problem can now be restated as that of sampling from the distribution $p(\mathbf{d}_{n,n-1}(\mathbf{x}_u), O_{n,n-1}(\mathbf{x}_u)|\mathbf{K})$. Employing Bayes rule results in an expression for the posterior distribution as follows, where $\mathbf{K}_{-\mathrm{i}}$ denotes all the parameters excepting I_{n-1}, I_{n+1}.

$$\begin{aligned} p(\mathbf{d}_{n,n-1}(\mathbf{x}_u), O_{n,n-1}(\mathbf{x}_u)|\mathbf{K}) &\propto p(I_{n-1}, I_{n+1}|\mathbf{d}_{n,n-1}(\mathbf{x}_u), O_{n,n-1}(\mathbf{x}), \mathbf{K}_{-\mathbf{i}}) \\ &\times p(\mathbf{d}_{n,n-1}(\mathbf{x}_u)|O_{n,n-1}(\mathbf{x}), \mathbf{K}_{-\mathbf{i}}) \\ &\times p(O_{n,n-1}(\mathbf{x}_u)|\mathbf{K}_{-\mathbf{i}}) \qquad (8.6) \end{aligned}$$

The second term contains the prior constraints on motion, described by the product of equations 8.2 and 8.3. The third term is the spatial occlusion smoothness prior in equation 8.4. The first term has not been discussed so far, it is the likelihood of the image data in the surrounding frames $n-1$ and $n+1$ given the various motion and occlusion parameters. It is described implicitly by equation 8.1 to yield

$$\begin{aligned} p(I_{n-1}, I_{n+1}|\mathbf{d}_{n,n-1}(\mathbf{x}_u), O_{n,n-1}(\mathbf{x}), \mathbf{K}_{-\mathbf{i}}) &= \\ \frac{1}{\sqrt{2\pi\sigma_e^2}} \exp -\frac{1}{2\sigma_e^2} [\mathcal{V}_\imath(I_{n+1}(\mathbf{x}_u + \mathbf{d}_{n,n+1}(\mathbf{x}_u)) &- I_{n-1}(\mathbf{x}_u + \mathbf{d}_{n,n-1}(\mathbf{x}_u)))]^2 \\ \text{where } \mathcal{V}_\imath(x) &= \begin{cases} x & \text{for} |x| \le i_t \\ i_t & \text{for} |x| > i_t \end{cases} \qquad (8.7) \end{aligned}$$

Note that this expression does not involve the occlusion variables, but again engages a similar ploy to that used for spatial motion smoothness. The potential function $\mathcal{V}_i(x)$ implicitly allows temporal discontinuities in the image data across frame n. In some sense it also encourages motion vector haloes around moving objects. As was illustrated previously, this does not adversely effect the correct configuration of occlusion provided that the motion prior in equation 8.3 is adopted. Although one may expect that the temporal image interaction should be turned off at sites of occlusion, it is found that this likelihood expression allows a configuration of an interpolated vector field given an initial motion estimate provided by simple multiresolution motion estimators. This is explained since the multiresolution motion estimator adopted is block based and is necessarily unable to properly estimate the motion of object boundaries.

Considering the interaction of the occlusion variables in the posterior distribution, an alarming feature arises. If the occlusion is set to 1 at any site, this effectively deletes the temporal motion prior from the expression for the posterior. Since the final goal is the maximization of the posterior with respect to the unknowns, it is possible that setting occlusion to 1 everywhere will automatically be advantageous. To avoid this, a term must be introduced to penalize the introduction of a 1 into the occlusion field. This was recognized by several authors, e.g. Morris [118], Li [104], Konrad et al [98, 157]. The third term in the posterior therefore becomes

$$\begin{aligned} p(O_{n,n-1}(\mathbf{x})|\mathbf{O}(\mathbf{x})) &= \frac{1}{Z_o} \exp - \left[\sum_{k=1}^{k=8} \lambda_k |O_{n,n-1}(\mathbf{x}) - O_{n,n-1}(\mathbf{x} + \mathbf{q}_k)| \right] \\ &\times \frac{1}{Z_p} \exp - [\alpha O_{n,n-1}(\mathbf{x})] \qquad (8.8) \end{aligned}$$

where α is the penalty introduced, and Z_p is the associated normalizing term. This penalty must balance the degree of violation of the mirror constraint and acts as a switch which employs occlusion when this violation becomes intolerable. Since the temporal motion smoothness is assumed to follow $\mathcal{N}(0, \sigma_v^2)$, it is possible to calculate normalized confidence limits on $|\mathbf{d}_{n,n-1}(\mathbf{x}) - \mathbf{d}_{n-1,n}(\mathbf{x} + \mathbf{d}_{n,n-1}(\mathbf{x}))|^2$. This immediately gives a reasonable value for α. For instance, to be 99% confident that a motion vector has violated the normal assumption as far as temporal smoothness is concerned, $\alpha = 2.576^2/2$.

8.5 Choosing the most probable motion field interpolant

The resulting optimization problem is now potentially huge, with up to four variables at every unknown pixel site. The unknown sites could occupy more than half of an image frame. A number of practical points facilitate better convergence of the final iterative method, at the expense of optimality.

1. A multiresolution motion estimator is used to configure the motion fields between frames $n-1, n-2$, $n+1, n+2$ and between the known data in frame n and frames $n-1$, $n+1$.
2. The vector fields between $n-1, n-2$ and $n+1, n+2$ can be backprojected onto frame n and a record kept of the nearest pixel site that was hit. At the end of this step most pixels in frame n have attached to them a series of possible temporal motion vector candidates for use as possible interpolants. This is called the *Temporal Hit list.*
3. The backprojected vectors can be manipulated simply to yield an initial configuration for the interpolated vector field. The simple rules explored in Section 8.1 can be used here. Sites which defy all rules are simply left unassigned.

After this stage it is usual that much of the required interpolation is achieved. It is only in the regions which show complicated or fast motion that the complete treatment of the problem (using the Gibbs sampler) improves upon the initial guess.

8.5.1 Modifying the Gibbs sampler

To yield further gains in computation, the use of a 'griddy' Gibbs sampler [168] for the motion vectors is advocated. In essence it is assumed that the best motion vector at a site is one of the vectors in a candidate list. This list is a subset of all the possible motion vectors. It is created by augmenting the Hit List initially set up as described earlier, with the spatial motion candidates. This list is assumed to cover the significant space of motion vectors and is employed in generating a sample for motion.

8.5.2 Composition sampling

In some sense the harder problem here is sampling for motion since those variables are continuous. The approximation of sampling from a subset of vectors does improve the speed of the algorithm, but the subset can be made up of as much as 20 vectors. Sampling for the binary occlusion variable is much simpler since it has only 2 states. However, the convergence of the sampler is known to improve in some cases [109, 151] if it is possible to sample jointly for many variables at the same time. In this case, because of the binary occlusion variable it is simple to derive the recipe for sampling jointly for occlusion *and* motion.

Employing composition sampling decomposes the required joint draw as follows:

$$\begin{aligned} p(\mathbf{d}_{n,n-1}(\mathbf{x}_u), O_{n,n-1}(\mathbf{x}_u)|\mathbf{K}) &= \\ & p(O_{n,n-1}(\mathbf{x}_u)|\mathbf{d}_{n,n-1}(\mathbf{x}_u), \mathbf{K}) \times p(\mathbf{d}_{n,n-1}(\mathbf{x}_u)|\mathbf{K}) \end{aligned} \tag{8.9}$$

The joint sample is generated by first drawing $\mathbf{d}_{n,n-1}$ from $p(\mathbf{d}_{n,n-1}(\mathbf{x}_u)|\mathbf{K})$ and then employing that sample in the draw for $O_{n,n-1}$ from the first distribution on the right hand side above. The required conditional for $\mathbf{d}_{n,n-1}(\mathbf{x}_u)$ can be derived as follows:

$$p(\mathbf{d}_{n,n-1}(\mathbf{x}_u)|\mathbf{K}) = \int p(\mathbf{d}_{n,n-1}(\mathbf{x}_u), O_{n,n-1}(\mathbf{x}_u)|\mathbf{K}) dO_{n,n-1} \tag{8.10}$$

Since $O_{n,n-1}$ is a binary variable this integral is derived by substituting $O_{n,n-1} = 0, 1$ into equation 8.6 to yield

$$\begin{aligned} p(\mathbf{d}_{n,n-1}(\mathbf{x}_u)|\mathbf{K}) &\propto p(I_{n-1}, I_{n+1}|\mathbf{d}_{n,n-1}(\mathbf{x}_u), \mathbf{K}_{-\mathrm{i}}) \\ & \times \sum_{O=0,1} [p_s(\mathbf{d}_{n,n-1}(\mathbf{x}_u)|O, \mathbf{K}_{-\mathrm{i}}) p_o(O|\mathbf{K}_{-\mathrm{i}})] \end{aligned} \tag{8.11}$$

A recipe for sampling for $d_{n,n-1}(\mathbf{x}_u)$ can then be proposed as follows:

1. For each candidate motion vector in the Hit List, evaluate the expression 8.11. Care must be taken in scaling the exponentials before adding the two terms to allow for finite machine precision. A simple mechanism involves finding the maximum argument in the exponential terms and then scaling all other terms accordingly.

2. Generate a sample from this numerically evaluated probability distribution by first calculating the cumulative distribution (scaled appropriately so that the maximum is 1.0). Using this function a uniform random number, drawn between 0 and 1, can then be mapped onto a particular candidate vector in the list. This is known as the transformation method for sampling from a random distribution. See page 287 of [142] (2nd edition) for further details.

Of course, one could employ other algorithms here for drawing a sample e.g. Metropolis-Hastings, Rejection method (see [142]). The direct method above is employed since the computation required for each evaluation is not great.

8.6 The algorithm

The steps required to sample configurations of the vector field can now be presented as follows. The algorithm concedes optimality to allow gains in implementation complexity.

1. As described above, the algorithm begins by first backprojecting both vector fields $\mathbf{d}_{n-1,n-2}$, $\mathbf{d}_{n+1,n+2}$ onto the current frame n. These vector fields are estimated here using the multiresolution WBME described in earlier chapters, although any motion estimator will do. As this is done a record is kept of the vectors which have hit locations in the current frame. The result is a list of $N_b(\mathbf{x}_u)$, $N_f(\mathbf{x}_u)$ vectors at each location of unknown data $\mathbf{x}_u$. Each vector is denoted $v_n^b(\mathbf{x}_u)\forall n = 1 \ldots N_b(\mathbf{x}_u)$, for the hits on the $n, n-1$ side of frame n and similarly $v_n^f(\mathbf{x}_u)\forall n = 1 \ldots N_f(\mathbf{x}_u)$ for the hits on the $n, n+1$ side of frame n.

2. An initial configuration for the interpolated vector field is set up. This is done by assigning locations with single hits, the relevant single hit vector. For locations with multiple hits one side and one on the other, the vector which creates the smoothest trajectory when compared with the vector on the single hit side, is chosen as the interpolated vector. All other circumstances e.g. holes or multiple hits on both sides are left undefined.

3. Measure σ_v^2 directly by assuming that the interpolated vectors which are at locations which have a single hit on both sides, are correct interpolants. Measure σ_e^2 similarly at these locations directly from known image values, I_{n-1}, I_{n+1}. This gives a suboptimal estimate for these parameters.

4. At each missing pixel site $\mathbf{x}_u$, DO (for $n, n-1$):

   (a) Consider all vectors $v_n^b(\mathbf{x}_u)$ to be candidate vectors.

   (b) Add to this list the list of vectors from the opposite side of the frame, $v_n^f(\mathbf{x}_u)$ as well as the assigned vector there (if one exists), suitably negated of course. Recall that assuming low acceleration, $d_{n,n-1}(\mathbf{x}_u) = -d_{n,n+1}(\mathbf{x}_u)$. In addition, add to this list any vectors in the surrounding 8 connected neighbourhood which have already been assigned to a pixel site. Denote the cth vector in this new list by $\mathbf{v}_c$, and define the length of the list to be C. For each of the C vectors in the list calculate

      i. The motion compensated vector error

      $$e_v = \frac{1}{2\sigma_v^2}|\mathbf{v}_c - \mathbf{d}_{n-1,n-2}(\mathbf{x}_u + \mathbf{v}_c)|^2 \tag{8.12}$$

      ii. The spatial motion smoothness error

      $$e_s = \sum_{k=1}^{8} w_k|\mathbf{v}_c - \mathbf{d}_{n,n-1}(\mathbf{x}_u + \mathbf{q}_k)| \tag{8.13}$$

iii. The occlusion smoothness error when occlusion is not set

$$e_{o=0} = \sum_{k=1}^{8} \lambda_k | - O_{n,n-1}(\mathbf{x}_u + \mathbf{q}_k)| \tag{8.14}$$

iv. The occlusion smoothness error when occlusion *is* set

$$e_{o=1} = \sum_{k=1}^{8} \lambda_k |1 - O_{n,n-1}(\mathbf{x}_u + \mathbf{q}_k)| \tag{8.15}$$

v. The motion compensated pixel difference.

$$e_i = \frac{1}{2\sigma_e^2} \left[I_{n-1}(\mathbf{x}_u + \mathbf{v}_c) - I_{n+1}(\mathbf{x}_u + \mathbf{d}_{n,n+1}(\mathbf{x}_u))\right]^2 \tag{8.16}$$

If the opposite vector $\mathbf{d}_{n,n+1}(\mathbf{x}_u)$ is not already set then fix this difference at zero. The image data will not influence the parameter selection process in this case.

vi. Calculate $E_{v0}(c) = -(e_v + e_s + e_{o=0} + e_i)$, and $E_{v1}(c) = -(e_v + e_s + e_{o=1} + e_i)$. The values are therefore stored in two arrays indexed by candidate vector number.

(c) For each of the two arrays E_{v0}, E_{v1} calculate the maximum value, M_0, M_1. Set $M = \max(M_0, M_1)$.

(d) For each of the vector candidates, calculate $p_o(c)$ and $p_1(c)$ as follows: $p_o(c) = E_{v0}(c) - M$, $p_1(c) = E_{v1}(c) - M$.

(e) Calculate $P(c) = exp(p_o(c)) + exp(p_1(c))$

(f) Select the interpolated vector as that which has the maximum $P(c)$ (Suboptimal ICM), or draw a sample from the calculated distribution.

(g) For that chosen vector, if $p_0(c) > p_1(c)$ set the occlusion parameter $O_{n,n-1}(\mathbf{x}_u)$ to 0, else set it to 1 (Suboptimal ICM). Alternatively, draw a sample from the calculated binary probability distribution. This is to be done with the proviso that the condition $O_{n,n-1}(\mathbf{x}_u) = O_{n,n+1}(\mathbf{x}_u) = 1$ must not be allowed to occur. Should setting the occlusion parameter cause this condition, it should be left unchanged.

5. This process is repeated at each pixel location that is detected as missing in the image (both sides). Also several iterations of the process may be necessary to fill all holes with interpolated motion information. In addition, pixels are treated in a checkerboard fashion as previously discussed.

After the motion field is completely configured, an algorithm may be chosen for image data interpolation at the missing sites. A simple interpolating function (allowing for occlusion) that gives samples of the interpolated data is

$$\hat{I}_n(\mathbf{x}_u) = \begin{cases} (I_{n-1}(\mathbf{x}_u + \mathbf{d}_{n,n-1}) + I_{n+1}(\mathbf{x}_u + \mathbf{d}_{n,n+1}))/2 + e \ \forall \mathbf{O} = [0\ 0] \\ (I_{n-1}(\mathbf{x}_u + \mathbf{d}_{n,n-1}) + e \ \forall \mathbf{O} = [0\ 1] \\ (I_{n+1}(\mathbf{x}_u + \mathbf{d}_{n,n+1}) + e \quad \text{otherwise} \end{cases} \tag{8.17}$$

where $\mathbf{O} = [O_{n,n-1}(\mathbf{x}_u), \ O_{n,n+1}(\mathbf{x}_u)]$, $e \sim \mathcal{N}(0, \frac{\sigma_e^2}{2})$ and the argument $\mathbf{x}_u$ for the displacement has been dropped for simplicity. This expression can be derived from a modification of the original image likelihood using occlusion, but the details are omitted here. The main term is the average of the two known image pixels along a motion trajectory through frame n whenever non-occluded motion occurs. The additional term is a Gaussian noise term with zero mean and variance σ_e^2. The value of this term has been discussed in the previous chapter. When occlusion occurs, the image data is essentially copied from either frame $n-1$ or $n+1$.

8.6.1 Suboptimal MAP estimation

The algorithm above will approximate samples from the joint distribution for the interpolated vector field. The sampling strategy is only approximate because a subset of the space of motion vectors is being searched. The subset is defined by the use of the candidate motion vector list. The approximation is reasonable because it is expected that the candidate list covers the significant part of the probability mass. The mean of these samples may be taken after some appropriate *burn in* period to yield the MMSE estimate. Alternatively the approximate mode of the distribution can be found using the ICM algorithm of Besag [15]. This latter is a simple modification: instead of drawing a random sample for the motion and occlusion at each site, the variable pair which maximizes the posterior (from among the candidates) is taken as the interpolant. The pixel grid is scanned in a checkerboard fashion so that the next site visited does not employ information from the 8-nearest neighbor spatial neighbourhood of the previous site that was processed. It is difficult to choose a scan which also does not overlap *temporal* neighbourhoods since the motion estimated may not be regular enough to preserve the motion compensated grid spacing in the surrounding frames. Doing so would incur some computational overhead and so is not employed here.

Also of interest, once a candidate set of motion vectors is found, is the possibility of solving jointly for all the variables $d_{n,n-1}(\mathbf{x}_u)$, $d_{n,n+1}(\mathbf{x}_u)$, $O_{n,n-1}(\mathbf{x}_u)$, $O_{n,n+1}(\mathbf{x}_u)$ at the same time at each site. This can be done by proposing all possible combinations of candidate vectors as pairs which can be associated with occlusion configurations of $[O_{n,n-1}, \ O_{n,n+1}] = [0 \ 0], \ [0 \ 1], \ [1 \ 0]$. Note that the combination $[1, \ 1]$ is now conveniently disallowed since this implies data in the central frame which cannot be found in the surrounding frames, disabling subsequent image reconstruction at that site. The complete posterior is then evaluated for each of these combinations and the best combination chosen as the local mode employed in later iterations for ICM. Alternatively, a *sample* may be drawn directly from the distribution evaluated, using the numerical transformation method outlined above. Of course, with as much as say 20 candidate vectors at one site for $d_{n,n-1}$, the number of combinations to be evaluated can be of the order $400 \times 3 = 1200$. Fortunately, this case of large numbers of candidates does not occur often unless there is complicated motion of some kind; so for most of the image the computational overhead of this modification is not great.

8.7 Considering more severe degradation

Figure 8.2 illustrates one case where it is not possible to rely on estimated vectors between frames $n-1, n-2$ etc. as being uncorrupted. While it is possible to propose techniques which estimate image data and missing vectors simultaneously, it is more convenient to propose a modification to the algorithm presented above for use with just 3 image frames in which 2 are relatively uncorrupted [97].

8.7.1 Generating motion candidates

First of all, if it is assumed that frames $n-1$, $n+1$ are relatively uncorrupted in the regions which are corrupted in frame n, then it is reasonable to assume that the vectors e.g. $\mathbf{d}_{n-1,n+1}$, $\mathbf{d}_{n+1,n-1}$ can be correctly estimated with a deterministic motion estimation stage, by dropping frame n altogether. Assuming low acceleration implies that $\mathbf{d}_{n,n-1} \approx \mathbf{d}_{n+1,n-1}/2$, suitably compensated for motion. Thus at any site $\mathbf{x}_u$ it is possible to propose a set of temporal motion vectors, derived from the outer two frames, which may be used to start the Gibbs sampler. The Temporal Hit List is assembled by using the half length vectors $\mathbf{d}_{n-1,n+1}/2$ for *backward* hits and $\mathbf{d}_{n+1,n-1}/2$ for *forward* hits. This set may be augmented by spatial candidates from the already configured motion estimates in frame n as the algorithm proceeds. Furthermore, it is again possible to split the process into a motion/occlusion configuration stage and an image interpolation stage. This can be done by employing equation 8.1 (and the assumption of low acceleration) to approximate $|I_n(\mathbf{x}_u) - I_{n-1}(\mathbf{x}_u + \mathbf{d}_{n,n-1}(\mathbf{x}_u))|$ by $|I_{n-1}(\mathbf{x}_u + \mathbf{d}_{n,n-1}(\mathbf{x}_u) - I_{n+1}(\mathbf{x}_u + \mathbf{d}_{n,n+1}(\mathbf{x}_u))|$. Thus the algorithms described above for the five frame case can now be employed for this case using just three frames.

8.8 Illustrating performance

An artificial sequence was generated by panning a subportion of the Lenna image across the Baboon image (called the Lenbab sequence). 128 lines were removed every 3rd frame in the sequence and the five frame reconstruction technique was employed to reconstruct the missing data. A portion of three frames from the sequence are shown at the top of Figure 8.8. The displacement was 5 pixels horizontally and 3 pixels vertically. The corrupted version of the middle frame is shown on the left of the 2nd row in Figure 8.8. The next image on that row shows the reconstruction using the vector interpolation technique described in section 8.6. This was followed by a least squares interpolation using equation 8.17 above, except without the random component (so that RMS errors are more meaningful). The final image shows the result of reconstructing the vector field using a temporal zero-order hold operation (VZO). This implies simply copying the vector fields in the previous and next frames into the missing location in the current frame and then using equation 8.17 with the occlusion terms all set to 0. This second process is the simplest default action in this circumstance. The bottom row shows the difference between the corresponding images in the previous row, and the original, clean data. The error is multiplied by 2.0 and offset by 128, thus mid-grey is zero error. The following values were used for the five frame technique $\lambda_k = 1.0$, $w_k = 5.0$, $i_t = 20.0$, $v_t = 5.0$. The MWBME was used to kick start the sampler, and 20 iterations were employed. The image shown is the result of employing the 20th sample from the sampler, no manipulation of the samples for motion or occlusion was employed.

The images show at first glance that the differences between the simplest and more complicated technique are not dramatic. In both cases a major improvement from the original results. However, closer inspection shows that the result of applying the sampling technique gives a much sharper result than the VZO image. The right hand eye of Lenna as well as the boundary are much better delineated. This is highlighted in the error images. This improved sharpness is due primarily to the attempt to estimate occlusion fields and use them in the interpolation. The sharpness is reflected in the lower RMS error through the whole sequence, as is illustrated in Figure 8.10. Note that Figure 8.10 only shows points every 3 frames, since these were the only ones corrupted in order to give the initial motion estimates some usefulness. When the restored sequence is viewed, however, both versions show a certain 'halo' of moving background which 'sticks' to Lenna. This is an artefact of the block based nature of the initial motion estimates as well as the lack of good definition of edges in the initial motion field provided by the MWBME.

A much more marked difference occurs at higher velocity, when occlusion becomes more important and motion haloes in a block based system such as the MWBME become more detrimental. This is illustrated in Figure 8.9. The top row again shows part of the original sequence, this time with motion $15, 10$ pixels horizontally and vertically. The central frame in the top row is frame 3 in the new sequence and the entire portion displayed was removed. A low threshold for motion was deliberately set for the MWBME to investigate the performance of the algorithm when there are increased motion haloes. The second row shows the five frame restoration as previously (same parameters, and 100 iterations) as well as the result of VZO. The final image on that row is the result when the true motion fields in the frames after and before the current one are known exactly to pixel resolution (so the algorithm still had to interpolate the missing motion field). The error images are not scaled this time. The restorations show reconstructions of the entire image shown as the second frame on the top row.

The images show that the quality of the initial motion estimates from frames $n-1, n-2$ and $n+1, n+2$ is the main factor which limits the performance of the algorithm presented. The left hand reconstruction, using estimated motion

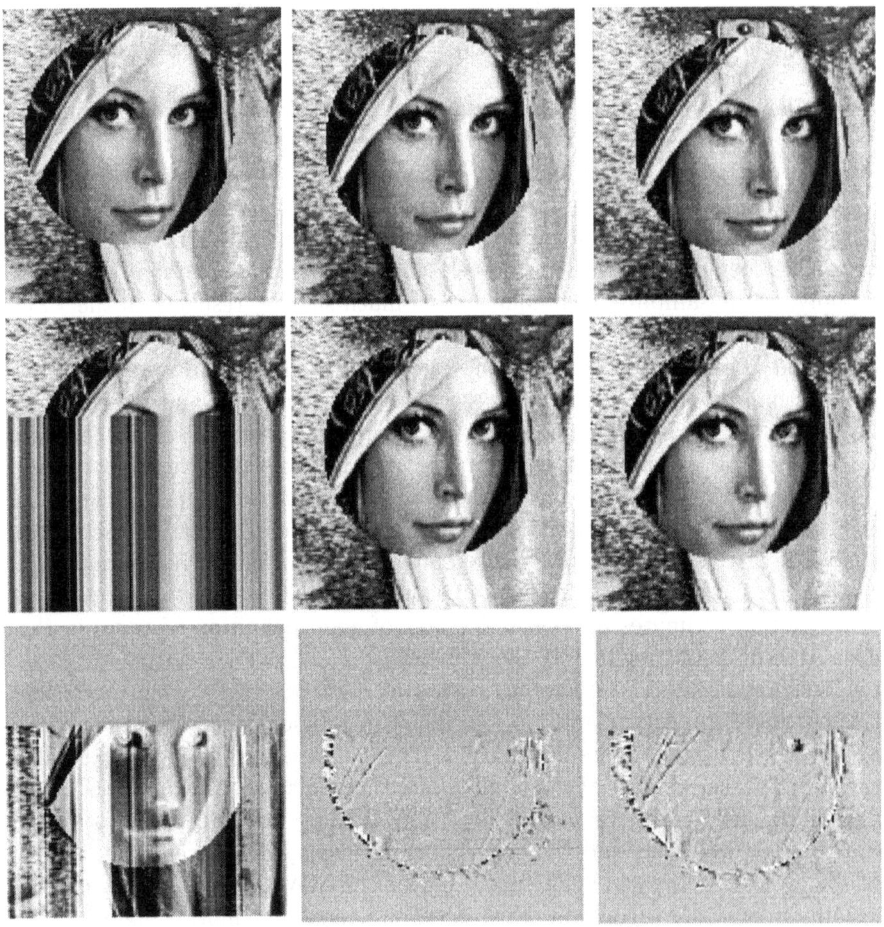

FIGURE 8.8. Top row to bottom row: Zoom on three frames from clean sequence; Degraded 2nd frame, Restored using 5 frame method, Restored using zero order hold; Corresponding difference images with original 2nd frame. Images are 128 × 128.

for candidates, is quite ragged compared to the reasonably pristine right hand reconstruction which used the ideal motion field in the outer frames for the same purpose. Nevertheless the restoration achieved by the five frame technique is much clearer than that resulting from VZO. Whether the reader prefers the smooth but obscure VZO interpolation to the somewhat ragged five frame result, is a separate issue. The RMS error again reflects the better quality of the five frame technique, 12.6 as opposed to 19.2 from the VZO for the Figure 8.9.

Figure 8.10 shows the RMS error for the two systems with perhaps a more realistic situation, the MOBCAL sequence. The results were generated by extracting a 256×256 portion of the image and then removing 40 lines in the region of the moving ball, in every 3rd frame of 25 frames. The result of using VZO is again shown to be inferior. Figure 8.11 allows a visual comparison of the original, degraded and reconstructed frame 5 of the sequence. It is interesting to note that, although the VZO result is more blurred in the region of moving objects and occluded regions, it is quite acceptable compared to the original. This is presumably because of the relatively slow motion of the objects involved. However, note in particular the better definition of the region between the train and the ball in the images shown. This is the main region of occlusion/uncovering. The edges of the ball are also better delineated, although this is perhaps better seen in the error images. It is encouraging that the five frame algorithm performed well in the region of the rotating ball despite the fact that the underlying assumption of translational motion and no acceleration was violated.

8.9 Real degradation

In a real situation the sizes of the degradation vary substantially. In some sense the tests above are harsh since the regular size of the distorted areas means that defects are easier for the human observer to detect. Also the temporal smoothness of the motion and the extremely sharp edges of the foreground object are not common in practice. Figure 8.12 illustrates the result of reconstructing the central portion of the 2nd frame in Figure 8.2 using the three frame technique outlined in section 8.7. The MWBME was again used to generate initial motion estimates. The result is quite good at this scale. Closer inspection of the edges of moving objects shows that the occlusion fields were not yet configured properly (10 iterations were used). The edge of the arm shows some small specks of raggedness. Nevertheless this is quite a useful result, achieved despite the fact that this sequence is corrupted in *every other field* with a dropout which is at least as large as shown.

Figure 8.1 shows the result of applying the five frame algorithm to a very badly degraded sequence. The center column shows the detection result using the ROD detector and the right hand column shows the result after 20 iterations of the algorithm in the locations detected as missing by the detector. Again the MWBME was used for initial motion estimates and the values of the various

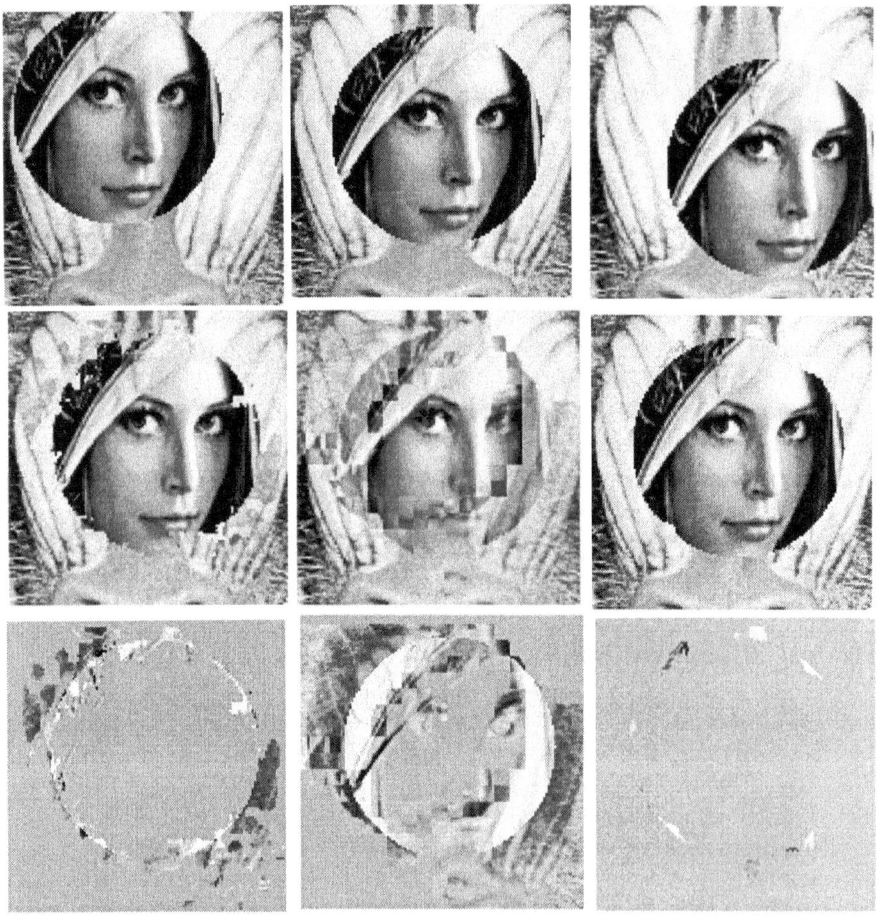

FIGURE 8.9. Top row to bottom row: Zoom on three frames from clean sequence (faster motion); Restored 2nd frame using 5 frame method, Restored 2nd frame using zero order hold, Restored 2nd frame with true motion; Corresponding difference images with original 2nd frame. Images are 128 $\times$ 128

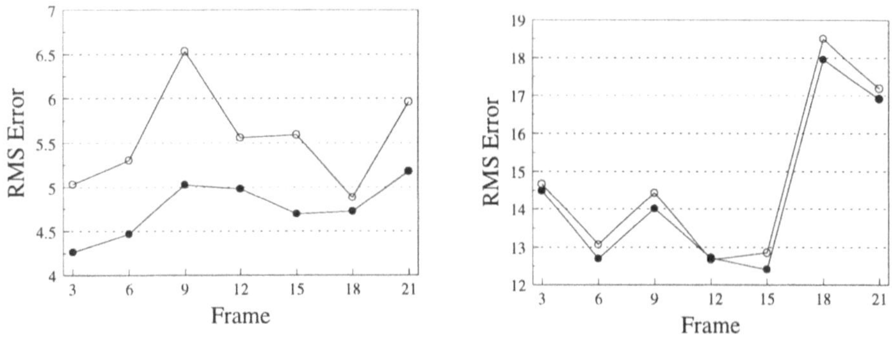

FIGURE 8.10. RMS error of reconstructions on degraded Lenbab sequence (left) and degraded MOBCAL sequence (right). Five frame reconstruction •, Zero Order Hold ○.

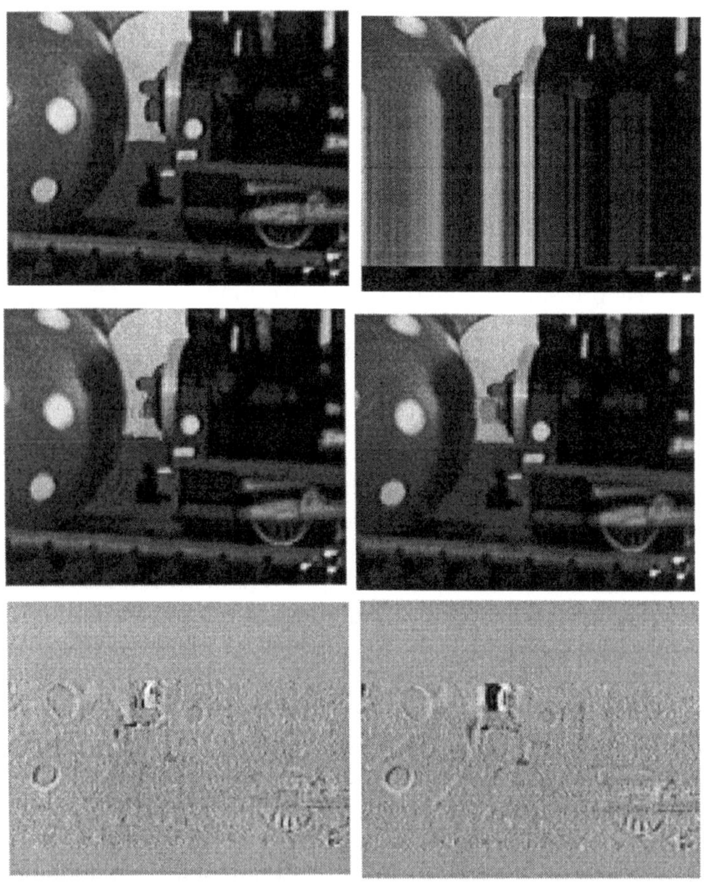

FIGURE 8.11. Top row to bottom row: Zoom on frame from clean/dirty sequence; Restored using 5 frame method, Restored using zero order hold; Corresponding difference images with original frame (magnified by 2.0). Images are 128 × 100.

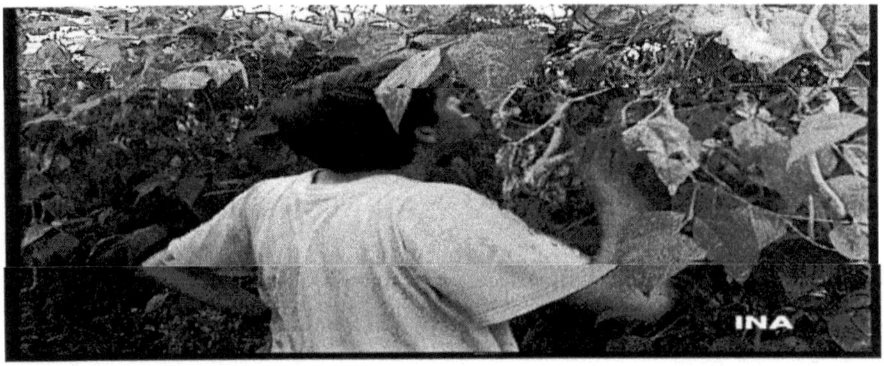

FIGURE 8.12. An example of reconstruction using 2 frames only $(n-1, n+1)$. Image is 720×288. White lines delineate reconstructed region.

motion field hyperparameters were kept the same as for the artificially degraded sequences. The result is very encouraging as all of the areas that have been detected as distorted, have been restored with good fidelity, especially when put in the context of the high noise level in the data.

8.10 Final comments

This chapter has presented a different approach to image interpolation for the case of very large missing regions. The underlying premise has been that the model described by equation 8.1 holds. Thus image reconstruction is completely defined if one could reconstruct the missing motion field in the corrupted areas. The approach was motivated by the lack of an efficient means to cope with both large and small missing areas without resorting to modifications to the somewhat complex joint estimation algorithm introduced in the previous chapter. The resulting algorithm is potentially very fast. It is based on the approximation that the correct motion field is contained within the motion estimates resulting from the use of a simple motion estimator in the known image regions around the frame treated. This approximation manifests in the design of a candidate selection process for the assignment of motion vectors at pixel resolution. Unfortunately, because of this assumption that the correct vector is among one of the candidates, the solution is suboptimal. It may not necessarily locate the global maximum of the posterior distribution for the unknown motion field. This may be overcome, at the expense of convergence perhaps, if the candidate selection process employed to draw samples of the motion field is replaced by a Metropolis sampling step [142], for instance.

It is the selection or generation of candidate vectors for examination which governs the performance of this algorithm. This was illustrated in figure 8.9. In that example, the true motion in the outer frames was used for candidate selection for the motion reconstruction process. That showed a much better performance than the use of the MWBME for generating the initial motion estimates. The algorithm is limited especially by the fact that the MWBME is block based and so there is some conflict between configuring a *pixel* accurate motion field in frame n when a block field exists at frames $n-1, n-2$; $n+1, n+2$. This point was overlooked in the search for a practical solution. To resolve this conflict, a pre-processing step could involve the manipulation of the initial estimates in the frames $n-1, n-2$; $n+1, n+2$ in order to generate a pixel accurate motion field there. Alternatively, the algorithm proposed here could be adapted for block based operation, and pixel accurate configuration of motion in the current field would then be sacrificed. This would impose an upper limit on the resolution of occluded regions. Further improvements may be made by incorporating a multiresolution strategy for candidate selection. In essence a more robust algorithm could be derived if the motion reconstruction process was allowed to operate from a coarse to fine resolution of the image.

The algorithm described here is a completely general purpose reconstruction process. It does not need to operate only on large areas, and in fact it could be used as the reconstruction step in a general detection/reconstruction system for missing data. In this mode it is able to achieve a more consistent result than any of the previous proposed systems, given the computational load. This is illustrated in Figure 8.13. This figure shows a comparison between the restoration shown in Figure 8.1 and the JOMBADI result after the same number of iterations. JOMBADI has not as yet removed all the missing data because the size of the missing patch is much larger than the influence of the 3DAR model used, whereas the algorithm discussed here has already achieved a reasonable result.

As a final note, it is possible to make modifications to the joint estimation system described in the previous chapter to allow joint estimation of image and motion information for this large case. To maintain practical algorithm speeds (minutes as opposed to hours per frame) the model in 8.1 can be adopted. This is really a 1 tap temporal 3DAR model with $a_1 = 1.0$. The spatial temporal smoothness prior used in the previous chapter would then need to be augmented with the temporal smoothness prior presented in equation 8.3. This would allow the algorithm to be able to propagate information temporally as well as spatially in order to solve the very ambiguous problem of data interpolation within a large area. The model statement would need to be modified to include occlusion, but the details are omitted here. The process for joint solution of the missing data and motion can then proceed exactly as described in the previous chapter, except at the pixel resolution. The main modification necessary is the incorporation of temporal motion smoothness. These ideas are the focus of current work.

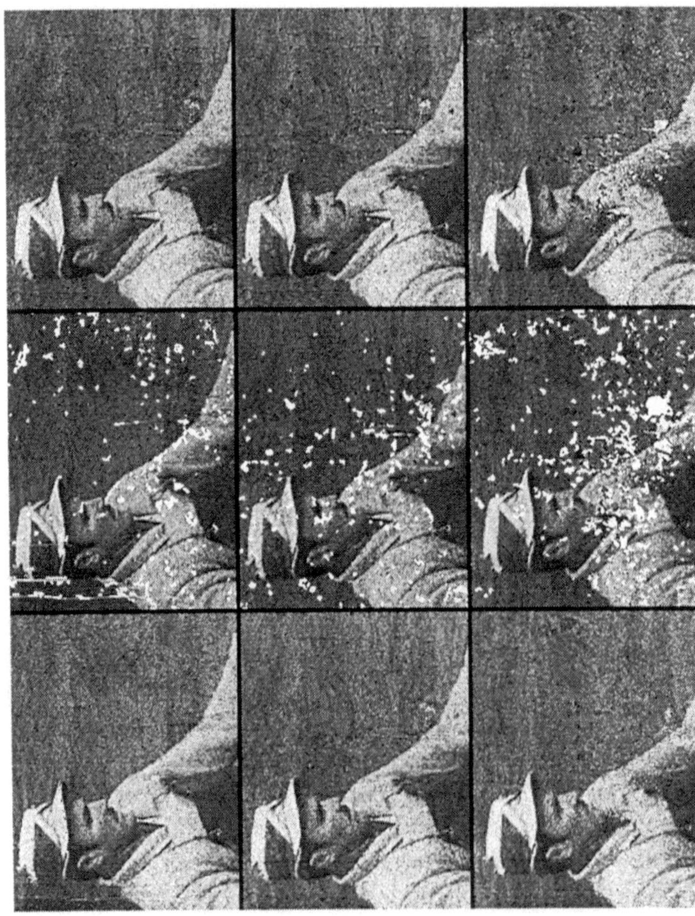

FIGURE 8.13. Comparing with the JOMBADI algorithm. Columns left to right: Result of restoration using algorithm discussed in this chapter (20 iterations), Detection result with JOMBADI (20 iterations), JOMBADI Reconstruction (20 iterations)

9

Line Scratch Detection and Removal

Line scratches are a common problem in archived film. The problem is transferred to video during the telecine transfer process. The artefact is easily visible as a vertical line of bright or dark intensity, oriented more or less vertically over much of the image. It may be caused when material from some particle is smeared vertically over the film material in the projector or by the abrasion of the film as it passes over some particle caught in the mechanism. The task is to propose a technique for the automatic detection and removal of the artefact. Detection is complicated by the fact that lines occur as natural phenomena in interesting scenes. Furthermore, the defect can occur in the same or nearly the same location in consecutive frames. Thus detection of line artefacts cannot rely on temporal discontinuity in image brightness, and the chapter concentrates on purely spatial line detection. Figures 9.1 and 9.2 show two examples of line scratches, illustrating also the persistence over frames.

It is difficult to propose a mathematical model for the effect of the abrasion or occlusion of the film (causing the scratch) on the intensity of the projected light. This would be the first step in the design of a detection algorithm. Instead it is possible to make some observations regarding the luminance cross-section of the line defect, from which a model useful for detection can be proposed. The use of this feature is explored in the first part of the chapter, which concludes by proposing a validation step for separating possible line defects from false alarms.

Since no reliable model of the degradation process is proposed, the removal of the line defect is treated as a missing data problem. A few varieties of reconstruction are explored. The region which is reconstructed persists across several frames and also persists across much of the vertical extent of the picture.

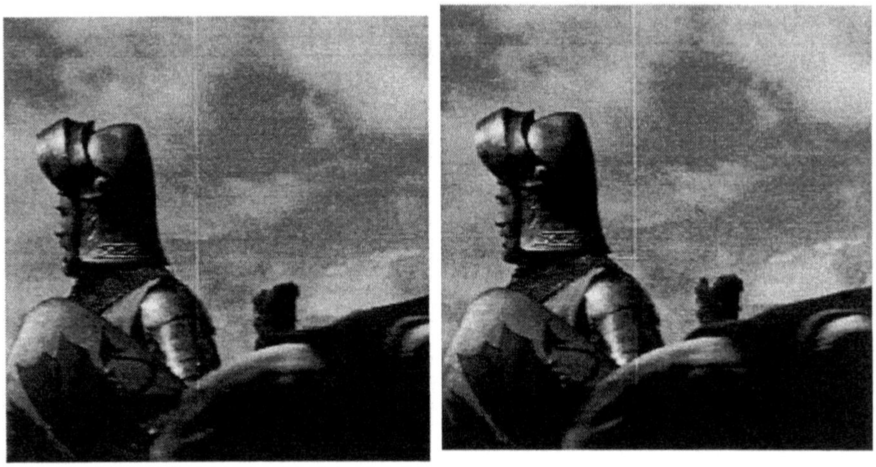

FIGURE 9.1. Two frames of the KNIGHT sequence showing line defects. Images are 256 × 256.

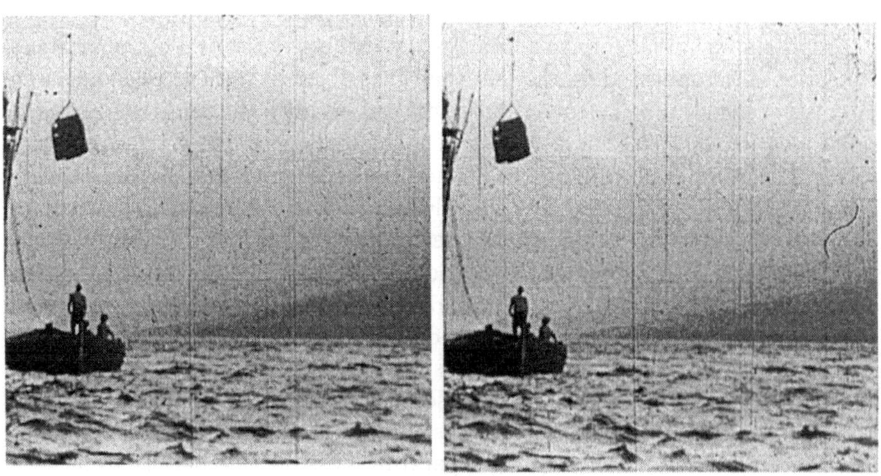

FIGURE 9.2. Two frames of the SITDOWN sequence showing line defects. Images are 512 × 512.

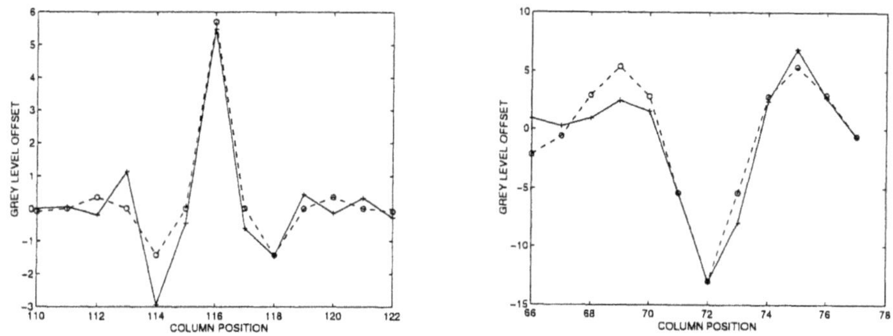

FIGURE 9.3. Cross sections of the lines (solid) at column 116 in figure 9.1 and at column 71 in Figure 9.2. The dashed lines show matched line profiles using $w = 3.0, k = 0.5, b = 5.5$ and $w = 4.8, k = 0.75, b = -13$ respectively.

It is interesting to note that errors in the reconstruction algorithm chosen are more visible than in the case of Blotches.

9.1 Line features and model for degradation

A horizontal cross-section through two of the line artefacts illustrated in figures 9.1 and 9.2 is shown in Figure 9.3. These sections were generated by taking the vertical mean of the columns after correcting for a space varying local mean image intensity. The shape of the sections is similar to a damped sinusoid and illustrates a characteristic of line defects as opposed to other horizontally impulsive features: *the line scratch is accompanied by sidelobes of appreciable intensity*. Dark fringes accompany lines of high intensity and vice versa.

In general lines which are defects tend to extend over most of the vertical length of the frame. Although lines can be slightly curved in some cases, the number of such occurrences are small. Therefore, to further limit the class of lines treated, it is assumed that the line is straight and will not deviate by large amounts from the vertical. Finally, the line features are narrow, with widths from about 3 to 10 pixels at most for video resolution.

The line profile may therefore be modelled by a damped sinusoid as below (where i, j are the horizontal and vertical abscissa)

$$L_n^{(p)}(i,j) = b_p k_p^{|i-(m_p j + c_p)|} \cos\left(\frac{3\pi|i-(m_p j + c_p)|}{2w_p}\right) \tag{9.1}$$

where $\mathbf{x} = [i, j]$, m_p, c_p are the slope and intercept with the horizontal edge of the image which defines the orientation of the pth straight line profile $L_n^{(p)}(i,j)$, w_p is defined as the line width, b_p is the brightness of the central portion of the line and k_p is the decay of the line profile so that $0.05 < k_p < 0.95$. The match between this form and the actual averaged line profile is shown in figure 9.3. This is only one model for the observed cross-section. In fact it may be more accurate to model the line profile formation as the pulse response of an appropriately chosen linear 2nd order system. However, to avoid problems with the estimation of further hidden system parameters the parametric form introduced above is adopted.

Denoting the intensity of the degraded image at position $\mathbf{x}$ as $G(\mathbf{x})$ and the original, clean image as $I(\mathbf{x})$, the distortion caused by P lines (whose maximum magnitude occur at horizontal positions $\mathbf{x}_p = [x_p,\ 0]))$ is assumed to be

$$G(\mathbf{x}) = I(\mathbf{x}) + \sum_{p=0}^{P-1} L_n^{(p)}(\mathbf{x} - \mathbf{x}_p) + e(\mathbf{x}) \tag{9.2}$$

where $e(\mathbf{x})$ is additive Gaussian noise $\sim \mathcal{N}(0, \sigma_e^2)$. Thus it is assumed here that the line artefact itself is an additive distortion.

Because P is not known beforehand, detection of lines using this model is complicated, in fact it is a type of model order selection problem. Morris [118] employs a *Reversible Jump* [56] technique which proposes lines for verification (as well as rejects proposals) as part of a Monte Carlo scheme for line detection. However, it is possible to take a much less compute intensive, but still successful route. A pessimistic estimate of the number and location of lines can be made by using a simple, deterministic process. This is then followed by a Bayesian refinement strategy which rejects or accepts proposals from the deterministic stage to avoid reliance on thresholds set in the deterministic process.

9.2 Finding suspect lines

Line features which are artefacts are vertically smooth but horizontally impulsive. A useful first step in processing is to *vertically subsample* the observed image (e.g. Figure 9.1) by a factor s, after low pass filtering using a vertically oriented Gaussian filter, thus yielding an image denoted $G_s(i,j)$. This procedure enhances vertical line features and suppresses noise. Figure 9.4 shows the result of applying this step to the image in Figure 9.1. Vertical low-pass filtering before subsampling also helps to smooth the line profile vertically, thus artificially reinforcing the assumption that the line cross-section does not change.

Having emphasized the artefacts of interest, the defect line pixels on each row can be detected as horizontally impulsive features. In a sense this is now a particular manifestation of the general problem of impulse detection in degraded signals. There are many approaches which may be employed here, and in fact this problem has been discussed in the earlier chapters of this book as missing data detection. The underlying feature of all the impulse detection work e.g. [183, 184, 54, 55] is that impulses represent significant deviation from an otherwise smooth underlying signal. Unfortunately, the problem is complicated here because step edges are a common feature in the image, thus the smoothness assumption can be violated. Paisan [138] introduced a detection scheme for impulses in 1-D signals which flagged suspect distortion when the observed signal deviated significantly from a median filtered version of the signal. In a sense, the underlying signal model employed was that of a MIN-MAX generating system, an idea currently being explored by Armstrong et al [9]. The approach is attractive since the underlying system model is tolerant to step features, being based on rank order structures. Furthermore it is computationally simple.

Thus detection of the line artefacts can proceed by generating a horizontally median filtered image $M_s(i,j)$, created by filtering $G_s(i,j)$. The line features may be detected by thresholding the signal $e(i,j) = G_s(i,j) - M_s(i,j)$, to yield $B(i,j)$ which is 1 when $e(i,j) > e_t(i,j)$ and 0 otherwise. $e(i,j)$ will be large at sites of horizontal impulses and it is vertically correlated at line scratch sites. This is because the median filter will remove lines with a width that is roughly less than its window. The result is a binary image which is vertically correlated primarily at the sites of short, horizontal impulses. The choice of median filter and threshold level is made less important because the purpose of this stage is to point out candidate line features. Figure 9.4 shows the result of applying this operation on using a 3 tap median filter and $e_t = 3$. The next problem is to find the lines which contribute most to the observed binary pattern.

The Hough transform is a geometric technique which is well suited to identifying the lines in $B(i,j)$. Good descriptions of the technique can be found in [153, 71]. Simply put, the idea is to map lines in (i,j) space onto points in m, c space, where m, c are the gradient and intercept on the $j-$axis respectively. This space is called the Hough space. For instance, a line $i = m_1 j + c_1$ will map into the point m_1, c_1. Conversely, the point (h,k) maps onto a line $c = -km + h$ in m, c space. This latter is the more useful observation. It implies that lines in the Hough space which correspond to points that belong to the same line in $B(i,j)$ in this case, will intersect at the value of m, c for that line in $B(i,j)$. Thus for every binary 1 in $B(i,j)$, a line is drawn in an array (called the accumulator array) which samples Hough space. When all lines are 'drawn', peaks in this space identify lines in $B(i,j)$. In simple terms, for each point $B(i,j) = 1$ a line of height 1 is drawn in an array so that $c = -jm + i$. The size of the bins in this array are chosen so that the space of m, c is adequately covered for the lines to be found. In this case suitable ranges are $m = -0.1 : .01 : 0.1$ and $c = 0 : 0.5 : N_1$ for an image with N_1 columns (using Matlab notation).

Note that the line equations take the gradient to be the slope compared to the vertical. This is a convenient representation since it avoids numerical problems which would occur if the axes were arranged in the usual manner with the horizontal axis being the independent variable. In that case the lines to be drawn would all have a close to infinite gradient. Using this axis arrangement, c gives directly the intercept of the detected line with the top row of the image, given the definition of coordinates of the top left hand pixel in the image as $(0,0)$. The narrow range for m is allowable because the lines are primarily vertical. This makes the application of the Hough transform very computationally attractive here, since only short lines are 'drawn' in Hough space. The result of applying this transform to Figure 9.1 is shown in Figure 9.4. The line at $c = 116$ stands out clearly as a peak in the transform. Note that since the accumulator array is effectively sampling impulsive features, i.e. lines which are certainly

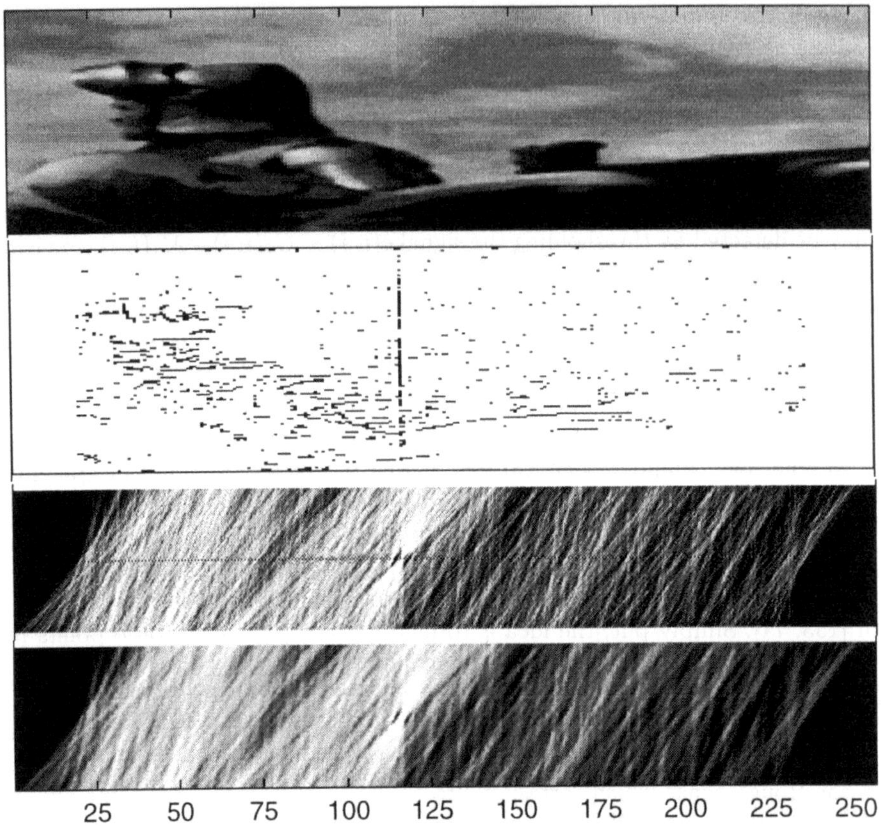

FIGURE 9.4. **Top to bottom**: Median filtered (1 × 3), Low passed and subsampled version of Figure 9.1; Thresholded median error (threshold = 3), $B(i, j)$; Hough transform of $B(i, j)$; Filtered Hough Transform using Gaussian window. Transform bounds: $m = -0.2 : 0.01 : 0.2, c = 0 : 0.5 : 255$, the top left corner is $m = -0.2,\ c = 0.0$.

not bandlimited, the process necessarily causes aliasing. For this application, the aliasing in Hough space is avoided by the crude, but effective, technique of filtering the raw accumulator array with a 2-D Gaussian filter with a variance of 0.5 and having 5 × 5 taps.

To select candidate lines, it is convenient to place an extremely conservative threshold on the height of the peaks selected in the Hough space, (H_t say) to avoid selecting peaks corresponding to lines which are certainly spurious. This is made possible because the height of the Hough bins is proportional to the number of pixels flagged as being part of a particular line. Given the assumption that line artefacts traverse a substantial portion of the image frame, it is reasonable to select Hough bins which have some substantial line contribution. An additional feature which may be incorporated in this stage is the number of lines which are required to be removed. Thus each peak is visited in turn from the maximum to the minimum, and the first N lines say, which exceed the bin threshold are selected for removal. This order allows the pixels in $B(i, j)$ which contribute to a particular selected line to be removed from the accumulator array before the next peak is selected. This is a very simple mechanism for avoiding the spread of the peaks in Hough space to affect the choice of the next line. Therefore, assuming that the maximum line width is w_t, after a line candidate is selected, the Hough contributions from all pixels which are set to 1 in $B(i,j)$ and which are within $w_t/2$ pixels from the candidate line in $B(i,j)$, are subtracted from the accumulator. $w_t = 6$ in the restorations here. The corresponding Hough transform for the image in Figure 9.2 using $e_t < -5.0$ and a median filter of length 7, is included on the CD-ROM. It gives evidence for not ignoring non-vertical lines, since at least one of the major lines has a significant gradient of -0.01.

9.3 Bayesian refinement

Having employed the Hough transform to find likely line candidates it is now necessary to evaluate them as line artefacts and choose those most likely to fit the line profile model defined in equation 9.1. Proceeding in a Bayesian fashion and denoting the parameter vector as $\mathbf{P} = [k_p, b_p, w_p, m_p, c_p, \sigma_e^2]$, the probability of the parameters for the pth line given the data may be written as

$$p(\mathbf{P}|G(\mathbf{x}), I(\mathbf{x})) \propto p(G(\mathbf{x})|\mathbf{P}, I(\mathbf{x}))p(\mathbf{P}|I(\mathbf{x})) \tag{9.3}$$

The likelihood can be written as a Normal distribution of errors, $\mathbf{E} = [e(\mathbf{x}_1), e(\mathbf{x}_2), \ldots]$,

$$p(G_n(\mathbf{x})|\mathbf{P}, I_n(\mathbf{x})) \propto \exp\left(-\frac{1}{\sigma_e^2}\sum_{\mathbf{x}\in\mathcal{P}}[G(\mathbf{x}) - L^{(p)}(\mathbf{x}) - I(\mathbf{x})]^2\right) \tag{9.4}$$

where $\mathcal{P}$ is an area which includes the pth line. These expressions may be solved to locate the MAP or ML solution for the parameter vector for each line. However, rather than take this approach, it is found to be useful to numerically investigate the marginal distribution for the brightness, $p(b_p)$. Examination of this distribution will then indicate the importance of the line. The line feature is assumed to be significant if $p(b_p \leq 1.0) < Rp(b_p > 1.0)$. In other words, when a significant proportion (measured by R) of the marginal probability distribution for the brightness is concentrated below the grey level 1.0, the line indicated is a false alarm. The use of the grey level 1.0 as a fixed point is due purely to considerations of perceptual visibility.

9.3.1 Gibbs sampling

Each line candidate provides an initial estimate of m_p, c_p. These values are kept fixed as it is found in practice that the peaks of the Hough transform correspond well with the line centres if a true line is found. Making the assumption that

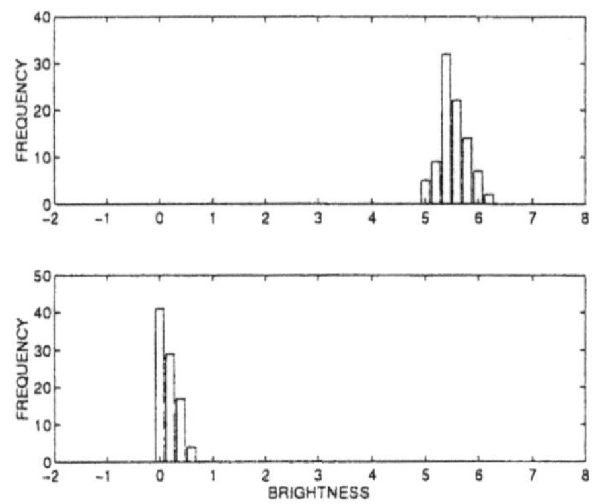

FIGURE 9.5. Histograms of the 1st 100 samples of b (after 10 sample *burn in*. Top: True line, c = 116; Bottom: False alarm, c = 56.

$M(i,j)$ is a good estimate of $I(i,j)$ for detection, Gibbs sampling [118, 151] can then be employed[1] to provide marginal probability densities for the remaining parameters, $b_p, w_p, k_p, \sigma_e^2$.

Gibbs sampling has been discussed in Chapter 3.1. In this case, it is executed by successively drawing samples from the conditional distributions $k_p \sim p(k_p|\mathbf{P}, I(\mathbf{x}))$, $w_p \sim p(w_p|\mathbf{P}, I(\mathbf{x}))$, $b_p \sim p(b_p|\mathbf{P}, I(\mathbf{x}))$, given the current values of the other parameters at each random draw. After some *burn in* period, the sequence of samples drawn converges to samples from the marginal distributions $p(k_p)$, $p(w_p)$, $p(b_p)$, $p(\sigma_e^2)$ respectively. It is possible to determine exactly the expressions for the conditional distributions for b_p, σ_e^2 since the required integrations are standard. The distributions are as follows

$$p(b_p|\mathbf{P}, I(\mathbf{x})) = \mathcal{N}(L_2/L_1, \sigma_e^2/L_1) \tag{9.5}$$

$$p(\sigma_e|\mathbf{P}, I(i,j)) \propto \sigma^{-N} \exp\left(-\frac{|\mathbf{E}|^2}{2\sigma_e^2}\right) \tag{9.6}$$

where

$$L_1 = \sum_{\mathbf{x} \in \mathcal{P}} L_p(\mathbf{x})^2$$

$$L_2 = \sum_{\mathbf{x} \in \mathcal{P}} [G(\mathbf{x}) - I(\mathbf{x})] L_p(\mathbf{x})$$

and $|\mathbf{E}|^2$ is the sum squared error in equation 9.4. Hence drawing samples for b_p is simple (See [151] or Chapter 3.1). However sampling for k_p, w_p is difficult. In order to avoid the computational cost of Metropolis-Hastings sampling, the allowed range for each of these parameters is quantized to 0.05 : 0.05 : 0.95 and 0.5 : 0.5 : 5.0 respectively, assuming a maximum w_p of 5 pixels. Then the conditional probability distribution over each of these ranges can be calculated

[1]With uniform priors, suitable parameter ranges, and a Chi Squared prior on σ_e^2.

numerically using equation 9.3. This can then be normalized and a sample for the quantized variable drawn using the Cumulative p.d.f. It is found that the sequences of samples converge very quickly to the marginal distributions, typically within the first 10 iterations.

Figure 9.5 shows two histograms of the two lines which were selected as candidates from the output of the deterministic process on the left hand image in Figure 9.1. The line at $c = 56$ is a false alarm and can be seen to be differentiated from the line at $c = 116$ (which is the true line) by the fact that its distribution is mostly concentrated below $b = 1.0$. Note that in the actual implementation it is assumed that $I(i,j)$ can be well approximated by $M(i,j)$, where $M(i,j)$ is a horizontally median filtered version of $G(i,j)$ at the full original scale.

9.4 Line removal

Equation 9.2 is not good enough for line removal. Figure 9.6 shows a zoom on a region showing the line artefact in Figure 9.1 before and after correction by subtracting the estimated line profile. The problem is that the line profile can change substantially as it traverses the image and subtraction of a constant profile can leave residual line traces. These traces are visible as annoying distractions in the restored image. Therefore, the conservative assumption is made that the line scratch has obliterated data in the region of the artefact (which does occur).

Interpolating data in this gap is not as simple as it may seem. The 2D Autoregressive (AR) model can be used as the basis for image interpolation in the same manner as that introduced in Chapter 3.1 for image sequences and the 3D AR model. However the region to be interpolated now extends over much of the image and so the underlying data is statistically non-stationary. Thus the interpolation ideas in Chapter 3.1 must be modified for use here. A simple scheme for dealing with non-stationarity is to overlap blocks of data for coefficient estimation and interpolation. In overlapping there are two alternatives. The first is introduced in Chapter 10, in which a Raised Cosine window can be used to blend (add) the results of blocks which overlap by a factor of 2 (only vertically in this case). The other method involves tiling blocks so that the data used for estimation of model parameters and interpolated output does overlap, but the interpolated output is tiled in continuous portions which fit together to create a smooth interpolant. These two alternatives are shown in Figure 9.7.

The choice of windowing scheme is determined by considering another more important problem that must be addressed. In any particular sub-block used for interpolation, there will exist an entire vertical strip which is not valid. It is assumed to be missing. The weighted AR coefficient estimation stage can deal with this problem since the weights can be set to disregard the assumed missing portion. However, the interpolation process requires boundary values to determine the interpolant. The problem is illustrated in Figure 9.8. Here a three

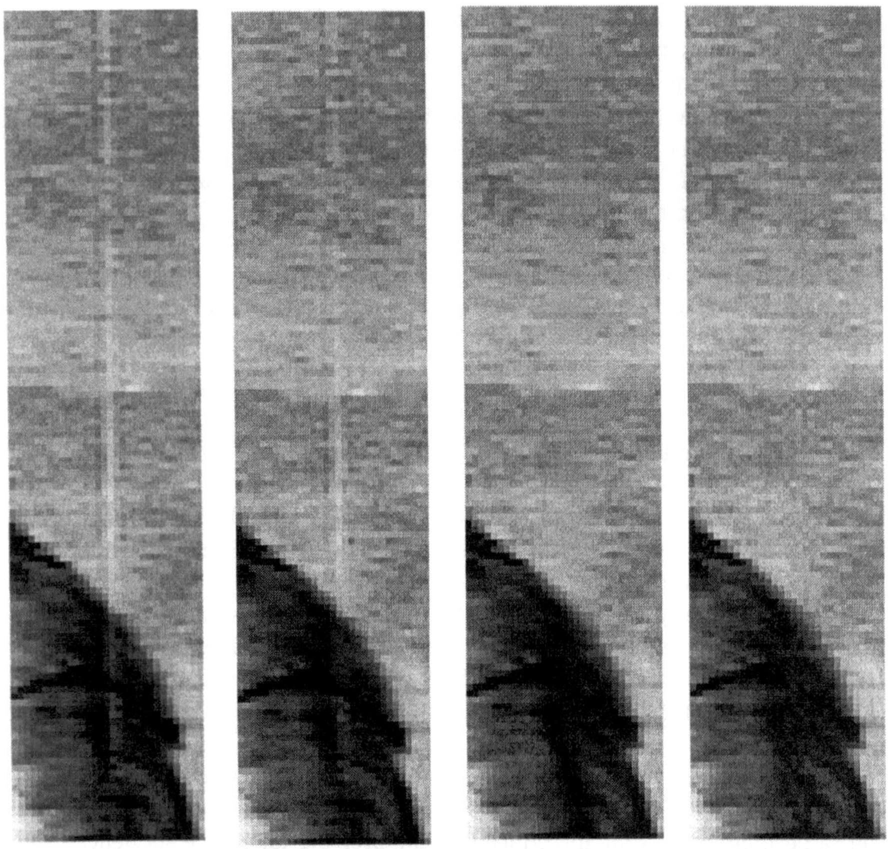

FIGURE 9.6. Left to right: Portion of original KNIGHT image; Result of subtracting estimated line profile; Least Squares Interpolation of line region; Sampled Interpolation in line region.

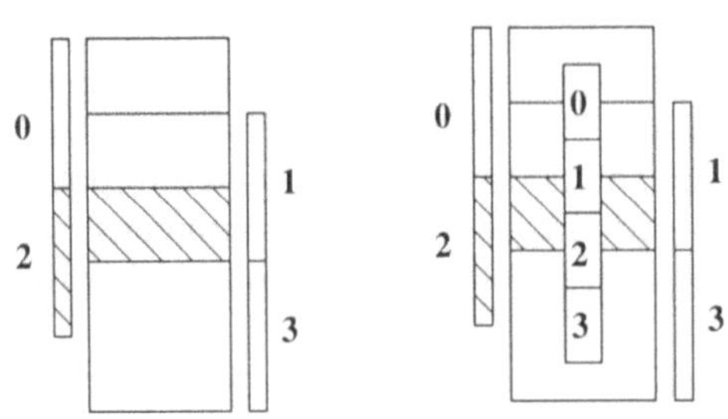

FIGURE 9.7. Left: Overlapping using windowing; Right: Overlapping with Tiling. The bars on each side show the vertical position of each overlapping block 0, 1, 2, 3. With tiling, the central bars show the vertical extent of the data which is used as the output. On the left hand side, no output information from the overlapping blocks is discarded.

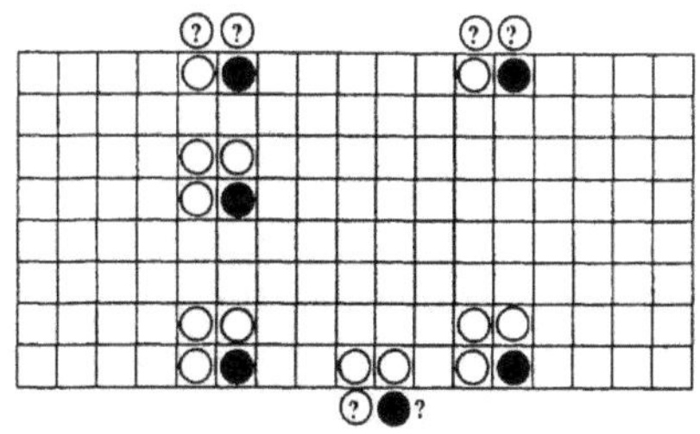

FIGURE 9.8. The three tap linear predictor does not have valid support at the boundaries of the missing regions (shaded). The solid dots indicate selected prediction error locations and the other circles show the corresponding support pixels. '?' indicates missing support. Each square represents one pixel and white squares are *known* data.

tap AR predictor is shown, having no known pixels in its support at the vertical extremities of the missing patch. One method of overcoming this problem is to choose purely horizontal models, but this yields rather poor interpolation since it is equivalent to performing many 1-D interpolation processes in each row, thus disregarding any 2-D correlation which certainly exists. The more complete alternative is to circularly extend the data block. Therefore, prediction equations at the block extremities can find support at the opposite side of the block. This does not solve the problem of lack of known boundary values, but it makes the interpolator more stable at the block extremities. It is possible to extend the data block using reflection at the block boundary. However this yields extremely poor results. Note further that as the interpolation process proceeds vertically down the missing region, previous interpolants can be used as known values for the next interpolation block. Of course no computational overhead need be caused by this data extension. In Chapter 3.1 a fast algorithm for AR interpolation was outlined and that need only be modified by a test which checks when the filters access out-of-block locations and the data lookup modified accordingly.

Now that the boundary problem has been identified, it may be thought that the better choice of overlapping process is tiling. Tiling avoids using unstable interpolations at the block boundaries which may occur at some instants despite the use of circular blocks. If such instabilities do occur and windowed overlapping is employed, this distortion is added into the next block. In practice, tiling tends to give slightly better detail preservation than windowing but it is difficult to find cases where the improvements are dramatic. In the restorations shown (figures 9.6, 9.9) tiling was used since it then becomes more convenient to interpolate non-vertical lines.

Ultimately, the interpolator is to be used on a real degraded image sequence. Unfortunately, because of the sometimes large gaps that need to be interpolated and the subsequent increase in non-stationarity over these large regions, the least squares interpolant is somewhat blurred. This is not easily noticed in a still image, but in a sequence of images the rather flat interpolant stands out against the otherwise grainy film texture, especially if there is motion. In fact the main contributor to the annoying visibility of the reconstructed region is that it remains in nearly the same place in each frame. In some respects this is a moot point since the line removal is one stage in a chain of restoration modules which will also involve noise reduction. After noise reduction it is harder to see the difference between the texture in the interpolated regions as compared to the surroundings. However, to affect a complete solution it is better to employ samples from the posterior distribution for the unknown data rather than employ the least squares interpolant. The samples tend to have a closer match with the textural behaviour of the film in the surrounding regions. This process is described in Chapter 3.1 for a similar problem in image sequences, and can be adopted unchanged here except that it is now a 2-D manipulation. Of course, the disadvantage is that this now incurs the additional expense of calculating the square root of a correlation matrix (See Chapter 3.1).

Figure 9.6 illustrates the decreased visibility of the interpolated patch when the interpolant is sampled rather than taking the least squares interpolant. The image is a portion of Figure 9.1. Figure 9.9 shows a portion of the original and restored real degraded images in Figure 9.2, after detection using the described technique and interpolation using sampled interpolants tiled vertically. Both interpolations used 3×3 (8 tap) causal model support and the detector used $H_t = 20, 40$ respectively for Figures 9.6 and 9.9. Full frames of these sequences are included on the accompanying CD-ROM. The reader may verify the decreased visibility of the sampled interpolant in the image sequence by comparing the restorations using least squares and sampled interpolants in the sequence. It should be noted that the visibility of the interpolated line is always further reduced after subsequent processing such as Dirt and Sparkle removal and in particular noise reduction. Both these subsequent stages would be necessary in most cases where line scratches occur.

9.5 Summary comments

This chapter has again outlined an algorithm which combined deterministic and stochastic methods. The overall concept is that a deterministic pre-processing algorithm can yield a very good starting point for a stochastic process, allowing the power of MCMC methods (for example) to be used in a practical, low cost solution by improving the convergence of the stochastic optimization stage.

The work has shown that the automatic detection of line scratches is complicated by the fact that they persist in nearly the same location in each frame. In some sense the deterministic process which was introduced can stand on its own as an effective detection system if the user is willing to identify the number of lines in the image. This kind of user interaction may be reasonably viable in the film post-processing industry, although much less viable for real-time television pre-processing. Nevertheless the deterministic stage is so computationally simple that it is conceivable that the selection of a 'suitable' threshold for detection of lines (H_t) can be coped with as part of a real time system.

The Bayesian refinement step was introduced solely in an effort to improve the hands-off operation of the algorithm. It suffers from one major disadvantage in that it assumes that the line traverses the entire image, which is usually the case but not always. It is possible to design a scheme which uses binary indicator

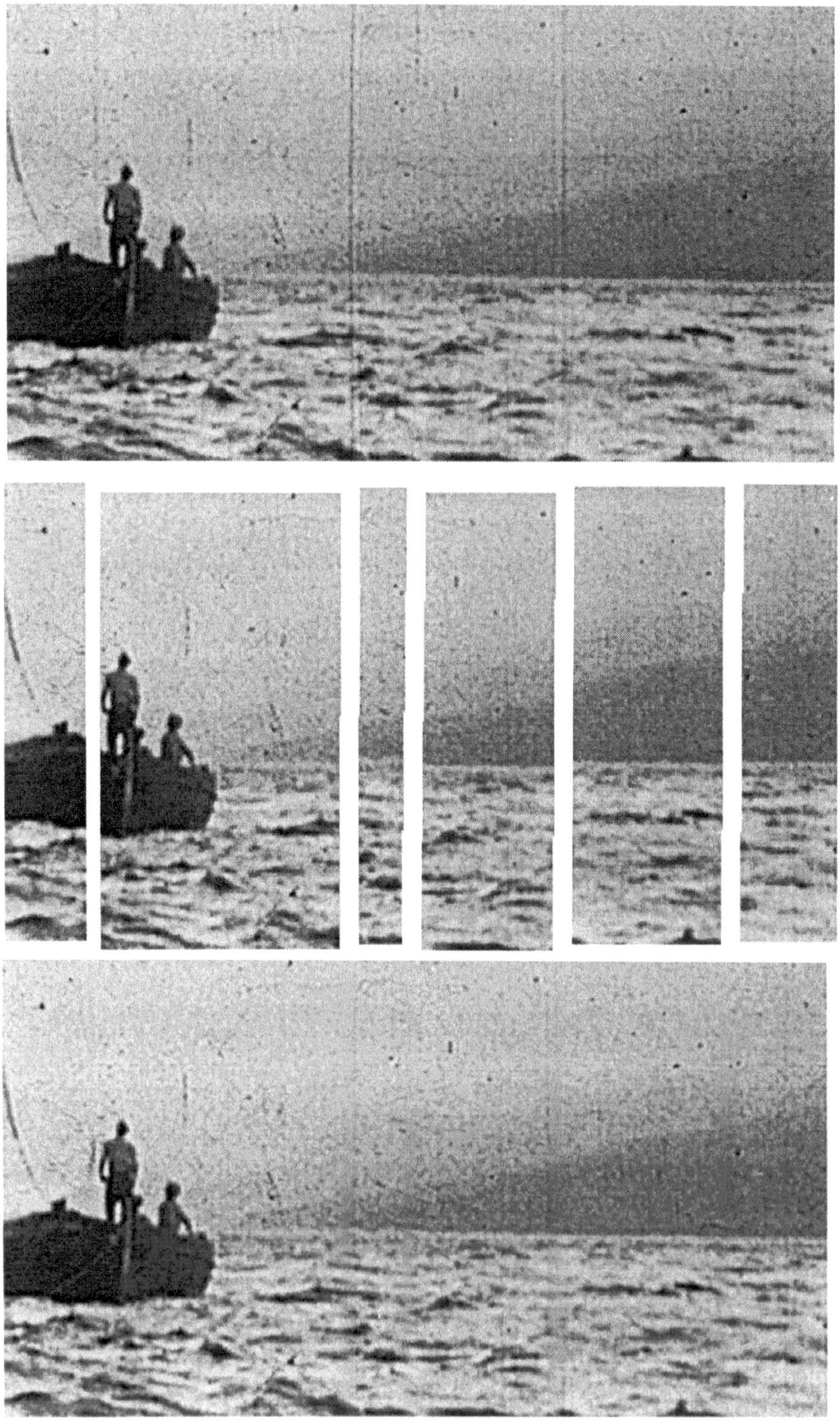

variables to switch on and off the introduction of a line profile at different points vertically along the line. The estimation of these variables could be incorporated as part of the refinement strategy.

The idea of treating the line as an area of missing data was necessary only because a good model of degradation could not be found. An alternative degradation model is that of the line being formed by a pulse passed through a second order system. This can be adopted in the refinement stage so that the coefficients of the system as well as the height and width of the pulse become model parameters. The estimation of these parameters may allow a more general shape to be fitted and so the evolution of the line vertically may be better tracked and so better removed. This is one focus of current work.

As a final note, it must be recognized that the techniques presented by Hirani et. al. [64] and Strohmer [166] are viable alternatives to the spatial AR interpolation process presented here. Hirani et al employ a POCS[2] [169] based method in the frequency domain for reconstructing missing patches by manually locating regions of similar texture in the image. This can be adapted for use here. Strohmer's technique interpolates regions using trigonometric polynomials, in a kind of weighted FFT formulation. This is a much more suitable approach than presented by Hirani, although the computational complexity is higher. However, in both cases it would be harder to deal with the visibility of the relatively 'smooth' interpolant.

[2]Projections on Convex Sets

10
Noise Reduction for Image Sequences

Noise is one of the most common forms of distortion observed in image sequences. It can be encountered both due to degradation of the original film material or due to transmission and receiver noise in the case of live video streams. A particularly severe level of noise is observed in video sources such as Electron Microscopes and cameras that are sensitive to non–visible radiation.

As outlined in the introduction to this book, there have been many proposed solutions to the problem of noise reduction in image sequences. The trend has been toward motion compensated temporal filtering techniques and in the last few years optimal (in the least squared error sense) 3D filtering has been applied. This chapter considers the use of three dimensional motion compensated Wiener filters both in the spatio–temporal domain and in the frequency domain. These filters are seen to be a viable alternative to the approach to Wiener filters discussed by previous authors. One version of the Wiener filter that was presented in [135] for image sequences has not been compared but is discussed in Appendix G.

This book does not give comprehensive treatment of possible noise reduction systems for film and video. Noise reduction has been an extremely well traversed area of research and there are quite comprehensive discussions given by Lajendijk, Kleihorst et al [102, 80, 169] for instance. In recent years there has been exponential activity in the use of Wavelets for noise reduction in images and image sequences [5, 185, 148]. It is hoped that this chapter will serve to illustrate the potential of motion compensated spatio-temporal noise reduction for image sequences. In addition the reader is exposed to some practical considerations which make possible the use of the more traditional Wiener filters for image sequence processing.

10.1 Motion compensated Wiener filtering

Wiener filtering is a well known technique for the reduction of noise in degraded signals. It has been used to good effect by Vaseghi [182] for reducing the noise in archived gramophone recordings. The filter can be applied both in the frequency domain, as an approximation to the IIR filter, or in the spatio–temporal domain as an FIR filter.

At this stage it should be clear to the reader that provided good motion estimates can be obtained, a 3-D implementation of a filter would provide substantial improvement over a 2-D implementation. Therefore, the approach considered here uses a 3-D filtering operation on data from three motion compensated frames in a sequence. The data in a block of $N \times N$ pixels in each frame is extracted allowing for motion, to give a data volume which is filtered to suppress noise in the central frame. As is typical for motion compensated filtering techniques, the algorithm begins by estimating the motion vectors that map the data in the current frame n into the next frame $n+1$ and the previous frame $n-1$. These vectors are then used to compensate the necessary data.

The observed image sequence model is defined by the following equation.

$$G(i,j,n) = I(i,j,n) + \eta(i,j,n) \tag{10.1}$$

Where $G(i,j,n)$ is the observed signal grey scale value at position (i,j) in the nth frame, $I(i,j,n)$ is the actual non-degraded signal and $\eta(i,j,n)$ the added Gaussian noise of variance $\sigma_{\eta\eta}$. Appendix F outlines the derivation of the Wiener filter for noise reduction in a signal which is degraded according to that equation. The filter is derived by first deciding on a filter structure, either FIR or IIR, and then finding the coefficients which minimize the expected value of the squared error between the filter output and the original, clean signal.

10.1.1 The 3D IIR/3D frequency domain filter

The IIR form of the filter can be expressed as a frequency domain operation which requires an estimate of the power spectrum of the original, clean data to yield the attenuation factor required for each frequency bin. The 3D Wiener filter in the frequency domain is then given as below:

$$A(\omega_1,\omega_2,\omega_3) = \frac{P_{gg}(\omega_1,\omega_2,\omega_3) - P_{\eta\eta}(\omega_1,\omega_2,\omega_3)}{P_{gg}(\omega_1,\omega_2,\omega_3)} \tag{10.2}$$

where $A(\omega_1,\omega_2,\omega_3)$ defines the frequency response of the filter, and $P_{gg}, P_{\eta\eta}$ refer to the Power spectral densities (PSD's) of the degraded and noise signals respectively. The arguments to the functions refer to *discrete* frequency bins as gained via the 3D DFT[1].

It is important to recognize that although the 3D IIR filter can be implemented in the 3D frequency domain using equation 10.2, the resulting filter is no

[1] The n–dimensional DFT is a separable operation and is implemented as a recursive DFT operation in orthogonal directions. See [71, 107]

longer IIR in practice. In practice, via the 3D DFT, the 3D Frequency domain Wiener filter operates only on a finite input data volume whereas a 3D IIR filter (as it is normally implemented) would additionally operate on previous outputs.

The noise is assumed to be uncorrelated with the original signal, and so the following equation applies:

$$P_{gg} = P_{ii} + P_{\eta\eta} \tag{10.3}$$

With this in mind it follows that the effect of the Wiener filter is to attenuate the frequency components of $G(i,j,k)$ according to the observed power of that particular component. When it is high, $P_{gg} >> P_{\eta\eta}$ less filtering is done, and when this power is low, the component is heavily attenuated. This is a useful property for images in particular [156, 42]. In regions of high image activity, such as highly textured areas and edges, there is less attenuation of the signal. In areas of low activity such as uniform areas, more attenuation is achieved. The signal detail is therefore less attenuated than in the uniform regions. The human visual system is known to be less sensitive to noise in regions of high activity. This is a useful bonus since the areas in which the filter will attenuate less noise corresponds to the areas in which noise is less easily observed.

The situation is exactly the same along the temporal axis. The activity along the temporal axis depends on the accuracy of motion estimation. When the motion estimation is not accurate, the filter will automatically reduce the contribution of the temporal information to the noise reduction. This property makes the filter robust to even bad errors in motion estimation. Of course the amount by which it reduces the attenuation when this problem occurs may not correspond to an optimal result in human visual terms. Nevertheless the filter has an advantage in this respect in comparison with motion compensated frame averaging [23] which tends to blur the image when the motion is not correctly estimated.

Practical considerations

The derivation of the 3-D frequency domain Wiener filter follows a simple extension to the standard 1D framework. However, in a practical implementation there are outstanding considerations.

1. Estimating the PSD:
   As indicated by equation 10.2, this Wiener filter can be defined in terms of the PSD of the observed degraded signal and the noise PSD. The work presented here does not automatically estimate the noise PSD, rather, it is a user selectable parameter which is altered to suit the tastes of the viewer. This is due to the fact that the noise PSD is assumed to be at a constant level across all the frequency components.

   The PSD of the degraded signal is estimated by the magnitude of the 3D DFT. To prevent spectral leakage effects, the signal must be windowed prior to the DFT operation . Many window functions exist for this purpose. The half–cosine window is chosen here, following [196]. For a block size of $N \times N$ the window is defined as below.

$$\begin{aligned} w(n_1, n_2) &= \cos(\frac{n_1 \pi}{2N}) \cos(\frac{n_2 \pi}{2N}) \\ \text{for } n_1, n_2 &= -(N - \frac{1}{2}), -(N - \frac{1}{2}) + 1, \ldots, (N - \frac{1}{2}) \end{aligned} \tag{10.4}$$

This window is called the *analysis* window since it is used to assist in acquiring an estimate of the Fourier transform of the image data.

The image information in each frame is therefore windowed with this 2D half–cosine window prior to taking the DFT. There is an argument for implementing a similar analysis window along the temporal axis since again there would be problems in taking the DFT in this direction. In practice the visible effect of using a half–cosine window across the 3 frames[2] is found to be small.

2. Overlapped processing:
The entire image cannot be treated with the same Wiener filter since the image information changes across the frame. Therefore, the image is broken into blocks of size $N \times N$. Each block is compensated for motion across the three frames used so that in restoring the information in a block in the current frame n a data volume of size $N \times N \times 3$ is used.

It is common in such a block based algorithm that the non–stationarity of the image causes blocking artefacts across the edges of the blocks. To suppress this effect the processed blocks can be overlapped [196]. In this implementation the blocks are overlapped by half their horizontal and vertical dimensions, an overlap of 2 : 1. If the processed blocks were merely placed so that one half of the next block replaced one half of the current block, artefacts could still occur. Overlapped processing implies windowing the output data so that when the overlapped blocks are summed, the effective signal gain is 1. Therefore any output pixel at the edge of a block has a contribution from several blocks around it.

The output or *synthesis* window must be chosen with regard to the analysis window. The two windows complement each other and taken together must not change the net gain through the noise reduction system. In this case, using a half–cosine window (and 2:1 overlap) as synthesis and analysis windows yields a net gain of unity as required. The net windowing effect is that of a raised cosine function defined below:

$$\begin{aligned} w(n_1, n_2) &= \cos^2(\frac{n_1 \pi}{2N}) \cos^2(\frac{n_2 \pi}{2N}) \\ \text{for } n_1, n_2 &= -(N - \frac{1}{2}), -(N - \frac{1}{2}) + 1, \ldots, (N - \frac{1}{2}) \end{aligned} \tag{10.5}$$

Overlapped processing has not been necessary so far in the work of the earlier chapters. This is primarily due to the fact that the solution to the

[2] Essentially, a 3 tap half–cosine window.

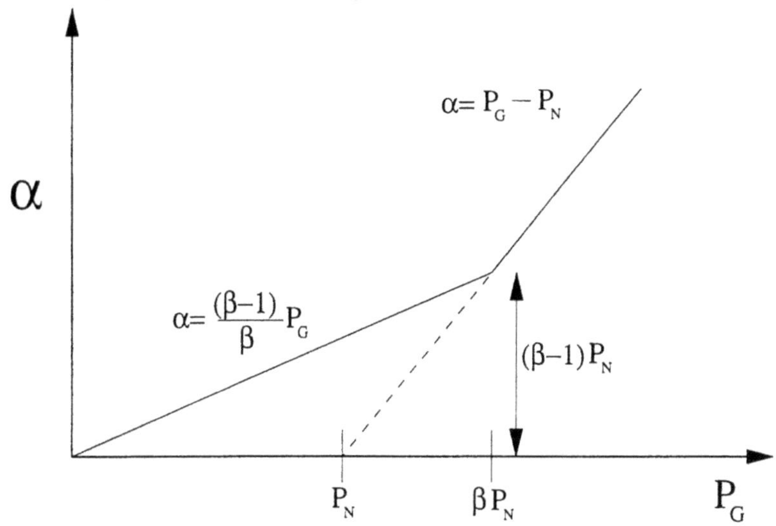

FIGURE 10.1. The mapping (solid line) for the numerator of the Wiener filter, α.

noise suppression problem involves filtering the entire image as opposed to selective treatment. Chapter 6 considered this point.

3. Noise margin:
The frequency domain Wiener filter essentially involves estimating an attenuation for each frequency component in the degraded signal. In the calculation of each factor it is possible that $P_{gg}(\omega_1, \omega_2, \omega_3) < P_{\eta\eta}(\omega_1, \omega_2, \omega_3)$ in the numerator of the expression 10.2. This would result in a negative attenuation which is impractical. The most common method for dealing with this problem is to set $P_G - P_N = 0$ when $P_G < P_N$ [156, 115]. Note that the frequency arguments have been dropped and the notation simplified.

However, this solution is somewhat drastic and can lead to ringing or patterned noise artefacts in the output frame. To avoid this difficulty, an alternative mapping may be used which reduces the attenuation of the filter for low values of P_G. This implies that more noise is left in the image, but this is preferable to the artefacts that would otherwise occur. The new mapping is as follows, where the Wiener filter is defined as $\frac{\alpha}{P_G}$.

$$\alpha = \begin{cases} P_G - P_N & \text{for } P_G > \beta P_N \\ \frac{\beta-1}{\beta} P_G & \text{otherwise} \end{cases} \tag{10.6}$$

When $\beta = 1$ the criterion becomes the same as that used previously in [156, 115]. β is a user defined parameter that governs the amount of noise left in the restoration, and is therefore called the Noise Margin. The mapping is illustrated in Figure 10.1. The modified Wiener filter attenuation is given below:

$$H(\omega_1, \omega_2, \omega_3) = \text{MAX}\left\{ (\frac{\beta - 1}{\beta}), \frac{P_{gg}(\omega_1, \omega_2, \omega_3) - P_{\eta\eta}(\omega_1, \omega_2, \omega_3)}{P_{gg}(\omega_1, \omega_2, \omega_3)} \right\} \tag{10.7}$$

Note that since $\left(\frac{\beta-1}{\beta}\right)$ is a constant, the effect is to define a minimum, non-zero, value of the filter coefficient at frequency bins where α would normally be set to zero.

10.1.2 The 3D FIR filter

This filter is defined in terms of the autocorrelation sequence of the degraded signal and the cross correlation of the degraded and original signals. If the spatial and temporal extent of the FIR filter is such that the filter solution includes all the significant terms in the relevant correlation sequences, then there is little difference between the FIR and IIR filters.

The filter solution is derived in Appendix F, and the result is stated here. The filter is of the form shown below:

$$\hat{I}(i, j, n) = \sum_{k_1=-N_1}^{N_1} \sum_{k_2=-N_2}^{N_2} \sum_{k_3=-N_3}^{N_3} a(k_1, k_2, k_3) G(i + k_1, j + k_2, n + k_3) \tag{10.8}$$

The filter can be defined with the same notation that was used to define the AR model equations in Chapter 3. The support of the filter can then be defined more generally as below:

$$\hat{I}(\mathbf{x}) = \sum_{k=0}^{N-1} a_k G(\mathbf{x} + \mathbf{q}_k) \tag{10.9}$$

Note that in this case a_0, which corresponds to the support vector $\mathbf{q}_k = [0\ 0\ 0]$ is not necessarily unity. The filter coefficients can be arranged into a vector $\mathbf{a}$, and the solution for the coefficients is given by the following equation.

$$\mathbf{a} = [\mathbf{R}_{gg}]^{-1} \mathbf{r}_{ig} \tag{10.10}$$

The correlation terms in the matrices used in the above expression are defined as below.

$$\begin{aligned} \mathbf{R}_{gg}(r, c) &= E[G(\mathbf{x} + \mathbf{q}_r) G(\mathbf{x} + \mathbf{q}_c)] \ r, c = 0 \ldots N - 1 \\ \mathbf{r}_{ig}(c) &= \begin{cases} E[(G(\mathbf{x}))^2] - \sigma_{\eta\eta}^2 & \text{for } c = 0 \\ E[G(\mathbf{x}) G(\mathbf{x} + \mathbf{q}_c)] & \text{otherwise} \end{cases} \end{aligned}$$

The filter therefore operates on a 3 block data volume to give a filtered output for the centre frame.

Practical considerations

In a similar manner to the previous discussion regarding the frequency domain Wiener filter, the following points are notable.

1. Data Windowing:
   No analysis window is used prior to estimating the correlation terms required since this would detrimentally affect the values measured. To prevent blocking artefacts, the output is windowed and overlapped with a 2:1 ratio in the same way as previously discussed. Because the analysis window is effectively a rectangular function, a raised cosine window is needed as the synthesis window prior to overlapping. This window is defined by equation 10.5.

2. The Noise Margin:
   In a similar fashion to the mapping used for the IIR Wiener filtering, the value for $r_{ig}(0) = \alpha = r_{gg}^2(0) - \sigma_{nn}^2$ that is used, is defined by a mapping shown below.

$$\alpha = \mathrm{MAX}\left(\left(\frac{\beta - 1}{\beta}\right) r_{gg}(0),\ \left(r_{gg}(0) - \sigma_{\eta\eta}^2\right) \right) \tag{10.11}$$

3. Ill Conditioning:
   It is possible that the solution to equation 10.10 in a real situation becomes ill conditioned. It is best to detect this condition using some form of eigen analysis. To keep computation low, a less effective yet satisfactory solution is adopted. After finding $\mathbf{a}$, equation 10.10 is evaluated. This yields a calculated value, $\hat{\mathbf{r}}_{ig}$ for the same observed quantity. If any term in this calculated vector differs from the observed term by more than 5% the solution is labelled as ill-conditioned. This situation does not occur often but when it does occur it is usually in uniform regions. Therefore, the filter coefficients are all set to $\frac{1}{N}$ when this condition is detected. The output of the filter is then the average of all the pixels in its support. It would be useful to consider the effect of using eigen analysis to detect ill-conditioning in future work.

10.2 Performance

Figures 10.2 and 10.3 illustrate the ringing phenomenon when Wiener filters are used for noise reduction. Three frames (256 × 256) containing a single corner feature were degraded with additive noise of variance 100.0. The corner was made stationary in the sequence in order to concentrate only on the ringing phenomenon. The images were processed with the IIR and FIR Wiener filters using a huge block size of 64 × 64 pixels so that the ringing was more easily seen. The top row of figure 10.2 shows a portion (64 × 64) of the clean original and

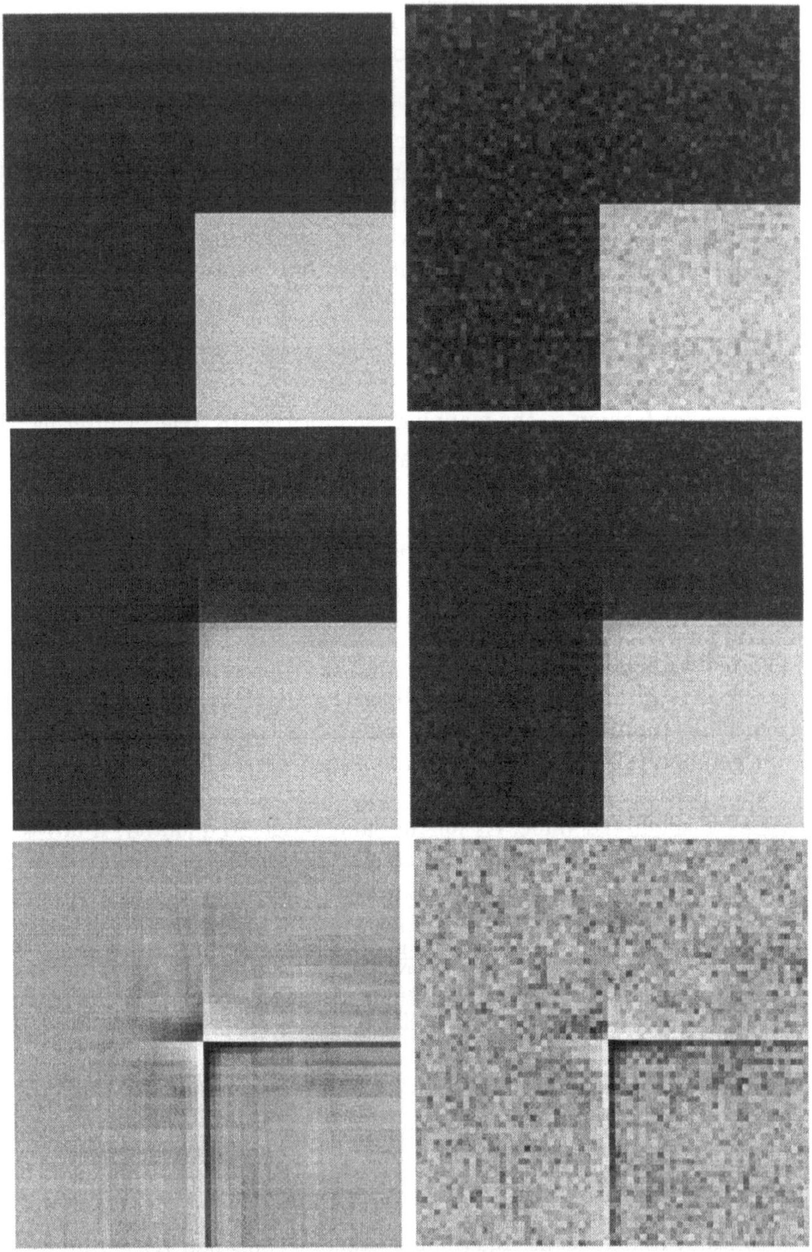

FIGURE 10.2. Illustration of the ringing phenomenon in Wiener filters. Top Row: Clean original (left) and degraded image (right). Middle Row: Wiener IIR result using $\beta = 1.0, 1.4$ (left and right respectively). Bottom Row: Corresponding difference images with clean original.

FIGURE 10.3. Illustration of the ringing phenomenon in Wiener filters. Left and right images are Wiener FIR result using $\beta = 1.0, 1.5$ respectively.

the dirty image processed. The middle row shows the result of applying the IIR Wiener filter with $\sigma_{nn} = 500.0, \beta = 1.0$ on the left and $\sigma_{nn} = 500.0, \beta = 1.4$ on the right. This overestimate of the noise variance increases the ringing effect. The left shows some ringing at the corner feature. In fact there is some ringing near to all the edges of the white square, but this is better seen with the actual image data on the CD-ROM. The picture immediately below this one shows the difference between the filtered output and the original clean signal (multiplied by a factor of 4). The ringing manifests as edge ghosts which can be more easily seen in the difference image.

The right hand image shows that the effect of increasing the noise margin is to increase the amount of residual noise left in the filtered image. This therefore reduces the ringing effect at the corner and along the edges. Again the effect on the actual image is better seen on the CD-ROM, but some appreciation for the increased noise and less ringing can be seen in the corresponding difference image. Obviously there is scope here to perform a more intelligent signal analysis at the edge feature so that the noise margin is only increased in a particular *direction*, thus maintaining some level of noise reduction *parallel* to the edge feature. This, coupled with some edge sensitivity in the filter, would allow a better noise reduction by being able to adaptively alter the noise margin depending on the proximity to edge features. This is discussed later. Note that the observed image after filtering is the result after overlapping block outputs, therefore the observed ringing effect is also influenced by the exact block position relative to the edge.

Figure 10.3 shows how the phenomenon manifests in the FIR Wiener filter using $\sigma_{nn} = 100$. The effect is visibly quite different. There is no identifiable ringing feature but there is an increased erroneous *textural* effect at the edges. This is because of the finite spatial extent of the FIR filter as well as the fact that the phenomenon observed is the result after blocks are overlapped. As explained previously, the net effect depends on the location of the edge feature relative to the placing of each of the overlapping blocks. When $\sigma_{nn} = 500$ (as used for the IIR filter) the effect is extremely pronounced and appears as a high level of noise added to the image. The FIR filter output is sometimes difficult to control because of ill-conditioning in the matrix solution. In fact this residual noise or textured noise problem at edge features can occur quite often, but because of the texturing the effect is sometimes more preferable to ringing from the IIR filter. Again the reader will observe that the increased noise margin does help to reduce the visibility of 'texturing' but the effect is still present even at $\beta = 1.5$.

The two filters presented above were compared with several filters that have been presented previously for noise reduction. The filters were compared with respect to noise reduction of 22 frames of the MOBCAL sequence degraded with Gaussian noise of variance 100. This gave an SNR of about 23db after quantization to 8 bits of grey scale and clipping between 0 and 255. A 3 level multiresolution BBM motion estimator (as described in Chapter 2) was used in all cases with the parameters kept the same as follows: Blocksize 17×17 at level 0, and 9×9 at levels 1,2; threshold MAE = 10 at level 1, and 5 otherwise, $r = 1.2,\ 1.1,\ 1.0$ for the highest to lowest resolution level. A variance of 1.0 was used for the Gaussian low pass filter (to create the pyramid) which was of size 9×9. The various filters that were compared are described below. The block based vector field that was generated by the algorithm was employed directly (using zero order hold) to provide the filters with motion vectors where required. This was found to yield better results than bilinear interpolation of the motion field.

1. The Temporal Wiener filter: Presented in [156], and discussed earlier in an implicit motion compensated scheme [115], this filter is essentially a one tap Wiener filter. The filter was reviewed in Chapter 4 and was defined as

$$\hat{I}(i,j,n) = \frac{\sigma_{gg}^2 - \sigma_{nn}^2}{\sigma_{gg}^2}(G(i,j,n) - \bar{g}(i,j,n)) + \bar{g}(i,j,n) \tag{10.12}$$

$\hat{I}(i,j,n)$ is the Wiener estimate of $I(i,j,n)$ and $\bar{g}$, σ_{gg}^2 are the mean and variance of the observed, degraded signal, $G(i,j,n)$. The filter does not explicitly incorporate information from more than one frame, in fact it operates by scaling the current pixel. The multi-frame information comes from the parameter estimates which are calculated from 3 motion compensated pixels[3]. The mean of the original signal, $\bar{I}(i,j,n)$ is assumed to be well approximated by the mean of the observed degraded signal, given that the noise is white. A noise margin β, can also be presented for this filter. The margin is set to unity for these experiments since the noise variance is known.

[3] Of course these estimates can be calculated using a data volume, but to avoid problems with blurring it was decided to use the temporal estimate.

2. Frame Averaging: The output of this filter is the average of the 3 motion compensated pixels from 3 frames. The particular implementation used was presented by Boyce in [23]. The output of the filter is as follows

$$\hat{I}(i,j,k) = \frac{1}{3}\left[G(i,j,k) + G(i + sx_{k,k-1}, j + sy_{k,k-1}, k-1) \right.$$
$$\left. + G(i + sx_{k,k+1}, j + sy_{k,k+1}, k+1)\right] \quad (10.13)$$

3. Temporally Recursive Filtering: Presented in [36], this filter is of the form

$$\hat{I}(i,j,k) = \alpha G(i,j,k) + (1-\alpha)\hat{I}(i + sx_{k,k-1}, j + sy_{k,k-1}, k-1) \quad (10.14)$$

It was discussed in Chapter 4. The attenuation of the filter is regulated by the scalar α. This is varied according to the magnitude of the backward displaced frame difference at the current pixel to be filtered. This difference is defined as

$$e = G(i,j,k) - G(i + sx_{k,k-1}, j + sy_{k,k-1}, k-1) \quad (10.15)$$

When this error is large, α is set closer to unity so that the filtering is reduced. The opposite is the case when the error is small. α is varied according to the piecewise characteristic defined below.

$$\alpha = \begin{cases} \alpha_b & \text{for } |e| \leq P_b \\ (\frac{\alpha_e - \alpha_b}{P_e - P_b})(|e| - P_b) + \alpha_b & \text{for } P_b < |e| \leq P_e \\ \alpha_e & \text{for } |e| > P_e \end{cases} \quad (10.16)$$

The parameters used were as follows: $\alpha_e = 1.0$, $\alpha_b = 0.5$, $P_e = 40.0$, $P_b = 20.0$.

4. Recursive Frame Averaging: This filter is the same as that discussed above except that $\alpha_b = \frac{1}{N}$, where N is the number of frames processed. Therefore, when $|e|$ is small, and the motion estimation has been effective, the output is the average of all the past frames. The implementation used here is rather primitive since the number of frames is not reset when $|e|$ is large nor is it allowed to vary across the image. Thus the attenuation persists when the filter is turned on after the temporally discontinuous region is past, rather than the attenuation being reset. This causes rather severe 'comet tail' effects. It is best to implement this kind of filter in the form presented in the previous filter description.

5. The 3D IIR/3D Frequency Domain Wiener Filter: As defined earlier, with $\beta = 1.0, \sigma_{nn}^2 = 100.0$. A Block size of 16×16 was used.

6. The 3D FIR Wiener filter: As defined earlier with $\beta = 1.0, \sigma_{nn}^2 = 100.0$. A Block size of 16×16 was used. The filter was non–causal using the support of a square of 3×3 pixels in each of 3 frames.

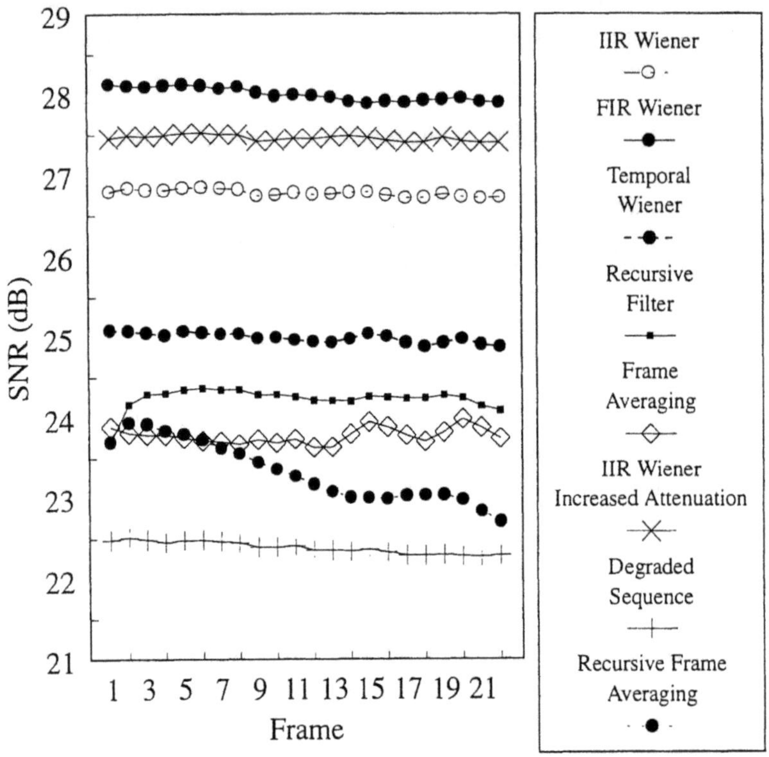

FIGURE 10.4. SNR of various restorations on MOBCAL. (Note that the IIR filter refers to the 3-D frequency domain implementation.)

Figure 10.4 shows the SNR (in db) of the restored frames with respect to the original frames for each of the filters used. The 3D Wiener filters are seen to give the best performance overall in terms of SNR. The frame averaging filters perform the worst. The 3 frame averaging filter is sensitive to erroneous motion estimation, causing blurring. The recursive frame averager does not perform well because it can consider too many frames in its current filter output again causing blurring. It is difficult to effectively control the operation of this filter with the piecewise linear characteristic used. The temporal recursive filter performs as expected, taking about 3 frames before its output stabilizes. Predictably, the filters with less spatial support are more sensitive to motion estimation error.

Note that it is possible to increase the noise attenuation of the Wiener filters by employing a higher value for σ_{nn}. This was done for the IIR Wiener filter (setting $\sigma_{nn} = 150.0$) and the improvement is observed to be about 1 db. The corresponding line in the plot is labelled at *IIR Wiener Increased Attenuation* in the associated legend.

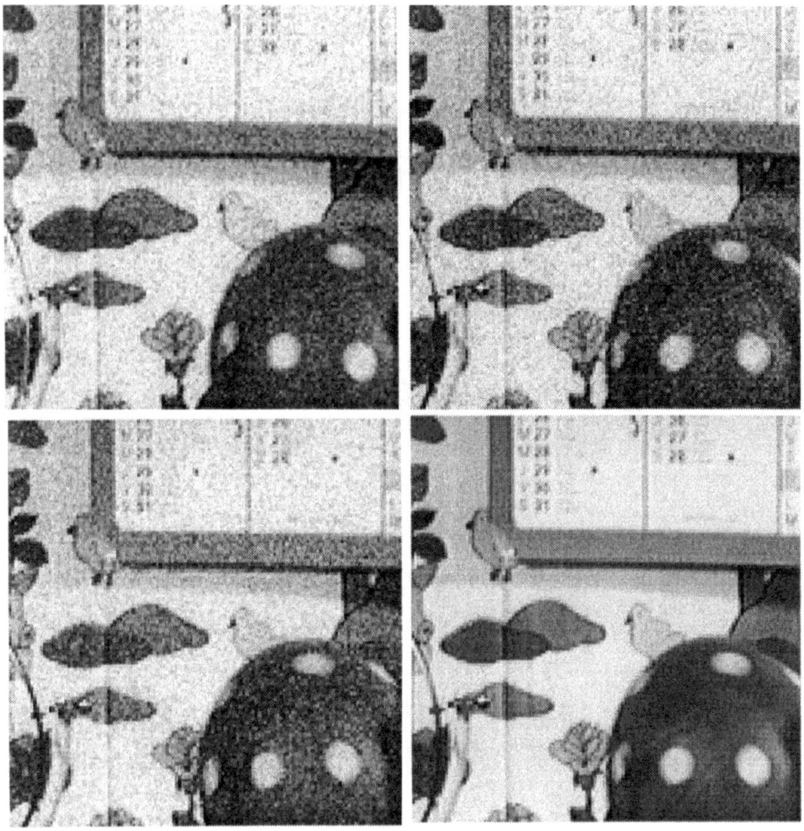

FIGURE 10.5. Clockwise from top left (zoom): Frames 9,10 of the degraded MOBCAL sequence, Original (clean) Frame 10, Degraded frame 11.

The SNR is, however, a bad measure of image quality and Figure 10.6 shows pictures of a zoom on the 10th frame of the restored sequences. The IIR Wiener result is actually the result using increased attenuation at $\sigma_{nn} = 150.0$. Figure 10.5 shows frames 9, 10, 11 of the degraded sequence as well as the clean original frame 10. The frames resulting from the temporal Wiener filter are sharper to some extent than the 3D DFT version, but less noise suppression is observed. The photographs illustrate well the compromise that must be struck between noise reduction and detail removal. Note that although the SNR plots indicate that the 3D FIR Wiener filter is better than the 3D IIR Wiener filter, the photographs show that visually the comparison is much closer. Figure 10.7 shows the difference between the filtered versions of MOBCAL and the original clean frames (added to 128 so that mid-grey is a difference of 0). Ideally these difference images would be constant at 128 showing no distortion. However edge features of the image can be observed, although not nearly as much as if motion compensation were omitted. (See later discussion on real data). The difference values have been magnified by a factor of 4. It is interesting the the 3D IIR

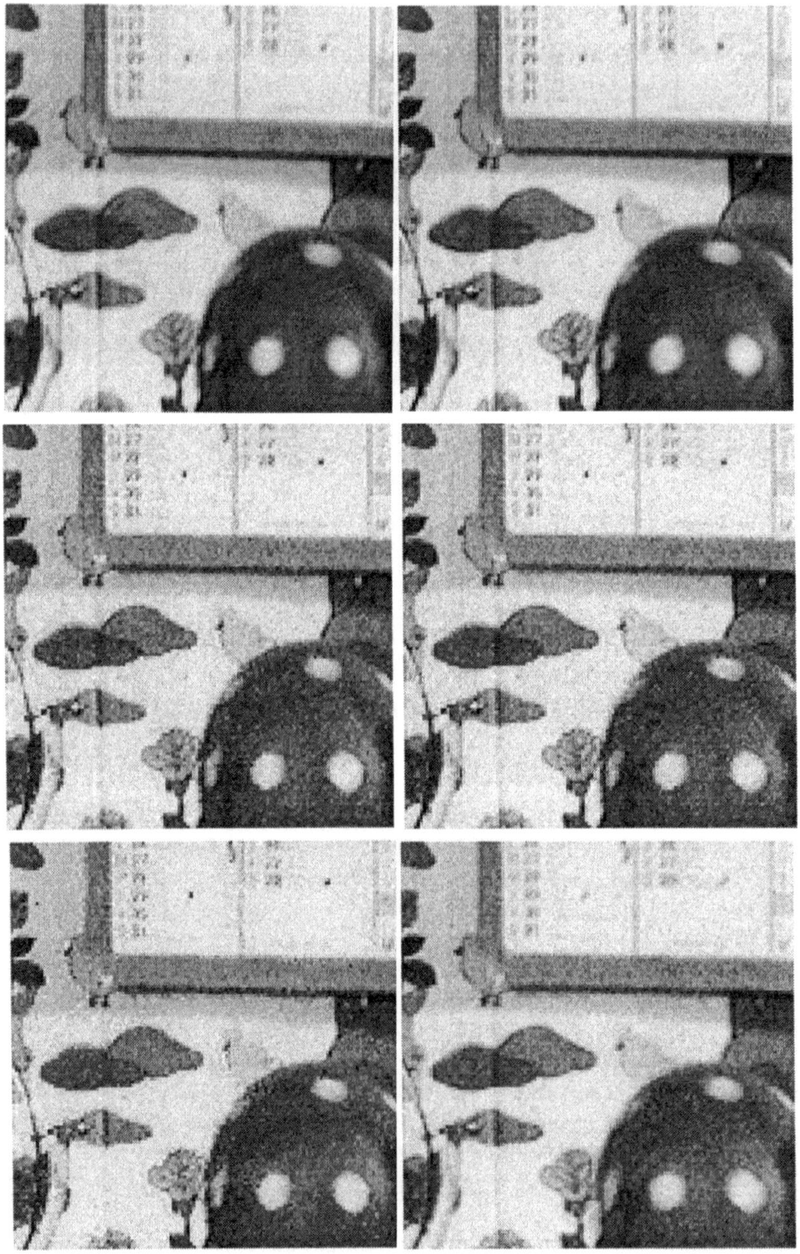

FIGURE 10.6. Restored versions of MOBCAL (zoom) using motion compensation. Top Row: IIR (left), FIR Wiener filters (right). Middle Row: Temporal Recursive Filtering (left), Temporal Wiener Filtering (right). Bottom Row: Recursive Frame Averaging (left), Frame Averaging (right).

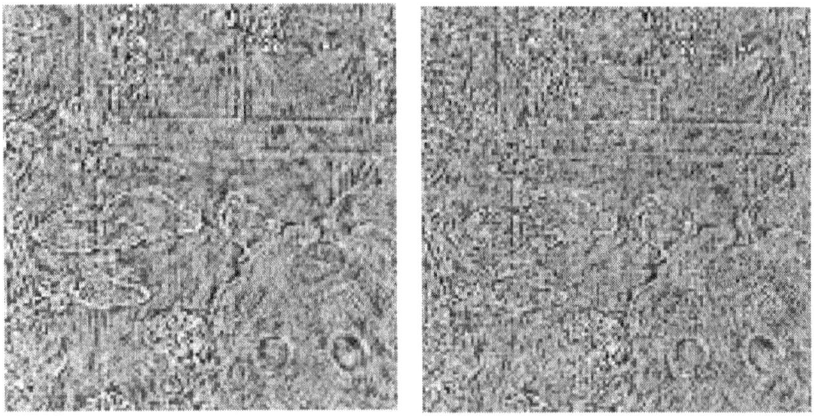

FIGURE 10.7. Difference between Wiener restored versions of MOBCAL (zoom) and the original clean image. IIR (left), FIR right.

filter appears to have a much less residual noise level in flat areas whereas the 3D FIR filter appears to have a higher level of noise overall. This is offset by the increased edge attenuation (though slight) of the IIR filter as compared to the FIR filter, hence yielding the measured SNR curves which show the IIR filter performing worse than the FIR filter. The reader is referred to the accompanying CD-ROM for a demonstration of this observation. The results show that the Wiener filters are consistently better than those previously introduced.

The Wiener filter works well because it is able to implicitly vary the attenuation of the filter with motion estimation accuracy. When there is an error in motion estimation, the DFT in the temporal direction would result in components of a high amplitude. The 3-D frequency domain filter would then reduce the attenuation of these components resulting in more residual noise at the output. This is a desirable effect which avoids signal degradation in regions where the noise is not usually noticeable since the motion estimation errors would typically occur in areas of high temporal activity e.g. fast motion.

For the FIR filter, motion estimation errors would reduce the strength of the temporal correlation of the blocks and so reduce the contribution of the corresponding taps, yielding the same behaviour. For the temporal one tap Wiener filter, the variance of the observed pixels would increase with the motion estimation error and so the attenuation would be reduced.

It is reasonable that the 3-D filter performances are better than the temporal filters because the temporal filter has no 2D fall back mode. It should be noted that the output of the FIR Wiener filter is somewhat more difficult to control than that of the IIR filter. In some regions, manipulation of the zeroth lag correlation coefficient as directed by equation 10.11 results in the addition of a high level of residual noise which is not well matched to the surrounding noise reduced area. It is thought that this phenomenon is related to ill-conditioning of the system of simultaneous equations for the filter coefficients.

10.3 Real sequences

Figures 10.8 and 10.9 show 3 frames of a cartoon sequence which are degraded by noise. The frames are of resolution 256×256 and 8 bit grey scale. The motion clearly poses problems to any of the standard motion estimators since some shapes change drastically from frame to frame. A 4 level pyramid was used with a 9×9 block size and ± 4 pixels search space at each level. Figure 10.8 also shows the result of noise reduction using the IIR Wiener filter on frame 43. The reader may have difficulty in observing the noise reduction at this scale. This is a rather good illustration of one of the main problems in designing noise reduction algorithms for motion picture film and video. In general, the visibility of noise in high activity regions of the image is less than in image areas with primarily low frequency information. Thus if the designer of an algorithm changes the size of the area used to display the results of the noise reduction process, inevitably the perceived noise reduction effect is different. If the display area is reduced, generally the noise reduction effect is reduced because it is then more difficult to see the noise on the original itself. Conversely, increasing the size of the display could make even the most tame noise reduction processes be perceived as dramatic since the noise (and associated artefacts in restoration) would then be much more visible. The reader is directed to the CD-ROM to assess independently the effectiveness of the noise reduction on this cartoon sequence.

Six filter outputs are shown in Figure 10.9. They show that the visual quality of the Wiener filters is superior. The performance of the 3-D frequency domain filter is better in uniform regions because it is easier to generate a uniform region by attenuating AC frequency components than by using a spatio–temporal domain weighted average approach[4]. Both Wiener filters used a 16×16 blocksize. The IIR and FIR filters used $\beta = 2.0, 1.1$ and $\sigma_{nn}^2 = 40.0, 20.0$ respectively. The temporal Wiener filter used $\sigma_{nn}^2 = 100.0, \beta = 1.1$. A border of 16 pels all around the image has been omitted in the restoration, hence the dark borders. The reader is reminded when viewing the images on CD-ROM that at the edges of the image the synthesis and analysis windows cannot overlap to give unity gain, hence an 8 pel border at the edge of the image is attenuated. A border of 16 pels is deleted from the restorations on the CD-ROM to remove the distraction of this region which can be mistaken for blurring.

From the images on the CD, it can be seen that the Wiener filter attenuation is reduced in regions of fast motion. Around the legs of the cow (for instance) there is a higher level of residual noise. The residual noise manifests itself in different ways depending on the filter used. The residual from the FIR filter appears more correlated than that left by the frequency domain filter. In the latter case it may be difficult to see this residual noise from the pictures in the figure. The reader is referred to the CD-ROM supplement for a more accurate reproduction. These effects are less visible in the case of the temporal Wiener filter because it leaves more noise in the image overall. The noise in the station-

[4] The FIR filter

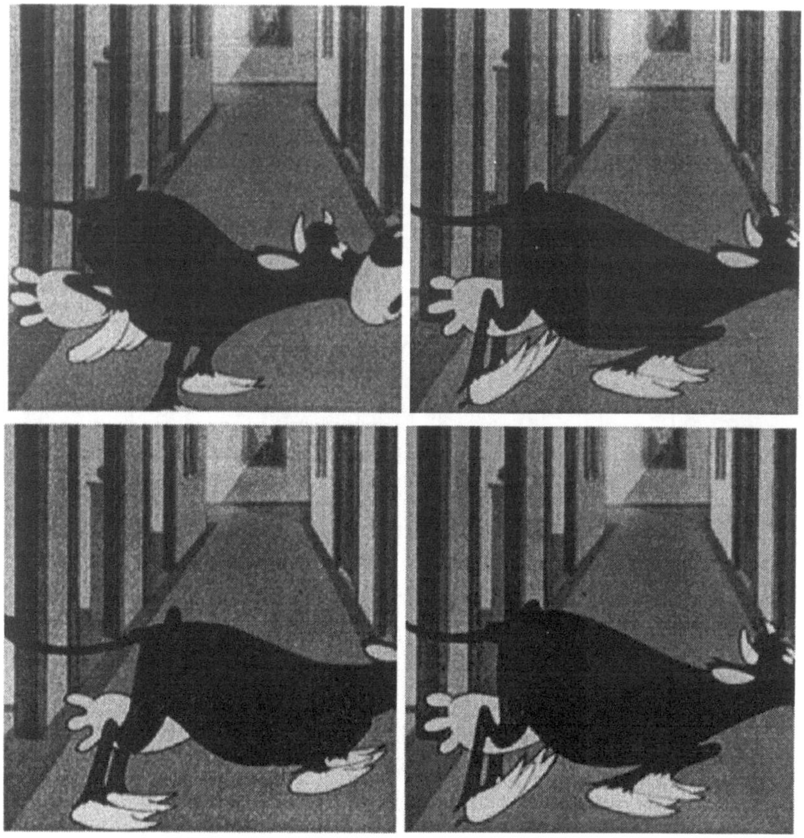

FIGURE 10.8. Clockwise from top left (zoom): Frames 42,43 of the cartoon sequence, Restored using Wiener IIR filter, Degraded frame 44.

ary regions of the image has been effectively reduced by all the Wiener filters, with the best performance being given by the frequency domain filter.

The averaging process shown in Figure 10.6 does not yield a good result. The restoration is distorted because of errors in motion estimation. This cartoon is an extreme case where it is impossible for a BBM algorithm to give an accurate estimate for motion all over the moving object. The example illustrates well the need for a noise reduction algorithm to be robust to motion estimation errors. Only in the case of a slow moving or purely translational sequence would the frame averaging process be successful. In order to reiterate how important motion estimation is in the correct application of these filters, the lower right hand image in Figure 10.6 shows the result using frame averaging if *no* motion compensation is used. As can be seen the result is quite poor, since regions which are completely different are being treated with the same filter.

It is important to note that the Wiener filters have been designed to be optimal with regard to mean squared error. This error measure is certainly not

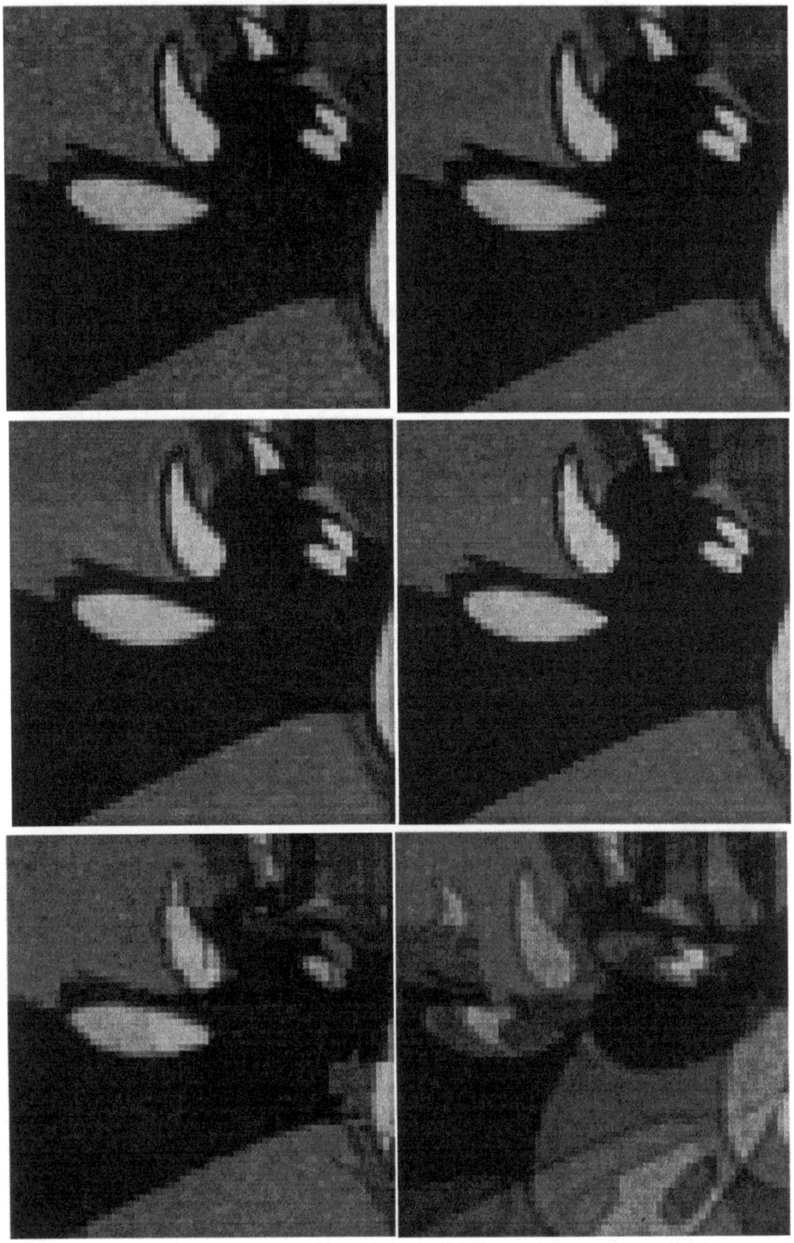

FIGURE 10.9. Restored versions of cartoon (zoom). Top Row: Original (dirty) frame 43 (left), IIR Wiener result (right). Middle Row: FIR Wiener result (left), Temporal Wiener result (right). Bottom Row: Frame averaging *with* (left) and *without* (right) motion compensation.

optimal as far as human visual characteristics are concerned and this explains why the filtered output may show artefacts.

These results are illustrated on the supplementary CD-ROM. The reader is directed to the sequences stored there to gain a visual appreciation of the relative performance of these filters.

10.4 A note on motion estimation

As in Chapter 6, which introduced the problem of missing data removal, it should be evident that the performance of all the filters discussed here is affected by the robustness of the motion estimator to noise. In general hierarchical motion estimators show better behaviour with respect to noise than single level estimators because at the coarse levels the noise present is much reduced. This chapter does not explore the robustness of motion estimators to noise except to note that it is an issue. The reader is referred to Kleihorst [80], Magarey et al. [110, 111] for some consideration of the subject. Note in addition that as the noise level increases there will come a point where no motion estimator can perform well. It is then that the improvement gained in noise reduction with motion compensated approaches will be diminished. The result could potentially become worse if the structures used for noise reduction have no spatial fallback mode. It is possible to propose schemes which estimate motion and perform noise reduction simultaneously. This can be done by altering the degradation model employed in Chapter 7 to include noise. The algorithm design could then proceed in an identical fashion to that discussed.

10.5 Final comments

The Wiener filter for spatio–temporal signals has been presented. It has been shown to compare favourably with other temporal only filters. Overlapped processing has been used to suppress blocking artefacts. The Wiener filter performs well because it is robust to motion estimation errors. When this error is large, the attenuation is reduced. The FIR Wiener filter is the most computationally intensive followed by the frequency domain filter, the temporal Wiener filter and the others which are roughly similar in this respect. This must be considered when choosing a filter for a particular application. Although the frequency domain filter can attenuate noise more effectively than the temporal filter, it requires more computation. Therefore, in cases where the noise levels are relatively low, the temporal filter may be the better choice. As far as output quality is concerned, the frequency domain filter performs best out of the those examined.

The performance of temporal filters can be improved by incorporating some spatial information in a cascade arrangement. Thus, for instance, the temporal filter could be acting on images which have already been spatially filtered. This kind of strategy was proposed by Martinez [117, 115] although he employed an implicit motion compensation strategy which was limited to small motion. It is this kind of strategy which is most useful for fast implementation of noise reduction systems in the computationally demanding area of digital film post production because of the large sizes of images employed.

11
Using Colour

The previous chapters have presented several processes for reconstructing degraded monochrome data. No real consideration has been given to the use of colour information. This is natural since much of the archived film and video material is monochrome. However, colour film and television have been in use for at least twenty years and so there is also a substantial amount of archived colour material. Furthermore modern film and video media are also subject to degradation, be it digital dropout or poor recording and transmission conditions. It is traditional to consider that the processing of colour images can proceed by manipulating each colour component separately [71, 107, 4]. This mode of colour algorithm implementation is quite successful and most commercially available video manipulation hardware and software currently process colour data in this way.

Despite this success, many authors [4, 45, 3, 167] have recognized that it would be best to use colour information in a multichannel framework for signal estimation. In this mode the data is treated as vector valued samples of a spatio-temporal signal. This chapter outlines the adjustments necessary to apply some of the algorithms discussed in this book to multichannel (colour) data. The work is still in progress and the pictures at the end of the chapter mainly illustrate system performance using separate processing of each channel.

11.1 Heuristics for detection

Figure 11.2 shows the full colour version of a very badly degraded image sequence which was used to demonstrate the performance of the large area reconstruction algorithm (see Figure 8.13). It shows clearly that in the locations of missing data (rips in the film) the colour is quite different from the surrounding, undegraded image. This extra information can be used by proposing a simple translational model for the *colour* sequence.

Defining $\mathbf{G}_n(\mathbf{x})$ as a 3-component vector containing observed image data from each colour component at location $\mathbf{x}$ in the image, the model can be stated as follows:

$$\mathbf{G}_n(\mathbf{x}) = \mathbf{G}_n(\mathbf{x} + \mathbf{d}_{n,n-1}(\mathbf{x})) + \mathbf{e}_n(\mathbf{x}) \tag{11.1}$$

where $\mathbf{e}_n$ is a 3-component error vector which ideally is due to observation noise and can be taken as $\sim \mathcal{N}(0, \mathbf{R}_{ee})$. Generally, the correlation matrix of $\mathbf{e}_n$, $\mathbf{R}_{ee}$, is *not* diagonal although the approximation to a diagonal is often made.

This simple model can be deployed in the same manner as was used in Chapter 6 for presenting a number of heuristics for missing data. Thus the *colour* SDIa detector can be defined as

$$b_{\text{SDIa}}(\mathbf{x}) = \begin{cases} 1 & \text{for } (|\mathbf{E}_b| > E_t) \text{ AND } (|\mathbf{E}_b| > E_t) \\ 0 & \text{otherwise} \end{cases} \tag{11.2}$$

where

$$\begin{aligned} \mathbf{E}_b &= \mathbf{I}_n(\mathbf{x}) - \mathbf{I}_{n-1}(\mathbf{x} + \mathbf{d}_{n,n-1}(\mathbf{x})) \\ \mathbf{E}_f &= \mathbf{I}_n(\mathbf{x}) - \mathbf{I}_{n+1}(\mathbf{x} + \mathbf{d}_{n,n+1}(\mathbf{x})) \end{aligned} \tag{11.3}$$

Thus the thresholding operation is now performed on the *magnitude* of the vector differences between motion compensated colour image data for example, $\mathbf{I}_n(\mathbf{x}), \mathbf{I}_{n-1}(\mathbf{x} + \mathbf{d}_{n,n-1}(\mathbf{x}))$ in frames $n, n-1$. However the threshold E_t now maps out a *volume* in colour space and it may be more difficult to quickly determine useful values in particular cases. The SDI and ROD detectors can be similarly extended to handle colour, with the ROD requiring the use of a vector median operation as described in the next section.

It is interesting to note that in the case of detection of missing data, the separate processing of the colour channels can lead to quite different behaviour, which might be undesirable. In the case of film for instance, data cannot normally be lost in one colour channel and not the other, so corrupted pixels which are missed in some channels by a separate channel detector may cause an annoying reconstruction. In the case of digital dropout however, this is quite possible since dropout can occur independently in the colour channels. Thus a separate processing of channels for digital dropout detection is more acceptable.

11.2 Vector median reconstruction

Chapter 6 introduces the use of 3D multilevel median filters as a first step toward designing structures for interpolation of missing data. The idea was to turn on the filter only where missing data was detected. The ML3Dex filter was introduced in order to allow the reconstruction of areas larger than the spatial window width of the filter. There is a natural extension of this filter into the colour domain by employing the vector median [10] operation which acts on vector data samples.

Given a set of P vector valued samples $\mathbf{z}_k$ $(k = 1 \dots P)$ the vector median of this set is defined as that sample $\mathbf{z}_n$ which minimizes the following expression.

$$\sum_{k=1}^{P} ||\mathbf{z}_n - \mathbf{z}_k|| \tag{11.4}$$

where $|| \cdot ||$ denotes the Euclidean norm. In other words, the vector median is that sample (from the set of P samples) which is in some sense in the 'middle' of the space occupied by the set of P vector samples. There are other definitions of the vector median, and Trahanias and Venetsanopolus [177] have proposed the use of a vector *directional* median which assesses the penalty expression above based on the *angle* as well as the length of the vector differences. Using either of these possibilities, the ML3Dex filter can be employed for colour images.

The straightforward vector median described above is computationally quite expensive. This is because the initial step is the calculation of the distances between all the pairwise vector combinations from the set of samples. Several alterations to the vector median have been proposed to reduce this complexity. It has been acknowledged that these altered vector median operations change the nature of the operation. A simple example is the marginal vector median [139]. The output of that filter is that vector resulting from applying a scalar median operation separately to each of the colour planes. The output of the marginal vector median is thus not guaranteed to be a sample from the input vector set. A very interesting fast algorithm for the 1-norm vector median filter (using the 1-norm instead of Euclidean norm above)was presented in [12].

11.3 Model based reconstruction

In a similar manner to that proposed above it is possible to extend the use of the 3D AR model employed in Chapter 7. The colour image sequence model can be assumed to be that of a multivariate autoregressive signal. Assuming all data is already compensated for motion, the model becomes

$$\mathbf{I}_n(\mathbf{x}) = \sum_{k=1}^{P} \mathbf{A}_k \mathbf{I}_{n+\mathbf{q}_k^t}(\mathbf{x} + \mathbf{q}_k^s) + \mathbf{e}(\mathbf{x}) \tag{11.5}$$

where the model coefficients, $\mathbf{A}_k$, are now 3×3 matrices. This new formulation (investigated exhaustively for time series in [143]) directly exploits correlation between the colour channels. Of course the computational load of subsequent algorithms using this model is much increased over the single channel or monochrome case. If it is assumed that the coefficients $\mathbf{A}_k$ are diagonal, then the subsequent algorithms derived using this model are equivalent to separate processing of each channel.

The reader may wish to verify that interpolation using this framework involves exactly the same manipulations as described in Chapter 7. Note this major alteration: where scalar coefficients are manipulated, matrix coefficients are substituted. This equivalence can more easily be seen by writing in full the multichannel error equations for the block of data as discussed in Figure 7.3 of Chapter 7. The spatial arrangement of missing and known pixels is the same as that shown in Figure 7.3 in Chapter 7. The equations can be written as follows:

$$
\begin{bmatrix} \mathbf{e}_{11} \\ \mathbf{e}_{12} \\ \mathbf{e}_{13} \\ \mathbf{e}_{21} \\ \mathbf{e}_{22} \\ \mathbf{e}_{23} \\ \mathbf{e}_{31} \\ \mathbf{e}_{32} \\ \mathbf{e}_{33} \end{bmatrix}
=
\begin{bmatrix}
 & & & & & * & * & & & * & * & & & & & \\
\mathbf{A}_2 & \mathbf{A}_3 & 0 & 0 & \mathbf{A}_1 & 1 & 0 & 0 & 0 & 0 & 0 & 0 & 0 & 0 & 0 & 0 \\
0 & \mathbf{A}_2 & \mathbf{A}_3 & 0 & 0 & \mathbf{A}_1 & 1 & 0 & 0 & 0 & 0 & 0 & 0 & 0 & 0 & 0 \\
0 & 0 & \mathbf{A}_2 & \mathbf{A}_3 & 0 & 0 & \mathbf{A}_1 & 1 & 0 & 0 & 0 & 0 & 0 & 0 & 0 & 0 \\
0 & 0 & 0 & 0 & \mathbf{A}_2 & \mathbf{A}_3 & 0 & 0 & \mathbf{A}_1 & 1 & 0 & 0 & 0 & 0 & 0 & 0 \\
0 & 0 & 0 & 0 & 0 & \mathbf{A}_2 & \mathbf{A}_3 & 0 & 0 & \mathbf{A}_1 & 1 & 0 & 0 & 0 & 0 & 0 \\
0 & 0 & 0 & 0 & 0 & 0 & \mathbf{A}_2 & \mathbf{A}_3 & 0 & 0 & \mathbf{A}_1 & 1 & 0 & 0 & 0 & 0 \\
0 & 0 & 0 & 0 & 0 & 0 & 0 & 0 & \mathbf{A}_2 & \mathbf{A}_3 & 0 & 0 & \mathbf{A}_1 & 1 & 0 & 0 \\
0 & 0 & 0 & 0 & 0 & 0 & 0 & 0 & 0 & \mathbf{A}_2 & \mathbf{A}_3 & 0 & 0 & \mathbf{A}_1 & 1 & 0 \\
0 & 0 & 0 & 0 & 0 & 0 & 0 & 0 & 0 & 0 & \mathbf{A}_2 & \mathbf{A}_3 & 0 & 0 & \mathbf{A}_1 & 1
\end{bmatrix}
\begin{bmatrix} \mathbf{i}_{00} \\ \mathbf{i}_{01} \\ \mathbf{i}_{02} \\ \mathbf{i}_{03} \\ \mathbf{i}_{10} \\ \mathbf{i}^*_{11} \\ \mathbf{i}^*_{12} \\ \mathbf{i}_{13} \\ \mathbf{i}_{20} \\ \mathbf{i}^*_{21} \\ \mathbf{i}^*_{22} \\ \mathbf{i}_{23} \\ \mathbf{i}_{30} \\ \mathbf{i}_{31} \\ \mathbf{i}_{32} \\ \mathbf{i}_{33} \end{bmatrix}
$$

Here, $\mathbf{e}, \mathbf{i}$ are *vector* samples of prediction error (or excitation) and colour image data respectively and $*$ denotes missing data. The derivation of the fast least squares solution for the missing data proceeds in the same way as discussed in Chapter 7.

11.4 A note on colour spaces

There are many alternative colour spaces which can be used to manipulate colour images [68]. For archived motion picture film and video the colour spaces most likely to be encountered are RGB and YUV. RGB (for red, green, blue) is the space of primary colours and most image acquisition/display takes place in this form [68].The colours are additive in that they superimpose to create a required colour. The YUV colour space was devised so that colour television signals could be viewed on standard black and white television sets [178]. The RGB signal is separated into a brightness or luminance component, Y, and two colour components U,V. The transformation from RGB to YUV colour space is as follows (see Jack [68] for further details).

$$
\begin{bmatrix} Y \\ U \\ V \end{bmatrix} = \begin{bmatrix} 0.299 & 0.587 & 0.114 \\ -0.147 & -0.289 & -0.436 \\ 0.615 & -0.515 & -0.100 \end{bmatrix} \begin{bmatrix} R \\ G \\ B \end{bmatrix} \tag{11.6}
$$

In a typical 8 bit digital system where R,G,B each take the range 0...255, Y,U,V each have ranges 0...255, −112...112, and −157...157 respectively. The mixing ratios of R,G,B to create the Y component were chosen with regard to human visual criteria as well as the constraint that those proportions of R,G,B must mix the 1 volt peak to peak signals to yield a 1 volt peak to peak luminance output (see Trundle [178]).

The various colour spaces are suited to different types of colour image manipulation. The RGB colour space is the dominant choice for display hardware since most display devices use red, green and blue phosphors. This may of course change as flat screen displays using active display technologies become more prevalent. The CMYK (cyan, magenta, yellow, black) space has not been discussed but it is used for printing devices since these are *subtractive* primary colours whose action is to delete colour components from white light. This action is similar to that of the mixing properties of ink. The YUV space is standard for broadcast digital video. By separating the colour components from the luminance the signal representation can take advantage of the reduced sensitivity to colour in the human eye. Therefore it is possible to employ a half resolution U and V frame with a full resolution Y frame and still maintain good viewing quality. This colour format is called 4:2:2 component video. When both U and V frames are at the same full resolution as the Y frame, the format is 4:4:4. The colour pictures used in this chapter are all 4:2:2 component video.

An additional advantage of using YUV instead of RGB is that in some sense, the colour information is more decorrelated since luminance and colour information is now separate. It should be recognized that maximum advantage in using colour is possible when the correlation between the channels used is small. In practice it is found that processing in the YUV space often yields visibly better results than processing in RGB.

Some authors (Magarey et al [110, 111] and Heeger et al [59]) have already used decorrelating transforms to good effect as a preprocessing step in the use of separate channel colour manipulation. By decorrelating the channel information prior to processing, there is less need for the kind of complete multivariate model presented in equation 11.5. Defining the observed correlation matrix of colour data to be the 3×3 matrix, $\mathbf{R}_{ii}$, it is possible to propose a transformation $\mathbf{T}$ which makes

$$\mathbf{R}_{I'} = \mathcal{I} \tag{11.7}$$

where $\mathcal{I}$ is the 3×3 identity matrix and $\mathbf{i}' = \mathbf{Ti}$. In the transformed colour space, $\mathbf{i}'$, each colour component contributes equally and independently to the final estimate, making the use of separate channel processing more beneficial. $\mathbf{i}$, $\mathbf{i}'$ are used here to represent a vector of image values in the original and transformed colour space respectively. This optimal transformation can be derived by writing the singular value decomposition (see [142]) of the symmetric matrix $\mathbf{R}_{ii}$ so that

$$\mathbf{V}^T \mathbf{R}_{ii} \mathbf{V} = \mathbf{D} \tag{11.8}$$

where $\mathbf{D}$ is a diagonal matrix of singular values and $\mathbf{V}$ is an orthogonal matrix. From this it follows that the required decorrelating transform is

$$\mathbf{T} = (\sqrt{\mathbf{D}})^{-1}\mathbf{V}^T \tag{11.9}$$

Any ill-conditioning in the inversion of $\mathbf{D}$, indicated by small singular values, can be solved by setting the reciprocal of the smallest singular value to zero [142]. This allows the use of a simple subspace method to determine when one colour component is completely redundant. Of course the correlation matrix $\mathbf{R}_{ll}$ may not be the same everywhere in the image. For missing data reconstruction an estimate of this matrix may be obtained locally for interpolating the patch being considered.

11.5 Pictures

The use of colour allows a rather good demonstration of the progression of ideas from Chapter 6 through Chapter 7. The visible difference between the various techniques is better seen in colour stills than in monochrome stills. Note that in order to reduce the cost of this book, the colour pages are not included in the page count and may not appear in the immediate vicinity of this text. Figure 11.1 shows several restorations of one frame (frame 2) of the VOITURE sequence (YUV 4:2:2 component video) artificially corrupted with regular sized missing blocks of data. More of the image can be seen on the accompanying CD-ROM. The first three frames were processed to interpolate these missing regions in the second frame. The MWBME as described in Chapters 2 and 3 was used for motion estimation employing 3 levels of the pyramid and 17×17 blocksize at the original resolution level.

The top right hand image in Figure 11.1 shows the result of using the vector median version of ML3Dex (VML3Dex) on the degraded image *without* using motion correction. The edge of the wall shows very bad reconstruction and the interpolated blocks in textured background behind the car are very 'bland' or faded. The left hand image on the next row shows the improvement *with* vector correction using VML3Dex and the motion correction step as outlined in Chapter 6. The interpolation of the headlamp in particular improves (it is 'dented' in the previous image), and the textured background is much better. The red ellipses highlight areas of interest. The next image shows interpolations generated by drawing samples from the posterior (MBI-2) as described in Chapter 7. The 3DAR support used for MBI-2 used 3×3 pixels in the previous frame and 3 causal pixels in the current. There is a marked improvement in the interpolation of texture, but at the scale shown in the pictures, the differences may be hard to appreciate. Some areas where the reader should notice improvements are highlighted, in particular the area highlighted with a red square. The last pair of images shows a zoom on four interpolated regions using VML3Dex on the left and MBI-2 on the right. It is hoped that the reader can now have a better feel for the improved interpolation using the sampled technique. These images are also included on the CD-ROM for comparison.

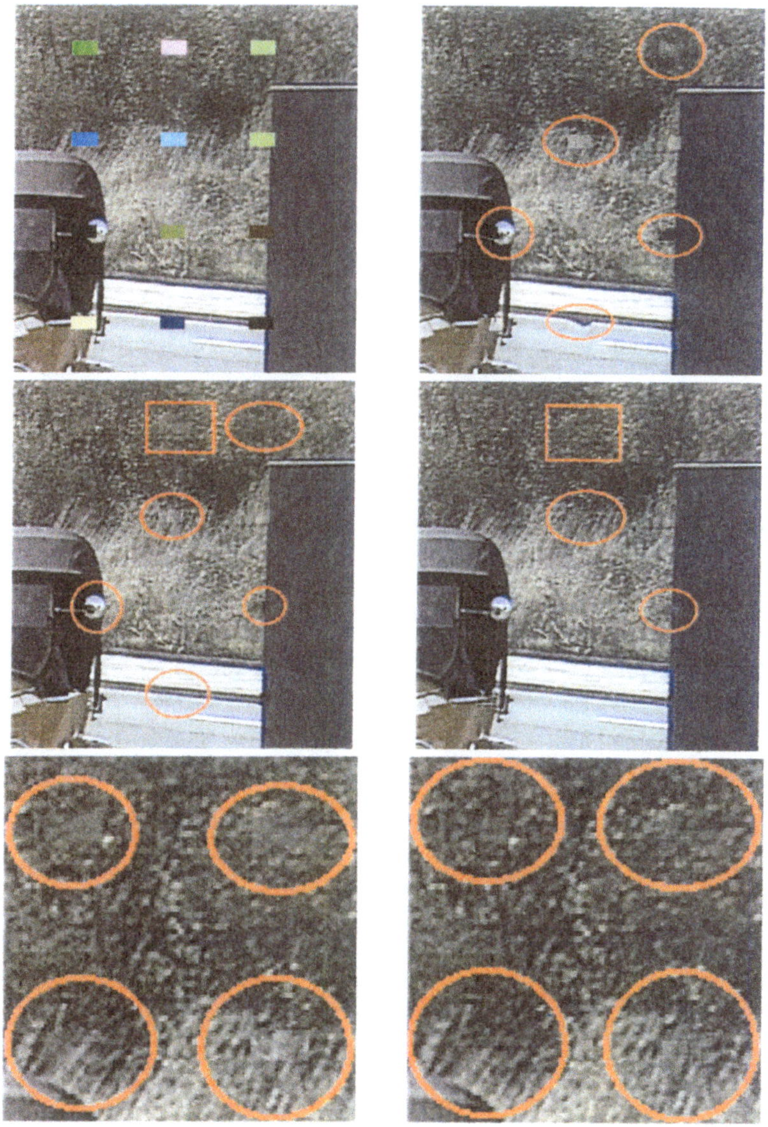

FIGURE 11.1. Top row : Portion of artificially degraded image (left), and restored using VML3Dex *without* motion correction. Middle Row : Reconstruction *with* motion correction using VML3Dex (left), MBI-2 (right). Bottom Row : Zoom on portions of images in middle row. Interesting areas are highlighted in red.

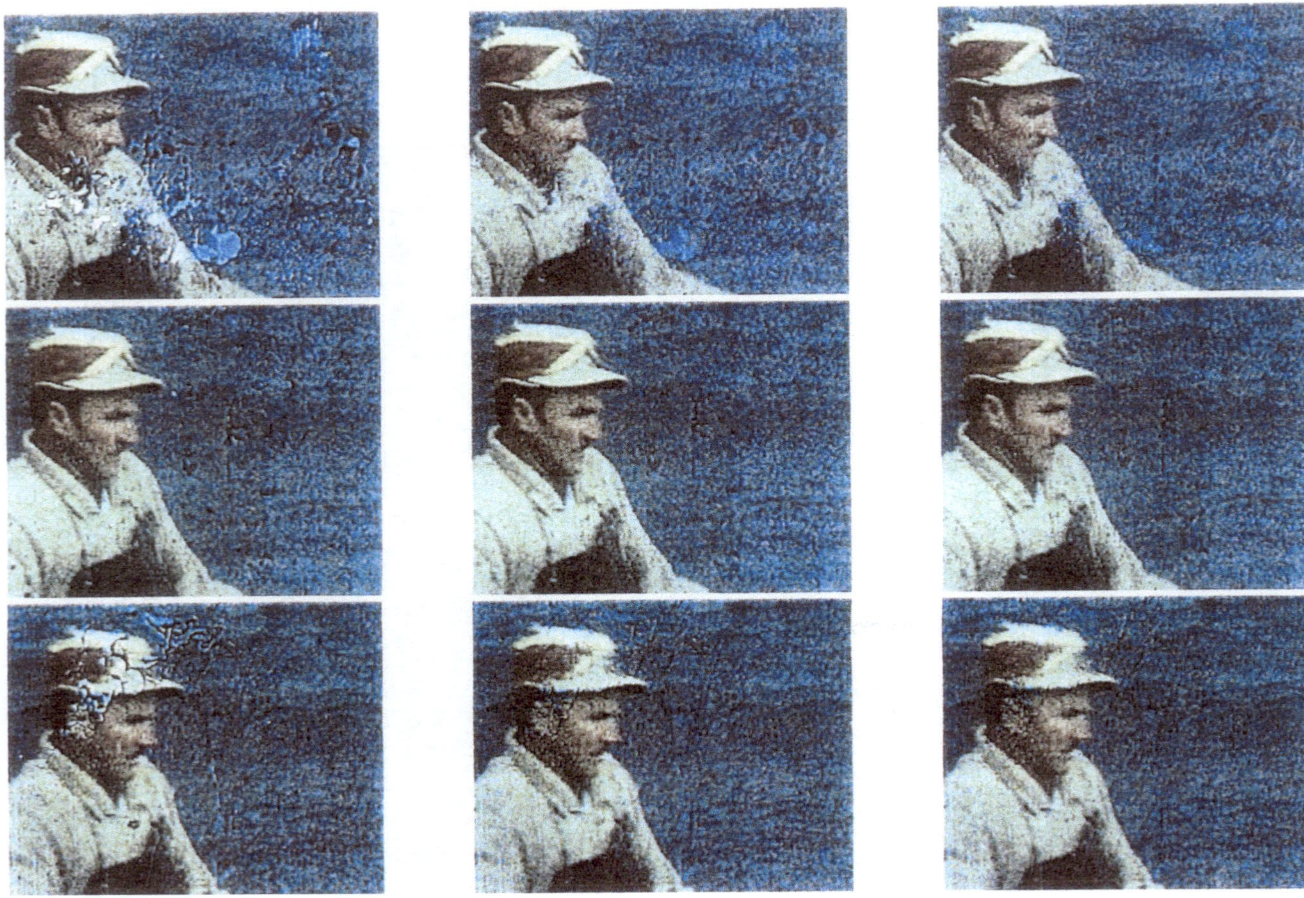

FIGURE 11.2. Left column: Portion of severely degraded image. Middle Column: Reconstruction using vector median ML3Dex, Right column: Separate reconstruction of Y,U,V fields as described in Chapter 8.

Figure 11.2 shows the result of reconstructing the colour version of the badly degraded sequence used in Chapter 8. Detection was performed using the ROD detector on the Y field only as in Chapter 8. The reconstructed vector fields employed only the luminance information in selecting vector candidates. Two processes for image interpolation (at the sites detected as missing by the ROD) are compared; both using the same interpolated motion vector fields. One uses simple temporal (vector) averaging with occlusion used to turn off the forward or backward temporal information. The other uses the VML3Dex to interpolate the image data. (Both techniques used motion compensation of course.) The VML3Dex output is much smoother overall. In fact it manages to reconstruct the brim of the fisherman's hat quite well whereas the averaging process has trouble at some pixels because the occlusion field is not quite configured properly in those locations. Conversely, although the averaging process does a good job of removing the cracking effect over the hat, VML3Dex is too conservative and leaves much of the effect behind.

It must be stressed that the images in Figure 11.2 are extremely severe cases of degradation. In addition the motion of the rope and hand are quite complicated. Therefore it is fortunate that both reconstructions are somewhat conservative and strike an appropriate balance between missing data reconstruction and damage to the underlying scene content. No noise reduction was performed on these images since by far the major distortion here is missing data due to physical degradation of the film material. In addition, the reader can observe in the images on CD-ROM that there are many fine 'cracks' on the image which have not been removed. There may be an argument in this case for a global application of a vector median process. As always, the subsequent loss in detail at the output of the system will have to be traded off against the increased level of artefact rejection.

11.6 Final comments

This chapter has outlined the steps required for extending the techniques introduced in previous chapters to colour images. The reader will observe that the separate processing of each channel appeared to yield very usable results. Although it is clear that a proper multichannel approach to colour manipulation is potentially much more beneficial than separate channel processing, the improvements may only be observed at high levels of degradation. The further development of multichannel extensions to the systems presented in this book is currently being pursued.

12
Parting Remarks

This book has presented algorithms for removing several different types of distortion, various forms of missing data and additive white noise. The distortions were treated as separate issues. The work has highlighted the usefulness of motion compensated algorithms in the treatment of image sequences. All the methods presented operate on 3 or 5 motion compensated frames with the goal being the restoration of the central frame. There exist many motion estimation algorithms, and a Multiresolution Gradient Based technique was chosen for use with the restoration algorithms that were developed. Multiresolution motion estimation was found to be invaluable for estimating the large displacements in a typical TV scene.

A solution was also presented for the problem of removing inter–line jitter in digitized TV frames. The method involves a 2DAR model which is used for estimating the relative displacement between the lines. The final algorithm was able to reduce the jitter because it incorporated a multiresolution image representation for displacement estimation. The image drift was successfully dealt with by removing the underlying low frequency drift component. It is a challenge to go beyond this algorithm and try to estimate jitter displacements which consist of low frequency components. To do this requires the use of an image model which contains more long range image information. Perhaps a wavelet approach will prove useful here since it would then be possible to de-jitter the image using several directionally oriented image decompositions. Of interest also is the design of algorithms which allow for *stretching* of the lines as well as displacement.

The reader will appreciate that the main body of the work discussed here deals with missing data of some kind. Chapters 6 to 9 illustrate the evolution of missing data treatment from the very simple to the more complex, comprehensive models. Through the use of a Bayesian framework for signal estimation it was possible to unify most of the known algorithms for missing data detection and removal. The presentation of algorithms which unify the treatment of the motion estimation and missing data problem jointly represents a significant step forward and was facilitated by the Bayesian approach.

For practical purposes, the simplest detectors and interpolators for missing data e.g. SDIa and ML3Dex remain the easiest to use in a situation where motion is relatively moderate and particularly if some user interactivity is allowed and the degradation is mild. However these detectors and interpolators behave quite erratically when the degradation becomes more serious. The JOMBADI algorithm is extremely good in most cases and represents the way forward for automatic reconstruction of missing data, but some more effort is required in improving the speed and efficiency of the basic steps involved. Current areas of research include a multiresolution approach to the problem and a better treatment of motion using more complex motion models.

There remains the problem of dealing effectively with large missing regions where there are many interacting objects. Chapter 8 takes the first steps toward a coherent solution to the problem and produces a good algorithm which is effective in many cases. Nevertheless, to deal with more complicated motion and even larger sizes of distortion it is necessary to increase the temporal aperture over which information is used. The layer models presented by Wang and Adelson [190] hold great promise here and provide an excellent direction for further work.

The chapter on line scratch removal has presented the reader with a good example of an algorithm which combines both deterministic and MCMC approaches to solving a problem. The initial heuristic is used to kick start a retrospective MCMC detector which is motivated by a Bayesian approach to the problem. The algorithm explicitly exploits cooperation between the experience of the researcher in using heuristics for 'almost' solving the problem and then using a more complete problem model but in a restricted space. It is this ethos which is employed to some extent (and to great advantage) in the previous chapters to improve the convergence of the MCMC algorithms proposed. It is possible to take advantage of the knowledge that line scratches persist in consecutive frames. This would increase the potential of the algorithm for detecting low contrast lines.

There have been several algorithms presented for noise reduction in image sequences, however none have considered the use of the standard 3D Wiener filter, be it IIR or FIR. The book presented these filters and showed that their performance compared favourably with some of the previously presented methods. Overlapped processing was used to prevent blocking artefacts in the output images. This was not an issue with the removal of Dirt and Sparkle in Chapter 6 since in that case the problem was a local one and not all the image was interpolated. Further, the blocking artefacts that may arise in measuring the displaced frame difference used for detecting missing data were much smaller than the discrepancies caused by the distortion itself. Therefore, for the detection of missing data, blocking is not as important and overlapped processing is unnecessary.

The noise reduction filters which are effective are those which are robust to motion estimation errors. The Wiener filter is implicitly robust to erroneous motion vectors because an error in motion estimation implies a large magnitude in the AC frequency components that are directed along the temporal axis. When the magnitude of a component is large, the attenuation is small. The filter is therefore easier to control than the recursive filter used in which a user-specified piecewise linear characteristic defines how the filter responds to motion estimation errors. However, the extent to which the attenuation is reduced by the Wiener filter when bad motion estimation occurs, is not governed by any human visual criterion [194]. If this were the case, one could be assured of a certain visual output quality. Future work in this area could involve consideration of such a constrained Wiener filter.

It is clear that the mean squared error criterion (which is the basis of Wiener filtering) does not necessarily reflect human visual criteria in the appreciation of image degradation. Future work in this area would benefit from the development of a tractable measure for image artefacts. Such a measure would allow the design of algorithms that are optimal with respect to the human visual system. Techniques such as those employed by Karunasekera and Yeh et al. [76, 194] could prove valuable for more effective image restoration algorithms.

There is also a need to develop an algorithm that can automatically generate an estimate for the noise variance of a degraded image as the sequence progresses. This is best done in stationary uniform regions and this involves detecting such regions. One approach may be to isolate blocks of small displaced frame difference and subtract the average image grey scale value to give a noise field from which the noise estimate may be found.

The last chapter on colour proposed a number of extensions to the algorithms presented in the rest of the book for using colour. It is the missing data algorithms, particularly JOMBADI, which stand to benefit most from the use of colour. It is clear that false alarms in the region of fast motion remain a problem for missing data detection and the use of colour would provide another feature useful for discrimination.

The algorithms presented here represent a useful set of techniques and they have been shown to be effective. The low computational cost of the algorithms in Chapter 6 and the Wiener filter in particular makes them suitable for a real time implementation. Because only local operations are required in JOMBADI which are each of low computational cost, it is possible to propose multiprocessor architectures which would run the algorithm in real time using technology available now. Although the algorithms discussed in the book can operate effectively in completely automatic mode, they are also well suited for semi-automatic processing of digital film. Semi-automatic processing is very popular for re-touching clips of film material. One can envisage a scenario where interactive missing data removal can be made much faster than a manual operation by allowing the user merely to touch a screen showing the degraded image in the *region* of a single distortion or group of distortions. An algorithm like JOMBADI for instance can be used to correct the distortions in that region without the user having to painstakingly delineate the exact blotch contour or 'paint' out the blotch manually. The line scratch removal algorithm presented here is particularly suited to interactive line removal in post production film houses [59, 64] where it is required to remove support wires which are visible when 'stunt scenes' are recorded. A good example of this is the making of *Terminator* [170].

The reader is referred to the CD-ROM to view some of the images included in this book. The images are presented in raw format so that they can be viewed on any device after some massaging by the reader. An HTML document gives a guide to the material on the CD. The colour illustrations in particular are more faithfully reproduced on the CD.

Appendix A

Estimating the AR Coefficients for the 3DAR Model

This appendix presents the least squared solution (the Maximum Likelihood solution may be found in [149]) for the coefficients of the three dimensional autoregressive model as outlined in Chapter 2. The model is best discussed in its prediction mode. The prediction equation is as below where $\hat{I}(i,j,n)$ is the predicted value of the pixel at (i,j,n).

$$\hat{I}(i,j,n) = \sum_{k=1}^{N} a_k I(i + q_k(x) + sx_{n,n+q_k(n)}, j + q_k(y) + sy_{n,n+q_k(n)}, n + q_k(n)) \tag{A.1}$$

The task then becomes to choose the parameters in order to minimize some function of the error, or residual,

$$\epsilon(i,j,n) = I(i,j,n) - \hat{I}(i,j,n) \tag{A.2}$$

The parameters of the model are both the AR coefficients $\mathbf{a} = [a_1, a_2, a_3..a_N]$, and the displacement $\mathbf{d}_{k,l} = [sx_{k,l}\ sy_{k,l}\ 0]$. This section is concerned only with coefficient estimation given an estimate for the displacement.

The coefficients are chosen to minimize the squared error, $\epsilon()$, above. This leads to the Normal equations [11, 32, 71]. The derivation is the same as the one dimensional case and the solution can be determined by invoking the principle of orthogonality. $E[\epsilon^2(i,j,n)]$ is minimized by making the error $\epsilon(i,j,n)$ orthogonal to the signal values used in its generation [71]. Therefore,

$$E[\epsilon(i,j,n) I(i + q_m(x) + sx_{n,n+q_m(n)}, j + q_m(y) + sy_{n,n+q_m(n)}, n + q_m(n))] = 0 \tag{A.3}$$

Where $m = 1 \ldots N$. Defining $\mathbf{q}_0 = [0, 0, 0]$ and $a_0 = 1.0$, then

$$\epsilon(i,j,n) = \sum_{k=0}^{N} a_k I(i + q_k(x) + sx_{n,n+q_k(n)}, j + q_k(y) + sy_{n,n+q_k(n)}, n + q_k(n)) \tag{A.4}$$

Note that the a_k are now reversed in sign to allow for this simpler formulation. To continue, the following notation is introduced.

$$\mathbf{x} = [i\ j\ n] \tag{A.5}$$
$$\mathbf{q}_k = [q_k(x)\ q_k(y)\ q_k(n)] \tag{A.6}$$
$$\mathbf{d}_{\mathbf{x},\mathbf{x}+\mathbf{q}_k} = [sx_{n,n+q_k(n)}\ sy_{n,n+q_k(n)}\ 0] \tag{A.7}$$

Substituting for $\epsilon()$ in equation A.3 gives,

$$\sum_{k=0}^{N} a_k E[I(\mathbf{x} + \mathbf{q}_k + \mathbf{d}_{\mathbf{x},\mathbf{x}+\mathbf{q}_k}) I(\mathbf{x} + \mathbf{q}_m + \mathbf{d}_{\mathbf{x},\mathbf{x}+\mathbf{q}_m})] = 0 \tag{A.8}$$
$$\forall m = 1..N$$

The expectation can be recognized as a term from the autocorrelation function of the 3-D signal $I(\mathbf{x})$. Matters may be simplified therefore by redefining the equation as

$$\sum_{k=0}^{N} a_k C(\mathbf{q}'_k, \mathbf{q}'_m) = 0 \tag{A.9}$$

Where $\mathbf{q}'_k, \mathbf{q}'_m$ are both motion compensated vector offsets as defined implicitly in the previous equation. However, a_0 has already been defined to be 1.0. Therefore, letting

$$\mathbf{a} = [a_1\ a_2\ ..\ a_N]^T \tag{A.10}$$

$$\mathbf{C} = \begin{bmatrix} C(\mathbf{q}'_1, \mathbf{q}'_1) & C(\mathbf{q}'_1, \mathbf{q}'_2) & \ldots & C(\mathbf{q}'_1, \mathbf{q}'_N) \\ C(\mathbf{q}'_2, \mathbf{q}'_1) & C(\mathbf{q}'_2, \mathbf{q}'_2) & \ldots & C(\mathbf{q}'_2, \mathbf{q}'_N) \\ C(\mathbf{q}'_3, \mathbf{q}_1{}') & C(\mathbf{q}'_3, \mathbf{q}'_2) & \ldots & C(\mathbf{q}'_3, \mathbf{q}'_N) \\ \vdots & \vdots & \vdots & \vdots \\ C(\mathbf{q}'_N, \mathbf{q}'_1) & C(\mathbf{q}'_N, \mathbf{q}'_2) & \ldots & C(\mathbf{q}'_N, \mathbf{q}'_N) \end{bmatrix} \tag{A.11}$$

$$\mathbf{c} = [C(\mathbf{q}_0, \mathbf{q}'_1)\ C(\mathbf{q}_0, \mathbf{q}'_2) \ldots C(\mathbf{q}_0, \mathbf{q}'_N)] \tag{A.12}$$
$$\mathbf{q}_0 = [0\ 0\ 0] \tag{A.13}$$

the parameters, $\mathbf{a}$ can be determined by solving

$$\mathbf{Ca} = -\mathbf{c} \tag{A.14}$$

It must be pointed out that although $\mathbf{C}$ is symmetric, it is not Toeplitz in the multidimensional case. This is due to the fact that along a diagonal, the differences between the offset vectors that define each correlation term are not necessarily parallel or the same magnitude. Consider the diagonal of matrix $\mathbf{C}$, consisting of terms at locations $[2,1][3,2][4,3]\ldots[N,N-1]$, where the top left element of $\mathbf{C}$ is at position $[1,1]$. Then vector $\mathbf{v}_1 = [q'_2 - q'_1]$ is not necessarily equal to $\mathbf{v}_2 = [q'_3 - q'_2]$ or $\mathbf{v}_3 = [q'_4 - q'_3]$ or any other such difference vector along the diagonal. The support vectors $\mathbf{q}$ may be chosen to allow this to occur by choosing vectors that lie along a line in the support volume. In general, however, when the support set delineates some volume, the vectors do not allow $\mathbf{C}$ to be Toeplitz. Therefore, it is difficult to exploit the structure of this matrix for computational purposes.

In the book, the equation A.14 is solved exactly. That is to say that no approximations about the autocorrelation function are made in estimating $\mathbf{C}$ or $\mathbf{c}$. The expectation operator in equation A.9 is taken to be the mean operation. Note that in order to calculate the required autocorrelation terms from a block of data of size $N \times N$ in the current frame n say, the offset vectors $\mathbf{q}$ require that data outside this block is necessary. The extent of this extra data is explained next.

Figure 2.5 shows a support set of 5 vectors. Calculation of $C(\mathbf{q}_0, \mathbf{q}_2)$, say, requires the following sum of products, where $\mathbf{q}_2 = [-1, 0, -1]$.

$$\sum_{\mathbf{x} \in \mathrm{B1}} I(\mathbf{x} + \mathbf{q}_0) I(\mathbf{x} + \mathbf{q}_2) \tag{A.15}$$

Block B1 is of size $N \times N$ as stated before, and this yields data for $I(\mathbf{x} + \mathbf{q}_0)$. The term $I(\mathbf{x} + \mathbf{q}_2)$ requires data from a block, B2, which is in the previous frame and the same size, but offset by $\mathbf{q}_2$ in that frame. In this case therefore, to solve for the AR coefficients exactly in blocks of size $N \times N$ involves data from a block of size $(N + 2) \times (N + 2)$ in the previous frame centred at the same position.

Appendix B

The Residual from a Non–Causal AR Model is not White

This section investigates the nature of the residual sequence from an AR model given a least squared estimate for the coefficients of the model. The analysis shows that unlike the causal AR model, the error or residual sequence of a non–causal model is not white but coloured (See [71, 149, 193, 192]). The model is considered in its 3D form as introduced in Chapter 2.

The model equation is as follows (see Chapter 2).

$$I(\mathbf{x}) = \sum_{k=1}^{N} a_k I(\mathbf{x} + \mathbf{q}_k) + \epsilon(\mathbf{x}) \tag{B.1}$$

This form of the model does not allow for any motion of objects between frames. Incorporation of this movement makes the expressions more cumbersome but does not affect the result. Typical support sets of $N = 9$ and $N = 1$ vectors defined by different $\mathbf{q}_k$ are shown in Figure 3.8.

In solving for the coefficients using the least squared approach (see Appendix A), the error, $\epsilon(\mathbf{x})$ is made orthogonal to the data at the locations pointed to by the support vectors, $\mathbf{q}_k$. This implies that

$$E[\epsilon(\mathbf{x}) I(\mathbf{x} + \mathbf{q}_n)] = 0 \text{ for } n = 1 \dots N \tag{B.2}$$

The goal of this analysis is to find an expression for the correlation function of $\epsilon(\mathbf{x})$. That is

$$R_{\epsilon\epsilon}(\mathbf{x}, \mathbf{q}_n) = E[\epsilon(\mathbf{x})\epsilon(\mathbf{x} + \mathbf{q}_n)] \tag{B.3}$$

Multiplying equation B.1 by $\epsilon(\mathbf{x} + \mathbf{q}_n)$ and taking expectations gives

$$E[I(\mathbf{x})\epsilon(\mathbf{x} + \mathbf{q}_n)] = \sum_{k=1}^{N} a_k E[I(\mathbf{x} + \mathbf{q}_k)\epsilon(\mathbf{x} + \mathbf{q}_n)] + E[\epsilon(\mathbf{x})\epsilon(\mathbf{x} + \mathbf{q}_n)] \tag{B.4}$$

Let the variance of $\epsilon(\mathbf{x})$ be $\sigma^2_{\epsilon\epsilon}$. Then from B.4, when $\mathbf{q}_n = [0\ 0\ 0]$,

$$E[I(\mathbf{x})\epsilon(\mathbf{x})] = \sigma^2_{\epsilon\epsilon} \tag{B.5}$$

The summation term disappears because of equation B.2, since $\mathbf{x} \neq (\mathbf{x} + \mathbf{q}_k)$. When the $\mathbf{q}_n$ refer to other positions within the support of the model then the following simplifications may be made

$$E[I(\mathbf{x})\epsilon(\mathbf{x} + \mathbf{q}_n)] = 0 \text{ by B.2} \tag{B.6}$$

$$\sum_{k=1}^{N} a_k E[I(\mathbf{x} + \mathbf{q}_k)\epsilon(\mathbf{x} + \mathbf{q}_n)] = a_n \sigma^2_{\epsilon\epsilon} \text{ by B.2 and B.5} \tag{B.7}$$

These simplifications can be substituted into B.4 to give the correlation term for non–zero vector lags. From this substitution it can be seen that the correlation structure of $\epsilon(\mathbf{x})$ is not white and it depends on the model coefficients. The final result then is

$$R_{\epsilon\epsilon}(\mathbf{x}, \mathbf{q}_n) = \begin{cases} \sigma^2_{\epsilon\epsilon} & \text{for } \mathbf{q}_n = [0\ 0\ 0] \\ a_n \sigma^2_{\epsilon\epsilon} & \text{for } n = 1 \ldots N \end{cases} \tag{B.8}$$

Appendix C

Estimating Displacement in the 3DAR Model

The three dimensional autoregressive model incorporating motion is defined as below (from Chapter 2).

$$I(i,j,n) =$$

$$\sum_{k=1}^{N} a_k I(i + q_k(x) + sx_{n,n+q_k(n)}, j + q_k(y) + sy_{n,n+q_z(n)}, n + q_k(n)) + \epsilon(i,j,n) \tag{C.1}$$

The parameters of the model are both the AR coefficients $\mathbf{a} = [a_1, a_2, a_3..a_N]$, and the displacement $\mathbf{d}_{k,l} = [sx_{k,l}\ sy_{k,l}\ 0]$. This section is concerned only with Least Squares displacement estimation given an estimate for the coefficients. For a Bayesian approach see Chapter 7.

In order to gain an explicit relation for $\epsilon(\cdot)$, in terms of $\mathbf{d}$, the approach used by Biemond [17] and Efstratiadis [37], was to expand the image function, $I(\cdot)$, in the previous frames, as a Taylor series about the current displacement guess. This effectively linearizes the equation for $\epsilon(\cdot)$ and allows a closed form estimate for $\mathbf{d}$. It is this solution that is used for estimating $\mathbf{d}$ in this work. The derivation given here is for a general non-causal model which involves support in both the past and future frames as well as the current frame.

It is necessary first of all to separate the support region for the AR model into three parts.

1. The support in the frames previous to the current one, i.e. $\mathbf{q}_k(n) < 0$. This is the temporally causal support.
2. The support in the current frame, $\mathbf{q}_k(n) = 0$.
3. The support in the frames to come, $\mathbf{q}_k(n) > 0$. The temporally anti-causal support.

Further, given the displacement $\mathbf{d}_{l,l+1}$ from frame l to frame $l+1$ and $\mathbf{d}_{l,l-1}$ defined similarly, the displacement $d_{l,l+k}$ is defined as the linear sum of the displacements from frame l through to frame $l+k$. That is

$$\mathbf{d}_{l,l+k} = \mathbf{d}_{l,l+1} + \sum_{m=l+1}^{k-1} \mathbf{d}_{m,m+1} \tag{C.2}$$

Similarly for $\mathbf{d}_{l,l-k}$.

The notation for the modelling equations is now improved slightly to allow a more condensed derivation of the estimation equations.

- $\mathbf{q}_k(f)$ is the spatial support vector in the (n+f)th frame
- n is the current frame
- $\mathbf{x}$ is a **spatial** position vector
- $I(\mathbf{x}, n)$ is the grey level at the position $\mathbf{x}$ in the nth frame
- $N(f)$ is the number of points in the support of the 3D AR model in frame $n + f$.
- N_0 is the number of points in the support of the 3D AR model in the current frame.
- F^- is the maximum frame offset in the causal support (a negative number for the number of causal frames)
- F^+ is the maximum frame offset in the anti causal support (a positive number for the number of anti causal frames)
- a^- the coefficients for the temporally causal support
- a the coefficients for the support in the current frame
- a^+ the coefficients for the temporally anti-causal support

The modelling equation can now be broken up into

$$\begin{aligned}
I(\mathbf{x}, n) &= \sum_{k=0}^{N_0} a_k I(\mathbf{x} + \mathbf{q}_k, n) \\
&+ \sum_{f=-1}^{F^-} \sum_{k=1}^{N(f)} a_k^- I(\mathbf{x} + \mathbf{q}_k(f) + \mathbf{d}_{n,n+f}, n + f) \\
&+ \sum_{f=1}^{F^+} \sum_{k=1}^{N(f)} a_k^+ I(\mathbf{x} + \mathbf{q}_k(f) + \mathbf{d}_{n,n+f}, n + f) \\
&+ \epsilon(\mathbf{x}, n)
\end{aligned} \tag{C.3}$$

If the various support is then expressed in terms of the displacement into the next and previous frames, the following equation results after using C.2.

$$\begin{aligned} I(\mathbf{x},n) &= \sum_{k=0}^{N_0} a_k I(\mathbf{x}+\mathbf{q}_k,n) \\ &+ \sum_{f=-1}^{F^-}\sum_{k=1}^{N(f)} a_k^- I(\mathbf{x}+\mathbf{q}_k(f)+\mathbf{d}_{n,n-1}+\sum_{m=n-1}^{n+f+1}\mathbf{d}_{m,m-1},n+f) \\ &+ \sum_{f=1}^{F^+}\sum_{k=1}^{N(f)} a_k^+ I(\mathbf{x}+\mathbf{q}_k(f)+\mathbf{d}_{n,n+1}+\sum_{m=n+1}^{n+f-1}\mathbf{d}_{m,m+1},n+f) \\ &+ \epsilon(\mathbf{x},n) \end{aligned} \tag{C.4}$$

It is assumed that there exist already estimates for $\mathbf{d}_{n,n+1}$ and $\mathbf{d}_{n,n-1}$. What is required is therefore an update for each value. Let the current estimates be $\mathbf{d}^0_{n,n+1}$ and $\mathbf{d}^0_{n,n-1}$. Further, let the updates required be such that

$$\mathbf{d}_{n,n+1} = \mathbf{d}^0_{n,n+1}+\mathbf{u}_{n,n+1} \tag{C.5}$$

$$\mathbf{d}_{n,n-1} = \mathbf{d}^0_{n,n-1}+\mathbf{u}_{n,n-1} \tag{C.6}$$

Where $\mathbf{u}$ represents the update to be found.

Equation C.4 can now be written as

$$\begin{aligned} I(\mathbf{x},n) &= \sum_{k=0}^{N_0} a_k I(\mathbf{x}+\mathbf{q}_k,n) \\ &+ \sum_{f=-1}^{F^-}\sum_{k=1}^{N(f)} a_k^- I(\mathbf{x}+\mathbf{q}_k(f)+\mathbf{d}^0_{n,n-1}+\mathbf{u}_{n,n-1}+\sum_{m=n-1}^{n+f+1}\mathbf{d}_{m,m-1},n+f) \\ &+ \sum_{f=1}^{F^+}\sum_{k=1}^{N(f)} a_k^+ I(\mathbf{x}+\mathbf{q}_k(f)+\mathbf{d}^0_{n,n+1}+\mathbf{u}_{n,n+1}+\sum_{m=n+1}^{n+f-1}\mathbf{d}_{m,m+1},n+f) \\ &+ \epsilon(\mathbf{x},n) \end{aligned} \tag{C.7}$$

The function for $I(\cdot)$ given in equation C.7 can then be linearized using a Taylor expansion[1] about $(\mathbf{x}+\mathbf{q}_k(f)+\mathbf{d}^0_{n,n+1}+\sum\mathbf{d}_{m,m+1},n+f)$, which represents the current displacement in both previous and next frames[2]. The form of the next expression is unwieldy unless the following definition is made.

$$\mathbf{D}(f,k,n) = \mathbf{x}+\mathbf{q}_k(f)+\mathbf{d}^0_{n,n+1}+\sum\mathbf{d}_{m,m+1} \tag{C.8}$$

---

[1]Note that this is not the only expansion that can be employed, an alternative is to use a Bilinear interpolation function. However, the first order Taylor expansion gives a simpler solution.

[2]The limits on the summation of displacement vectors are intentionally left out to allow the same expression to be used for the forward and backward displacement depending on the equation context.

The Taylor series expansion then yields the following expression.

$$\begin{aligned}
I(\mathbf{x},n) &= \sum_{k=0}^{N_0} a_k I(\mathbf{x}+\mathbf{q}_k,n) \\
&+ \sum_{f=-1}^{F^-}\sum_{k=1}^{N(f)} a_k^- I(\mathbf{D}(f,k,n),n+f) \\
&+ \mathbf{u}_{n,n-1}^T \sum_{f=-1}^{F^-}\sum_{k=1}^{N(f)} a_k^- \nabla I(\mathbf{D}(f,k,n),n+f) \\
&+ \sum_{f=-1}^{F^-}\sum_{k=1}^{N(f)} a_k^- \nu(\mathbf{D}(f,k,n),n+f) \\
&+ \sum_{f=1}^{F^+}\sum_{k=1}^{N(f)} a_k^+ I(\mathbf{D}(f,k,n),n+f) \\
&+ \sum_{f=1}^{F^+}\sum_{k=1}^{N(f)} a_k^+ \nu(\mathbf{D}(f,k,n),n+f) \\
&+ \mathbf{u}_{n,n+1}^T \sum_{f=1}^{F^+}\sum_{k=1}^{N(f)} a_k^+ \nabla I(\mathbf{D}(f,k,n),n+f) \\
&+ \epsilon(\mathbf{x},n)
\end{aligned} \tag{C.9}$$

$\nu(\cdot)$ represents the higher order terms in the Taylor series expansions.

For the current set of estimated parameters, $\mathbf{a}$ and $\mathbf{d}_0$ there will be some observed error ϵ_0. This error is defined as (following A.1),

$$\begin{aligned}
\epsilon_0(\mathbf{x},n) &= I(\mathbf{x},n) - \sum_{k=0}^{N_0} a_k I(\mathbf{x}+\mathbf{q}_k,n) \qquad \text{(C.10)} \\
&- \sum_{f=-1}^{F^-}\sum_{k=1}^{N(f)} a_k^- I(\mathbf{D}(f,k,n),n+f) \\
&- \sum_{f=1}^{F^+}\sum_{k=1}^{N(f)} a_k^+ I(\mathbf{D}(f,k,n),n+f)
\end{aligned}$$

Therefore substituting C.9, an expression involving $\mathbf{u}$ in terms of observables follows (where the limits on the sums have been dropped).

$$\begin{aligned}
\epsilon_0(\mathbf{x}, n) &= \mathbf{u}_{n,n-1}^T \sum\sum a_k^- \nabla I(\mathbf{D}(f,k,n), n+f) \\
&+ \sum\sum a_k^- \nu(\mathbf{D}(f,k,n), n+f) \\
&+ \mathbf{u}_{n,n+1}^T \sum\sum a_k^+ \nabla I(\mathbf{D}(f,k,n), n+f) \\
&+ \sum\sum a_k^+ \nu(\mathbf{D}(f,k,n), n+f) \\
&+ \epsilon(\mathbf{x}, n)
\end{aligned} \tag{C.11}$$

The spatial, two component, update vectors, $\mathbf{u}$ are now required, but there is only one equation. Collecting observations of $\epsilon_0(\mathbf{x}, n)$, $\nabla I(\cdot)$ at each position in some predefined region, an overdetermined system of equations results. These equations can be written as follows:

$$\mathbf{z}_w = \mathbf{G}_w \mathbf{u} + \mathbf{v}_w \tag{C.12}$$

The quantities are defined as follows, given a set of equations made by observing a block of $N_1 \times N_2$ pixels.

- $\mathbf{z}_w$, $(N_1 \times N_2 \times 1)$ is a column vector of current errors at all the points in the region used for estimation.

$$\mathbf{z}_w = \begin{bmatrix} \epsilon_0(\mathbf{x}_1, n) \\ \epsilon_0(\mathbf{x}_2, n) \\ \vdots \\ \epsilon_0(\mathbf{x}_{(N_1 N_2)}, n) \end{bmatrix} \tag{C.13}$$

- $\mathbf{G}_w$, $(N_1 \times N_2 \times 4)$ is a matrix of gradients at the past and future support positions.

$$\mathbf{G}_w = \begin{bmatrix}
\sum\sum a_k^- \frac{\partial I_1(\mathbf{D}(f,k,n))}{\partial x} & \sum\sum a_k^- \frac{\partial I_1(\mathbf{D}(f,k,n))}{\partial y} & \sum\sum a_k^+ \frac{\partial I_1(\mathbf{D}(f,k,n))}{\partial x} & \sum\sum a_k^+ \frac{\partial I_1(\mathbf{D}(f,k,n))}{\partial y} \\
\sum\sum a_k^- \frac{\partial I_2(\mathbf{D}(f,k,n))}{\partial x} & \sum\sum a_k^- \frac{\partial I_2(\mathbf{D}(f,k,n))}{\partial y} & \sum\sum a_k^+ \frac{\partial I_2(\mathbf{D}(f,k,n))}{\partial x} & \sum\sum a_k^+ \frac{\partial I_2(\mathbf{D}(f,k,n))}{\partial y} \\
\vdots & \vdots & \vdots & \vdots \\
\sum\sum a_k^- \frac{\partial I_{(N_1 N_2)}(\mathbf{D}(f,k,n))}{\partial x} & \sum\sum a_k^- \frac{\partial I_{(N_1 N_2)}(\mathbf{D}(f,k,n))}{\partial y} & \sum\sum a_k^+ \frac{\partial I_{(N_1 N_2)}(\mathbf{D}(f,k,n))}{\partial x} & \sum\sum a_k^+ \frac{\partial I_{(N_1 N_2)}(\mathbf{D}(f,k,n))}{\partial y}
\end{bmatrix} \tag{C.14}$$

- $\mathbf{u}$ is the (4×1) vector of updates defined as

$$\mathbf{u} = \begin{bmatrix} \mathbf{u}_{n,n-1} \\ \mathbf{u}_{n,n+1} \end{bmatrix} \tag{C.15}$$

- $\mathbf{v}_w$ is the collection of all the error terms ϵ and ν.

$$\mathbf{v}_w = \begin{bmatrix} \sum\sum a_k^- \nu_1(\ldots) + \sum\sum a_k^+ \nu_1(\ldots) + \epsilon(\mathbf{x}_1, n) \\ \sum\sum a_k^- \nu_2(\ldots) + \sum\sum a_k^+ \nu_2(\ldots) + \epsilon(\mathbf{x}_2, n) \\ \vdots \\ \sum\sum a_k^- \nu_{(N_1N_2)}(\ldots) + \sum\sum a_k^+ \nu_{(N_1N_2)}(\ldots) + \epsilon(\mathbf{x}_{(N_1N_2)}, n) \end{bmatrix}$$

So far the derivation for the parameter estimates has placed no restriction on the spatial or temporal nature of the model support. However, the work in the book is concerned with causal modelling primarily due to the decreased computation necessary.

Solving for the updates.

It is possible to estimate the displacement update vector in C.12 directly via the pseudo inverse of $\mathbf{G}_w$ as follows:

$$\begin{aligned} \mathbf{G}_w^T \mathbf{z}_w &= \mathbf{G}_w^T \mathbf{G}_w \mathbf{u} + \mathbf{G}_w^T \mathbf{v}_w \\ \mathbf{u} &= [\mathbf{G}_w^T \mathbf{G}_w]^{-1} [\mathbf{G}_w^T \mathbf{z}_w - \mathbf{G}_w^T \mathbf{v}_w] \end{aligned} \quad \text{(C.16)}$$

To arrive at a more robust solution, the approach adopted by [37] has been to derive a Wiener estimate for $\mathbf{u}$ [37, 17]. The method was initially presented by Biemond, and it attempts to find the estimate $\hat{\mathbf{u}}$ for $\mathbf{u}$ which minimizes the error $E[|\mathbf{u} - \hat{\mathbf{u}}|^2]$. Therefore,

$$\begin{aligned} E[|\mathbf{u} - \hat{\mathbf{u}}|^2] &= E[(\mathbf{u}^T - \hat{\mathbf{u}}^T)(\mathbf{u} - \hat{\mathbf{u}})] \\ &= E[\mathbf{u}^T\mathbf{u} - \mathbf{u}^T\hat{\mathbf{u}} - \hat{\mathbf{u}}^T\mathbf{u} + \hat{\mathbf{u}}^T\hat{\mathbf{u}}] \end{aligned} \quad \text{(C.17)}$$

The estimate, $\hat{\mathbf{u}}$, is found from a linear transformation of the observed error vector $\mathbf{z}$ such that

$$\hat{\mathbf{u}} = \mathbf{L}\mathbf{z}_w \quad \text{(C.18)}$$

Substituting this expression for $\hat{\mathbf{u}}$ in C.17 and differentiating with respect to the required unknown, $\mathbf{L}$, to find the minimum squared error, yields the following equation.

$$\mathbf{L}E[(\mathbf{G}_w\mathbf{u} + \mathbf{v})(\mathbf{G}_w\mathbf{u} + \mathbf{v}_w)^T] = E[\mathbf{u}(\mathbf{G}_w\mathbf{u} + \mathbf{v}_w)^T] \quad \text{(C.19)}$$

Therefore, assuming that, $\mathbf{v}_w$, which involves higher order terms, is uncorrelated with the actual update, $\mathbf{u}$, an explicit expression for $\mathbf{L}$ results.

$$\mathbf{L} = \mathbf{R}_{uu}\mathbf{G}_w^T[\mathbf{G}_w\mathbf{R}_{uu}\mathbf{G}_w^T + \mathbf{R}_{vv}]^{-1} \quad \text{(C.20)}$$

This solution for $\mathbf{L}$ involves the inverse of a large matrix. If the number of positions at which observations are taken is P, then it involves the inverse of a $P^2 \times P^2$ matrix. Biemond [17] has employed a matrix identity which simplifies this solution considerably.

$$\mathbf{R}_{uu}\mathbf{G}_w^T[\mathbf{G}_w\mathbf{R}_{uu}\mathbf{G}_w^T + \mathbf{R}_{vv}]^{-1} = [\mathbf{G}_w^T\mathbf{R}_{vv}^{-1}\mathbf{G}_w + \mathbf{R}_{uu}^{-1}]^{-1}\mathbf{G}_w^T\mathbf{R}_{vv}^{-1} \quad \text{(C.21)}$$

Using C.21, therefore,

$$\mathbf{L} = [\mathbf{G}_w^T \mathbf{R}_{vv}^{-1} \mathbf{G}_w + \mathbf{R}_{uu}^{-1}]^{-1} \mathbf{G}_w^T \mathbf{R}_{vv}^{-1} \tag{C.22}$$

Assuming that the vector $\mathbf{v}$ represents white noise and that the components of $\mathbf{u}$ are uncorrelated, i.e. $\mathbf{R}_{vv} = \sigma_{vv}^2 \mathbf{I}$ and $\mathbf{R}_{uu} = \sigma_{uu}^2 \mathbf{I}$, $\mathbf{L}$ is given by

$$\mathbf{L} = [\mathbf{G}_w^T \mathbf{G}_w + \mu \mathbf{I}]^{-1} \mathbf{G}_w^T \tag{C.23}$$

where $\mu = \frac{\sigma_{vv}^2}{\sigma_{uu}^2}$. Due to the identity in C.21 and this assumption, the matrix inverse is reduced to the inverse of a 2×2 matrix , regardless of the number of equations.

It is important to recognize that the validity of the assumption regarding $\mathbf{R}_{vv}$ is affected by the causality of the model support. This is because part of $\mathbf{v}$ consists of the model error $\epsilon(\cdot)$. It has been shown in [107, 192, 193, 71], that this error is not white when the model support is non–causal. This implies that if the support for the model consists of points in the current frame that represent a non–causal region in that frame, the assumption is not valid. To ensure the validity of the white noise assumption, the support for the AR model in the current frame must be limited to a causal region, i.e. to the left and above the predicted location.

The Wiener estimate for $\mathbf{u}$ is therefore given by

$$\hat{\mathbf{u}} = [\mathbf{G}_w^T \mathbf{G}_w + \mu \mathbf{I}]^{-1} \mathbf{G}_w^T \mathbf{z}_w \tag{C.24}$$

This solution for the update for the current displacement is incorporated into an iterative refinement scheme. A guess for the displacement, which may be zero, is iteratively refined using the above equation until some convergence criterion is satisfied. Two main criterion are used in this book, a threshold on the magnitude of the error vector, $|\mathbf{z}_w|_t$ and a threshold on the size of the update vector, $|\mathbf{u}|_t$. The iterative refinement process is halted if the magnitude of the current error is less than $|\mathbf{z}_w|_t$ or the magnitude of the update is less than $|\mathbf{u}|_t$. A final criterion is the most harsh, and simply halts iteration if no other criterion has been fulfilled when a certain number of iterations have completed. These criterion are necessary to limit the computational load of the algorithm.

C.1 Summary

The approach taken by Biemond [17] can be used to generate a solution for the motion update in the case of the general 3DAR model. The solution is linear only if the model coefficients are known beforehand. These are not available in practice but it is possible to estimate the coefficients and displacement successively in the iterative process. The motion equations reduce to approximately the same form as the standard WBME.

It is important to recognize that the Taylor series expansion is not the only expansion which can be used to linearize the model equation. The purpose of the expansion is to make the displacement parameter available explicitly. To this end any interpolator would suffice. The compromise is one of interpolator quality versus computation. Sinc interpolation is a possibility but it would yield a non–linear solution. Bilinear interpolation is also an alternative which may prove better than the Taylor expansion. This book uses the Taylor expansion to facilitate a simple linear solution but other interpolators would be useful to consider in further work.

Appendix D
Joint Sampling in the JOMBADI Algorithm

This Appendix presents the details for the derivation of the various conditional distributions required for the JOMBADI algorithm. The discussion begins with a restatement of the basic models as follows:

$$\text{The observation model is } G_n(\mathbf{x}) = (1 - b(\mathbf{x}))I_n(\mathbf{x}) + b(\mathbf{x})c(\mathbf{x}) \tag{D.1}$$

$$\text{The (clean) image model is } I_n(\mathbf{x}) = \sum_{k=1}^{P} a_k I_{n+\mathbf{q}_k^t}(\mathbf{x} + \mathbf{q}_k^s) + e(\mathbf{x}) \tag{D.2}$$

where all the required data is compensated for motion, G_n is the nth degraded frame and I_n is the nth original, clean frame. The parameters to be estimated are $b(\mathbf{x})$ (set to 1 when site $\mathbf{x}$ is corrupted and zero otherwise), the P 3DAR model coefficients $\mathbf{a}$ ($a_0 = 1.0$), the clean original data $I_n(\mathbf{x})$ at sites where $b(\mathbf{x}) = 1$, and the motion vector fields $\mathbf{d}_{n,n-1}$, $\mathbf{d}_{n,n+1}$. The motion information is left out of the image model for simplicity. The parameter vector containing all the variables is denoted θ. When it is required to define a parameter vector which contains a subset of the parameters, this vector is denoted $\theta_{-(\mathbf{a},\sigma_e^2)}$ for instance, for a subset which does not contain the 3DAR model parameters. There are two main sampling steps, sampling jointly for $b(\mathbf{x}), c(\mathbf{x}), I(\mathbf{x})$ and sampling jointly for $\mathbf{a}, \sigma_e^2, \mathbf{d}_{n,n-1}, \mathbf{d}_{n,n+1}$. The latter sampling strategy is more straightforward to derive and it is considered first.

D.1 Sampling for $\mathbf{a}(\mathbf{x}), \sigma_e^2(\mathbf{x}), \mathbf{d}_{n,n-1}(\mathbf{x})$

Recall that the 3DAR model parameters and motion vectors are block based. In this section although $\mathbf{x}$ is the position vector of a particular site in a frame, the values for $\mathbf{a}, \sigma_e^2, \mathbf{d}_{n,n-1}(\mathbf{x}), \mathbf{d}_{n,n+1}(\mathbf{x})$ are the same over each $B \times B$ block of pixels in the image. This section drops the $\mathbf{x}$ argument for the various block based parameters to keep the notation simple. Consider a single block of pixels. The joint sample required is drawn from the distribution

$$p(\mathbf{a}, \sigma_e^2, \mathbf{d}_{n,n-1} | I_n, I_{n-1}, I_{n+1}, \mathbf{d}_{n,n+1}, D) \tag{D.3}$$

in which D denotes the block based motion vector neighbourhood around the current block position. Raster scanning all the necessary image data into the column vector $\mathbf{i}$, this distribution may be decomposed as follows:

$$\begin{aligned} &p(\mathbf{a}, \sigma_e^2, \mathbf{d}_{n,n-1} | \mathbf{i}, \mathbf{d}_{n,n+1}, D) = \\ &p(\mathbf{a}|\sigma_e^2, \mathbf{d}_{n,n-1}, \mathbf{i}, \mathbf{d}_{n,n+1}, D) p(\sigma_e^2 | \mathbf{d}_{n,n-1}, \mathbf{i}, \mathbf{d}_{n,n+1}, D) p(\mathbf{d}_{n,n-1} | \mathbf{i}, \mathbf{d}_{n,n+1}, D) \end{aligned} \tag{D.4}$$

The first conditional distribution on the right hand side results from

$$p(\mathbf{a}|\sigma_e^2, \mathbf{d}_{n,n-1}, \mathbf{i}, \mathbf{d}_{n,n+1}, D) = \frac{p(\mathbf{a}, \sigma_e^2, \mathbf{d}_{n,n-1}, \mathbf{i}, \mathbf{d}_{n,n+1}, D)}{\int p(\mathbf{a}, \sigma_e^2, \mathbf{d}_{n,n-1}, \mathbf{i}, \mathbf{d}_{n,n+1}, D) d\mathbf{a}} \tag{D.5}$$

The joint distribution in the numerator of the expression above is the joint posterior distribution in equation 7.4. To avoid some unnecessary algebraic manipulation, the following analysis is useful.

Suppose the conditional $p(a|b)$ is required, where a and b are some random variables. Proceeding in the usual way,

$$\begin{aligned} p(a|b) &= \frac{p(a,b)}{p(b)} \\ &= \frac{p(a,b)}{\int_a p(a,b) da} \end{aligned} \tag{D.6}$$

The denominator is independent of a and is just a normalizing term which ensures the derived expression, $p(a|b)$, integrates to 1, if b is treated as a *given* constant. This extremely important result shows that to derive the conditional distribution for a random variable given the joint distribution, it is only necessary to collect together the terms which involve that particular random variable, and then derive the normalizing constant (if required). This rearrangement may not be straightforward in some cases, but fortunately, in equation 7.4, the only term which involves $\mathbf{a}$ is the image likelihood and this is a multivariate Gaussian. This makes it simple to perform the necessary manipulations.

Returning then to the original problem of deriving the conditional for $\mathbf{a}$, note that the prediction error $\mathbf{e}$ in a volume of data including the current block of pixels can be written as $\mathbf{e} = \mathbf{i} - \mathbf{I}\mathbf{a}$. The image data is scanned into the matrix $\mathbf{I}$

so that $\mathbf{Ia}$ is the prediction of a pixel at a particular site in the volume. $\mathbf{e}, \mathbf{i}$ are column vectors that are B^2 elements long and the image data in $\mathbf{I}$ may need to come from outside the current block to provide the necessary support for the prediction required. Since each element of $\mathbf{e}$ is drawn from $\mathcal{N}(0, \sigma_e^2)$, and the only term involving $\mathbf{a}$ is the likelihood,

$$p(\mathbf{a}, \sigma_e^2, \mathbf{d}_{n,n-1}, \mathbf{i}, \mathbf{d}_{n,n-1}, D) \propto \frac{1}{\sqrt{2\sigma_e^2}} \exp\left(-\frac{[\mathbf{i} - \mathbf{Ia}]^T[\mathbf{i} - \mathbf{Ia}]}{2\sigma_e^2}\right) \tag{D.7}$$

where the proportionality symbol $\propto$ indicates that there are other terms which do not involve $\mathbf{a}$, such as the priors $p(\mathbf{d}_{n,n-1}|D)p(\sigma_e^2)$. As explained before, these terms are not important for the conditional $p(\mathbf{a}|\ldots)$.

This expression can be rearranged by completing the square in the argument of the exponent as follows:

$$\begin{aligned}[\mathbf{i} - \mathbf{Ia}]^T[\mathbf{i} - \mathbf{Ia}] &= \mathbf{i}^T\mathbf{i} - 2\mathbf{i}^T\mathbf{I}^T\mathbf{a} + \mathbf{a}^T\mathbf{I}\mathbf{I}^T\mathbf{a} \\ &= [\mathbf{a} - [\mathbf{I}\mathbf{I}^T]^{-1}\mathbf{Ii}]^T[\mathbf{I}\mathbf{I}^T][\mathbf{a} - [\mathbf{I}\mathbf{I}^T]^{-1}\mathbf{Ii}] \\ &\quad + \mathbf{i}^T\mathbf{i} - \mathbf{i}^T\mathbf{I}^T[\mathbf{I}\mathbf{I}^T]^{-1}\mathbf{Ii}\end{aligned} \tag{D.8}$$

Again ignoring the terms which do not involve $\mathbf{a}$, the argument of the exponent can be seen to have the form of a multivariate Gaussian and so the conditional may be written as

$$p(\mathbf{a}|\theta_{-\mathbf{a}}, \mathbf{i}, D) = \frac{|\mathbf{I}\mathbf{I}^T|^{\frac{1}{2}}}{(2\pi\sigma_e^2)^{P/2}} \exp\left(-\frac{[\mathbf{a} - [\mathbf{I}\mathbf{I}^T]^{-1}\mathbf{Ii}]^T[\mathbf{I}\mathbf{I}^T][\mathbf{a} - [\mathbf{I}\mathbf{I}^T]^{-1}\mathbf{Ii}]}{2\sigma_e^2}\right)$$

This results in the expression as required (framed for emphasis):

$$\boxed{p(\mathbf{a}|\theta_{-\mathbf{a}}, \mathbf{i}, D) \sim \mathcal{N}(\hat{\mathbf{a}}, \frac{\sigma_e^2}{[\mathbf{I}\mathbf{I}^T]})} \tag{D.9}$$

where $\hat{\mathbf{a}} = [\mathbf{I}\mathbf{I}^T]^{-1}\mathbf{Ii}$ is the least squares estimate for the P 3DAR coefficients, and the matrices $\mathbf{I}\mathbf{I}^T$ and $\mathbf{Ii}$ can be recognized as the required covariance matrix and vector as described in Appendix A. Note that the multivariate distribution is of dimension P, the order of the 3DAR model, hence the variance term in the normalizing factor is raised to this power, not N, the number of pixel sites used for prediction error equations in the motion compensated data volume.

To derive the conditional $p(\sigma_e^2|\theta_{-(\mathbf{a},\sigma_e^2)}, \mathbf{i})$, the coefficients need to be integrated out of the posterior since

$$p(\sigma_e^2|\theta_{-(\mathbf{a},\sigma_e^2)}, \mathbf{i}, D) = \frac{p(\theta_{-\mathbf{a}}, \mathbf{i}, D)}{\int_{\sigma_e^2} p(\theta_{-\mathbf{a}}, \mathbf{i}, D) d\sigma_e^2}$$

where

$$p(\theta_{-\mathbf{a}}, \mathbf{i}, D) = \int_{\mathbf{a}} p(\theta_{-\mathbf{a}}, \mathbf{i}, D) d\mathbf{a} \tag{D.10}$$

Having already completed the square to find the conditional for $\mathbf{a}$ the integration of the posterior is made simpler by the important observation that

$$\int_{\mathbf{x}} \frac{|\mathbf{K}|^{\frac{1}{2}}}{\sqrt{2\pi\sigma^2}^P} \exp\left(-\frac{[\mathbf{x}-\bar{\mathbf{x}}]^T\mathbf{K}[\mathbf{x}-\bar{\mathbf{x}}]}{2\sigma^2}\right) d\mathbf{x} = 1 \tag{D.11}$$

for a general multivariate Gaussian distribution $\mathcal{N}(\bar{\mathbf{x}}, \frac{\sigma^2}{|\mathbf{K}|})$, in which $\mathbf{x}$ is of dimension P. See [172] for some basic information about the multivariate Gaussian.

The analysis can now proceed as follows:

$$\begin{aligned}
\int_{\mathbf{a}} p(\theta_{-\mathbf{a}}, \mathbf{i}, D) d\mathbf{a} \propto \\
&\frac{1}{(2\pi\sigma_e^2)^{N/2}} \int_{\mathbf{a}} \exp\left(-\frac{[\mathbf{a}-[\mathbf{I}\mathbf{I}^T]^{-1}\mathbf{I}\mathbf{i}]^T[\mathbf{I}\mathbf{I}^T][\mathbf{a}-[\mathbf{I}\mathbf{I}^T]^{-1}\mathbf{I}\mathbf{i}]}{2\sigma_e^2}\right) d\mathbf{a} \\
&\quad \times \exp\left(-\frac{\mathbf{i}^T\mathbf{i}-\mathbf{i}^T\mathbf{I}^T[\mathbf{I}\mathbf{I}^T]^{-1}\mathbf{I}\mathbf{i}}{2\sigma_e^2}\right) \\
&= \frac{|\mathbf{I}^T\mathbf{I}|^{\frac{1}{2}}}{(2\pi\sigma_e^2)^{P/2}} \int_{\mathbf{a}} \exp\left(-\frac{[\mathbf{a}-[\mathbf{I}\mathbf{I}^T]^{-1}\mathbf{I}\mathbf{i}]^T[\mathbf{I}\mathbf{I}^T][\mathbf{a}-[\mathbf{I}\mathbf{I}^T]^{-1}\mathbf{I}\mathbf{i}]}{2\sigma_e^2}\right) d\mathbf{a} \\
&\quad \times \frac{|\mathbf{I}^T\mathbf{I}|^{-\frac{1}{2}}}{(2\pi\sigma_e^2)^{(N-P)/2}} \exp\left(-\frac{\mathbf{i}^T\mathbf{i}-\mathbf{i}^T\mathbf{I}^T[\mathbf{I}\mathbf{I}^T]^{-1}\mathbf{I}\mathbf{i}}{2\sigma_e^2}\right) \\
&= \frac{|\mathbf{I}^T\mathbf{I}|^{-\frac{1}{2}}}{(2\pi\sigma_e^2)^{(N-P)/2}} \exp\left(-\frac{\mathbf{i}^T\mathbf{i}-\mathbf{i}^T\mathbf{I}^T[\mathbf{I}\mathbf{I}^T]^{-1}\mathbf{I}\mathbf{i}}{2\sigma_e^2}\right)
\end{aligned} \tag{D.12}$$

In the final expression above, the reader is reminded that the motion vectors $\mathbf{d}_{n,n-1}, \mathbf{d}_{n,n+1}$ are implicit arguments to all the image data since they are needed to compensate for motion. There is only one other term in the posterior which involves σ_e^2 and that is the prior $p(\sigma_e^2)$. As stated earlier, this prior is suitably non-informative : $p(\sigma_e^2) \propto 1/\sigma_e^2$. The derivation of the required conditional is similar to that presented in [151]. Note that if $\mathbf{a} = \hat{\mathbf{a}}$ in equation D.8 above, then it is easy to rewrite the last term in equation D.12 as follows in the conditional.

$$\begin{aligned}
p(\sigma_e^2|\theta_{-(\mathbf{a},\sigma_e^2)}, \mathbf{i}) &\propto \frac{p(\sigma_e^2)}{\sqrt{2\pi\sigma_e^2}^{N-P}} \exp\left(-\frac{\hat{\mathbf{e}}^T\hat{\mathbf{e}}}{2\sigma_e^2}\right) \\
&\text{where } \hat{\mathbf{e}} = \mathbf{i} - \mathbf{I}\hat{\mathbf{a}}
\end{aligned}$$

substituting for the prior, (D.13)

$$\propto \frac{1}{\sigma_e^2\sqrt{2\pi\sigma_e^2}^{N-P}} \exp\left(-\frac{\hat{\mathbf{e}}^T\hat{\mathbf{e}}}{2\sigma_e^2}\right)$$

The Inverted Gamma distribution [14] with parameters α, β is defined as

$$IG(x|\alpha,\beta) = \frac{\beta^\alpha}{\Gamma(\alpha)} x^{-(\alpha+1)} \exp^{-(\beta/x)} \tag{D.14}$$

Comparing this with the expression for $p(\sigma_e^2|\ldots)$ results in the required expression as follows (setting $x = \sigma_e^2$):

$$p(\sigma_e^2|\theta_{-(\mathbf{a},\sigma_e^2)}, \mathbf{i}) = \mathrm{IG}(\sigma_e^2|\frac{N-P}{2}, \frac{\hat{\mathbf{e}}^T\hat{\mathbf{e}}}{2}) \tag{D.15}$$

The final conditional required is the conditional for the motion given the image data. This is derived by integrating out the variance from the expression D.12 for the same reasons as previously stated in deriving the conditional for the variance. Recall that the motion is an implicit variable in the assembly of the image data $\mathbf{i}, \mathbf{I}$ (and of course in $\hat{\mathbf{e}}$). Because the motion is involved in the data $\mathbf{I}$ the factor outside the exponential cannot be ignored. The integration can be performed by using equation D.14 and $\int \mathrm{IG}(x|\alpha, \beta)dx = 1$ to yield

$$\int_{\sigma_e^2} \frac{|\mathbf{I}\mathbf{I}^T|^{-\frac{1}{2}}}{(2\pi)^{\frac{N-P}{2}} \sigma_e^{2(\frac{N-P}{2}+1)}} \exp\left(-\frac{\mathbf{e}^T\mathbf{e}/2}{\sigma_e^2}\right) d(\sigma_e^2) = |\mathbf{I}\mathbf{I}^T|^{-\frac{1}{2}} \frac{\Gamma((N-P)/2)}{(\hat{\mathbf{e}}^T\hat{\mathbf{e}}/2)^{\frac{(N-P)}{2}}}$$

Therefore

$$p(\mathbf{d}_{n,n-1}|D, \mathbf{i}) \propto \frac{|\mathbf{I}\mathbf{I}^T|^{-\frac{1}{2}}}{(\hat{\mathbf{e}}^T\hat{\mathbf{e}})^{\frac{N-P}{2}}} p(\mathbf{d}_{n,n-1}|D) \tag{D.16}$$

where $p(\mathbf{d}_{n,n-1}|D)$ is the prior for motion, given the block based motion neighbourhood. A similar situation exists for $p(\mathbf{d}_{n,n+1}|D, \mathbf{i})$ and the relevant expression can be part of a separate joint sampling step for motion in this direction.

D.2 Sampling for $b(\mathbf{x}), c(\mathbf{x}), I_n(\mathbf{x})$

The derivation of this joint sample is somewhat unusual because the observed data likelihood is a delta function. In fact this makes the required integration steps quite simple, but it is unlikely that the reader would have previously encountered the manipulation of distributions like these in this context. As with the derivation of the previous joint sampling strategy, the heart of the analysis is the successive integration of two of the three variables out of the joint posterior. The order in which this integration is done does affect the complexity of the resulting expressions, although it does not affect the computational complexity of the final algorithm. Recall that these sampling steps are performed on the *pixel* grid. In the discussion which follows, $i_n(\mathbf{x}), g_n(\mathbf{x})$ refer to the value of a pixel at site $\mathbf{x}$ in the original, clean image frame n and the dirty observed frame n, respectively. Similarly $b(\mathbf{x})$, $c(\mathbf{x})$ refer to the values of the detection field and corruption field at site $\mathbf{x}$, respectively. To indicate that data has been compensated for motion, the superscript m is used. Thus $\mathbf{I}_{n-1}^m$ is shorthand for motion compensated data in frame $n-1$. The notation $\theta_{-(b,i,c)}$ is used to refer to the parameter vector *without* the variables $b(\mathbf{x}), i(\mathbf{x}), c(\mathbf{x})$. The sampling order adopted is $b(\mathbf{x})$ followed by $i_n(\mathbf{x})$ then $c(\mathbf{x})$. These steps require integrating out $c(\mathbf{x})$ then $i_n(\mathbf{x})$ in that order. The joint posterior can be written

$$\begin{aligned} p(b(\mathbf{x}), i_n(\mathbf{x}), c(\mathbf{x})|g_n(\mathbf{x}), \mathbf{I}_{n-1}^m, \mathbf{I}_{n+1}^m, B, C, \theta_{-b,-i,-c}) \propto \\ p(g_n(\mathbf{x})|\theta, \mathbf{I}_{n-1}^m, \mathbf{I}_{n+1}^m, B, C) \\ \times\, p_i(i_n(\mathbf{x})|\theta_{-b,-c}, \mathbf{I}_{n-1}^m, \mathbf{I}_{n+1}^m, B, C)\; p_c(c(\mathbf{x})|C)\; p_b(b(\mathbf{x})|B) \end{aligned} \quad \text{(D.17)}$$

where B, C refer to the current values of the 8 connected neighbourhood around the site $\mathbf{x}$ in the detection and corruption fields.

Each term in the posterior is as below

$$\begin{aligned} p(g_n(\mathbf{x})|\theta, \ldots) &= \delta(g_n(\mathbf{x}) - (1 - b(\mathbf{x}))i_n(\mathbf{x}) - b(\mathbf{x})c(\mathbf{x})) \\ &= \begin{cases} \delta(g_n(\mathbf{x}) - i_n(\mathbf{x})) & \text{for } b(\mathbf{x}) = 0 \\ \delta(g_n(\mathbf{x}) - c(\mathbf{x})) & \text{for } b(\mathbf{x}) = 1 \end{cases} \end{aligned} \quad \text{(D.18)}$$

$$\begin{aligned} p_i(i_n(\mathbf{x})|\theta_{-(b,c)}, \ldots) &= \frac{|\mathbf{a}_u^T \mathbf{a}_u|^{\frac{1}{2}}}{\sqrt{2\pi\sigma_e^2}} \exp\left(-\frac{1}{2\sigma_e^2}[i_n(\mathbf{x}) - \hat{i}](\mathbf{a}_u^T\mathbf{a}_u)[i_n(\mathbf{x}) - \hat{i}]\right) \\ &= \mathcal{N}(\hat{i}, \frac{\sigma_e^2}{\mathbf{a}_u^T \mathbf{a}_u}) \end{aligned} \quad \text{(D.19)}$$

$$\text{where } \hat{i} = \frac{\mathbf{a}_u^T \mathbf{A}_k \mathbf{i}_k}{\mathbf{a}_u^T \mathbf{a}_u}$$

$$p_c(c(\mathbf{x})|C) \propto \exp\left(-\sum_{z \in C} \lambda_z^c (1 - u(z, \mathbf{x}))|c(\mathbf{x}) - z|\right) \quad \text{(D.20)}$$

$$p_b(b(\mathbf{x})|B) \propto \exp\left(-\sum_{z \in B} \lambda_z^b (1 - u(z, \mathbf{x}))|b(\mathbf{x}) - z|\right) \quad \text{(D.21)}$$

The expressions D.20 and D.21 are both taken from their original definitions found in equations 7.9, 7.10. The expression for the conditional for $i_n(\mathbf{x})$ is derived following the same arguments that led to equation 7.35, except considering *one* missing pixel only, $i_n(\mathbf{x})$ in this case. Therefore $\mathbf{a}_u$ is now a column vector as opposed to a matrix. Further details about the composition of $\mathbf{A}_k$ etc. can be found in the discussion on fast algorithms in section 7.8 of Chapter 7. The manipulations required to yield the final form of equation D.19 are the same as those leading up to equation D.9 in this appendix.

Integrating out $c(\mathbf{x})$ from the posterior yields

$$\int_c p(b(\mathbf{x}), i_n(\mathbf{x}), c(\mathbf{x})|\theta_{-(b,i,c)}), \ldots)\, dc(\mathbf{x}) =$$

$$\begin{cases} p_i(i_n(\mathbf{x})|\theta_{-b,-c}, \ldots)p_c(c(\mathbf{x}){=}g_n(\mathbf{x})|C)p_b(b(\mathbf{x}){=}1|B) & \text{for } b(\mathbf{x}) = 1 \\ p_i(i_n(\mathbf{x}){=}g_n(\mathbf{x})|\theta_{-b,-c}, \ldots)p_b(b(\mathbf{x}){=}0|B) & \text{for } b(\mathbf{x}) = 0 \end{cases} \quad \text{(D.22)}$$

because

$$\delta(g_n(\mathbf{x}) - i_n(\mathbf{x}))p_i(i_n(\mathbf{x})|\theta_{-(b,c)}, \ldots) = p_i(i_n(\mathbf{x}){=}g_n(\mathbf{x})|\theta_{-(b,c)}, \ldots) \tag{D.23}$$

$$\delta(g_n(\mathbf{x}) - c(\mathbf{x}))p_c(c(\mathbf{x})|C) = p_c(c(\mathbf{x}){=}g_n(\mathbf{x})|C) \tag{D.24}$$

Integrating $i_n(\mathbf{x})$ out of the expression for $b(\mathbf{x}) = 1$ yields

$$\int_{i_n(\mathbf{x})} \left[p_i(i_n(\mathbf{x})|\theta_{-b,-c}, \ldots)p_c(c(\mathbf{x}){=}g_n(\mathbf{x})|C)p_b(b(\mathbf{x}){=}1|B)\right] di_n(\mathbf{x}) = p_c(c(\mathbf{x}){=}g_n(\mathbf{x})|C)p_b(b(\mathbf{x}){=}1|B) \tag{D.25}$$

because, by definition

$$\int_{i_n(\mathbf{x})} p_i(i_n(\mathbf{x})|\theta_{-(b,c)}, \ldots)\, di_n(\mathbf{x}) = 1 \tag{D.26}$$

Integrating $i_n(\mathbf{x})$ out of the expression for $b(\mathbf{x}) = 0$ yields

$$\int_{i_n(\mathbf{x})} \left[p_i(i_n(\mathbf{x}){=}g_n(\mathbf{x})|\theta_{-(b,c)}, \ldots)p_b(b(\mathbf{x}){=}0|B)\right] di_n(\mathbf{x}) = p_i(i_n(\mathbf{x}){=}g_n(\mathbf{x})|\ldots)p_b(b(\mathbf{x}){=}0|B) \tag{D.27}$$

because,

$$\int_{i_n(\mathbf{x})} p_i(i_n(\mathbf{x}){=}g_n(\mathbf{x})|\theta_{-(b,c)}, \ldots)\, di_n(\mathbf{x}) = p_i(i_n(\mathbf{x}){=}g_n(\mathbf{x})|\theta_{-(b,c)}, \ldots) \tag{D.28}$$

Therefore the required conditional $p(b(\mathbf{x})|B, \ldots)$ is

$$\boxed{p(b(\mathbf{x})|B, \ldots) = \begin{cases} p_c(c{=}g_n|C)p_b(b{=}1|B) & \text{for } b(\mathbf{x}) = 1 \\ p_i(i{=}g_n|\ldots)p_b(b{=}0|B) & \text{for } b(\mathbf{x}) = 0 \end{cases}} \tag{D.29}$$

where $\mathbf{x}$ has been omitted for brevity. This equation can be used to generate the required sample for the binary $b(\mathbf{x})$. Note that drawing a sample for $b(\mathbf{x})$ requires knowledge of the normalizing constants for the distribution $p_c(c(\mathbf{x})|C)$ and $p_i(i_n(\mathbf{x})|\ldots)$.

The form of the conditional for $i_n(\mathbf{x})$ can be seen to be

$$\boxed{p(i_n(\mathbf{x})|b(\mathbf{x}), \ldots) = \begin{cases} \mathcal{N}(\hat{i}, \frac{\sigma_e^2}{\mathbf{a}_u^T\mathbf{a}_u}) & \text{for } b(\mathbf{x}) = 1 \\ \delta(g_n(\mathbf{x}) - i_n(\mathbf{x})) & \text{for } b(\mathbf{x}) = 0 \end{cases}} \tag{D.30}$$

The delta function simply means that when $b(\mathbf{x}) = 0$ there is no corruption and so the underlying clean image is the observed image. Otherwise some interpolation needs to be done and the image is interpolated using a sample from the normal distribution as specified.

Similarly, the form of the conditional for $c(\mathbf{x})$ can be seen to be

$$\boxed{p(c(\mathbf{x})|b(\mathbf{x}), i_n(\mathbf{x}), \dots) = \begin{cases} \delta(g_n(\mathbf{x}) - i_n(\mathbf{x})) & \text{for } b(\mathbf{x}) = 1 \\ p_c(c(\mathbf{x})|C) & \text{for } b(\mathbf{x}) = 0 \end{cases}} \tag{D.31}$$

with delta function meaning that when $b(\mathbf{x}) = 1$ there is corruption and so the corruption field is the observed image. Otherwise the corruption field is a sample from the prior alone.

Sampling for the data in this order makes the samples for $i_n(\mathbf{x})$ and $c(\mathbf{x})$ easy to generate since $b(\mathbf{x})$ is already sampled first.

Appendix E

Examining Ill–Conditioning in $\mathbf{G}^T\mathbf{G}$

Several workers[1] all recognized that when gradient based algorithms failed they did so primarily because of the ill–conditioning of the Gradient matrix that was set up. This ill–conditioning can be quantified through the eigen values of the matrix. The analysis presented here considers this phenomerron and gives some basis for the final algorithm presented by Martinez [117] and reviewed in Chapter 2.

E.1 Condition for singularity

In the standard WBME the solution for the update vector is given as

$$\hat{\mathbf{u}} = [\mathbf{G}^T\mathbf{G} + \mu]^{-1}\mathbf{G}^T\mathbf{z} \tag{E.1}$$

(see Chapter 2). The gradient matrix $\mathbf{G}$ contains elements as follows:

$$\mathbf{G} = \begin{bmatrix} \frac{\partial I_{n-1}(\mathbf{x}_1)}{\partial x} & \frac{\partial I_{n-1}(\mathbf{x}_1)}{\partial y} \\ \frac{\partial I_{n-1}(\mathbf{x}_2)}{\partial x} & \frac{\partial I_{n-1}(\mathbf{x}_2)}{\partial y} \\ \vdots & \vdots \\ \frac{\partial I_{n-1}(\mathbf{x}_{N_1 \times N_2})}{\partial x} & \frac{\partial I_{n-1}(\mathbf{x}_{N_1 \times N_2})}{\partial y} \end{bmatrix} \tag{E.2}$$

Figure 2.3 illustrates the situation when solving for the motion update using this technique. Note how this solution follows from using a one point temporal

[1] For instance, Martinez [117], Kearney et al. [78] and Böröczky et al. [20, 21, 35, 34].

AR model as described in Appendix C, when the single coefficient used has a value of unity.

To facilitate further analysis, $\mathbf{G}$ can be split into two vector components, $\mathbf{G} = [\mathbf{g}_x, \ \mathbf{g}_y]$. The matrix of importance, $\mathbf{G}^T\mathbf{G}$ can then be written

$$\mathbf{G}^T\mathbf{G} = \begin{bmatrix} \mathbf{g}_x^T\mathbf{g}_x & \mathbf{g}_x^T\mathbf{g}_y \\ \mathbf{g}_y^T\mathbf{g}_x & \mathbf{g}_y^T\mathbf{g}_y \end{bmatrix} \quad \text{(E.3)}$$

This matrix is singular iff

$$(\mathbf{g}_x^T\mathbf{g}_x)(\mathbf{g}_y^T\mathbf{g}_y) = (\mathbf{g}_y^T\mathbf{g}_x)(\mathbf{g}_x^T\mathbf{g}_y) \quad \text{(E.4)}$$

The form of this equation is $(\mathbf{a}^T\mathbf{a})(\mathbf{b}^T\mathbf{b}) = (\mathbf{a}^T\mathbf{b})^2$. According to the Schwartz inequality, this is satisfied iff $\mathbf{a} = \alpha\mathbf{b}$, where α is some scalar constant. The equality therefore implies that $\mathbf{g}_x = \alpha\mathbf{g}_y$. This situation,

$$\frac{\partial I_{n-1}(\mathbf{x}_n)}{\partial x} = \alpha\frac{\partial I_{n-1}(\mathbf{x}_n)}{\partial y} \quad \text{(E.5)}$$

occurs when all the points used to set up the gradient matrix lie along an edge.

E.2 Relating ill–conditioning to the spatial contrast

Consider the problem of finding the directions of maximum and minimum average contrast. First of all, define a directional gradient at position $\mathbf{x}$ in a frame of the sequence with the following expression, $\mathbf{v}^T\nabla I(\mathbf{x})$, where $I(\mathbf{x})$ represents the grey level, and the direction is along the vector $\mathbf{v}$. The average magnitude of the spatial gradient (contrast) along $\mathbf{v}$ can then be defined as

$$\bar{C}(\mathbf{x}) = \frac{1}{N}\sum_{i=1}^{N}[\mathbf{v}^T\nabla I(\mathbf{x})]^2 \quad \text{(E.6)}$$

Minimizing and maximizing $C(\mathbf{x})$ with respect to $\mathbf{v}$ will yield the direction of minimum and maximum contrast. Of course, to yield a direction only, an additional constraint is needed to ensure the resulting vector is a unit vector. The constraint is

$$\mathbf{v}^T\mathbf{v} = 1 \quad \text{(E.7)}$$

Note that the definition of directional gradient introduced, which is intuitively reasonable, now allows the sum[2] magnitude of contrast to be expressed in terms of $\mathbf{G}$. It becomes easier to see how $\mathbf{G}^T\mathbf{G}$ is related to a measure of directional contrast.

$$\begin{aligned} C(\mathbf{x}) &= \mathbf{v}^T\mathbf{G}^T\mathbf{G}\mathbf{v} \quad \text{(E.8)} \\ &= \mathbf{v}^T\mathbf{W}\mathbf{v} \end{aligned}$$

[2]The average value differs only by a constant which is not of significance for this analysis.

Another simplification is introduced, letting $\mathbf{W} = \mathbf{G}^T\mathbf{G}$.

Combining equation E.7 and E.9 and a Lagrange multiplier λ, gives the following expression to optimize.

$$O(\lambda, \mathbf{v}) = \mathbf{v}^T[\mathbf{W} - \lambda\mathbf{I}]\mathbf{v} + \lambda \tag{E.9}$$

Differentiating E.9 with respect to $\mathbf{v}$ and setting the result to zero, yields E.10.

$$[\mathbf{W} - \lambda\mathbf{I}]\mathbf{v} + [\mathbf{W} - \lambda\mathbf{I}]^T\mathbf{v} = 0 \tag{E.10}$$

But $\mathbf{W}$ is symmetric, therefore,

$$[\mathbf{W} - \lambda\mathbf{I}]\mathbf{v} \quad = \quad 0 \tag{E.11}$$

This is an eigenvalue problem. Therefore, the directions of maximum and minimum contrast lie along the eigen vectors of $\mathbf{W}$. Similarly, the magnitudes of maximum and minimum contrast are given by eigenvalues of $\mathbf{W}$. The larger eigenvalue and corresponding vector refer to the magnitude and direction of maximum contrast.

When $\mathbf{W}$ is ill–conditioned, $\lambda_{\max} \gg \lambda_{\min}$. Hence $\mathbf{G}^T\mathbf{G}$ is ill–conditioned when the directions of averaged spatial gradient are not of equal importance; at an edge for instance.The solution for the motion estimate is therefore well conditioned when the image portion consists of two clearly defined gradient directions. A good example is at a corner [131, 125].

This agrees well with intuition. As discussed in Chapter 2, vertical motion at a vertical (step) edge cannot be estimated. In that case all the information in the vertical direction would be the same. Any vertical motion would yield no change in information. Horizontal motion would of course yield a change in information because of travel across the vertical step. Therefore this motion can be readily estimated. This phenomenon can be summed up by stating that the motion estimate is most accurate in the direction of maximum contrast. This observation would reflect itself in the ill–conditioning of $\mathbf{W}$ by giving meaning to only one of the two equations for motion in equation E.1 when the equations are set up at an edge. This cannot be unraveled in the straightforward inverse operation required in equation E.1, the only clue is that the solution becomes ill–conditioned.

In the light of this analysis it seems natural to detect ill–conditioning in the usual way, by thresholding the eigenvalue ratio, then proceed to stabilize the solution using eigen analysis. This analysis can then yield a direction in which the motion estimation is more accurate. This is the solution that Martinez proposed, and his solution is repeated here.

$$\mathbf{d} \quad = \quad \begin{cases} \alpha_{\max}\,\mathbf{e}_{\max} & \text{if } \lambda_{\max} \gg \lambda_{\min} \\ \left[\mathbf{G}^T\mathbf{G}\right]^{-1}\mathbf{G}^T\mathbf{z} & \textit{otherwise} \end{cases} \tag{E.12}$$

$$\alpha_{\max} \quad = \quad \frac{\mathbf{e}_{\max}^T}{\lambda_{\max}}\mathbf{G}^T\mathbf{z}$$

Here, λ and $\mathbf{e}$ refer to the eigen values and eigen vectors of $\mathbf{G}^T\mathbf{G}$, and α is a scalar variable introduced to simplify the form of the final solution.

Note that in the work of Martinez [117], the solution was neither pel–recursive nor was any consideration given to the use of the regularizing μ used in E.1.

E.3 Ill–conditioning in the general 3DAR solution.

Following a similar approach to that of Martinez, it is possible to examine the relation between the WBME and the general AR estimator with respect to conditioning of the solution.

Consider a temporally causal AR model in which the support region consists of points in the previous frame only, for example, AR9 in Figure 3.8 in Chapter 3. Then, using the definition of $\mathbf{G}_w$ in equation 3.7, it is possible to express the general matrix in terms of $\mathbf{g}_x$ and $\mathbf{g}_y$, defined above, as follows:

$$\mathbf{G}_w = [\mathbf{A}\mathbf{g}_x \ \mathbf{A}\mathbf{g}_y] \tag{E.13}$$

$\mathbf{A}$ is a matrix of AR coefficients that would weight the relevant gradient terms of $\mathbf{g}$ to yield the general form indicated by equation 3.7.

From E.13 $\mathbf{G}_w^T\mathbf{G}_w$ can be expanded to

$$\mathbf{G}_w^T\mathbf{G}_w = \begin{bmatrix} \mathbf{g}_x^T\mathbf{A}^T\mathbf{A}\mathbf{g}_x & \mathbf{g}_x^T\mathbf{A}^T\mathbf{A}\mathbf{g}_y \\ \mathbf{g}_y^T\mathbf{A}^T\mathbf{A}\mathbf{g}_x & \mathbf{g}_y^T\mathbf{A}^T\mathbf{A}\mathbf{g}_y \end{bmatrix} \tag{E.14}$$

The matrix is therefore singular when

$$(\mathbf{g}_x^T\mathbf{A}^T\mathbf{A}\mathbf{g}_x)(\mathbf{g}_y^T\mathbf{A}^T\mathbf{A}\mathbf{g}_y) = (\mathbf{g}_y^T\mathbf{A}^T\mathbf{A}\mathbf{g}_x)(\mathbf{g}_x^T\mathbf{A}^T\mathbf{A}\mathbf{g}_y) \tag{E.15}$$

As before, the Schwartz inequality may be applied to yield the condition for singularity as

$$\mathbf{A}\mathbf{g}_x = \alpha\mathbf{A}\mathbf{g}_y \tag{E.16}$$

Provided that $\mathbf{A}$ has an inverse, the condition for a singular weighted gradient matrix, $\mathbf{G}_w^T\mathbf{G}_w$, is the same as for the matrix $\mathbf{G}^T\mathbf{G}$, as discussed previously. Therefore, the general 3DAR solution for motion is also ill–conditioned at an edge in the image.

An eigen analysis would yield similar conclusions to that discussed previously. The weighting matrix, $\mathbf{A}$ would change the eigen values and vectors of the un–weighted matrix, $\mathbf{G}^T\mathbf{G}$. Nevertheless, in view of the similar conditions for singularity, it would appear that 3DAR solution is not any better conditioned in this respect. In practice however, it is noticed that the solution is better behaved, and this must be due to the noise reducing effect of the gradient weighting.

E.4 Summary

An eigen analysis of the simple gradient based motion solution has related ill–conditioning to the presence of an edge in the image. The eigen values and vectors of $\mathbf{G}^T\mathbf{G}$ have been shown to represent the value and directions of maximum and minimum contrast in the image. Hence, the ill–conditioned solution can be related to the contrast orientation in the image, a standard problem.

In this respect, the 3DAR solution carries no advantage. It is also singular at edges in the image. However, it uses a weighted gradient matrix and each term therefore represents a weighted combination of gradient observations in a small region. This results in a less noisy gradient observation and so the solution is more stable.

The eigen analysis of both $\mathbf{G}^T\mathbf{G}$ and $\mathbf{G}_w^T\mathbf{G}_w$ can be used to propose a strategy when the solution is detected as ill–conditioned. This strategy is used to good effect in Chapter 3.

Appendix F

The Wiener Filter for Image Sequence Restoration

The Wiener filter has been used frequently in the past as an effective method for noise and blur reduction in degraded signals. The form of the filter for image sequences is a direct extension of the 1–D result, however the limited amount of temporal information available implies different considerations in the implementation of the method. This Appendix makes explicit the various forms of the filter that can be used and in so doing lays the foundation for the ideas presented in appendix G. The theory presented here ignores motion between frames. Motion is considered in the main body of the text.

This book concerns itself with the problem of noise reduction in image sequences, blur is not considered. The observed signal model is then

$$g(i,j,n) = I(i,j,n) + \eta(i,j,n) \tag{F.1}$$

where $g(i,j,n)$ is the observed signal grey scale value at position (i,j) in the nth frame, $I(i,j,n)$ is the actual non-degraded signal and $\eta(i,j,n)$ the added white noise of variance $\sigma_{\eta\eta}$.

F.1 The 3D frequency domain/3D IIR Wiener filter

The Wiener filter attempts to produce the best estimate for $I(i,j,n)$, denoted $\hat{I}(i,j,n)$. It does so by minimizing the expected squared error given by

$$E[(e(i,j,n))^2] = E[(I(i,j,n) - \hat{I}(i,j,n))^2] \tag{F.2}$$

This estimate may be achieved using either an IIR or FIR Wiener filter, which operates on the observed noisy signal. The following two expressions show respectively IIR and FIR estimators.

$$\hat{I}(i,j,n) = \sum_{k_1}\sum_{k_2}\sum_{k_3} a(k_1,k_2,k_3)g(i+k_1,j+k_2,n+k_3) \quad \text{(F.3)}$$

$$\hat{I}(i,j,n) = \sum_{k_1=-N_1}^{N_1}\sum_{k_2=-N_2}^{N_2}\sum_{k_3=-N_3}^{N_3} a(k_1,k_2,k_3)g(i+k_1,j+k_2,n+k_3) \quad \text{(F.4)}$$

In equation F.3 there are no limits on the summations, hence IIR, and in equation F.4 the filter mask (or support) is a symmetric volume around the filtered location of size $(2N_1+1)\times(2N_2+1)\times(2N_3+1)$.

Proceeding with the IIR filter leads here to a frequency domain expression for the Wiener filter. Using the principle of orthogonality to bypass some lines of differential algebra, $E[(e(\dots))^2]$ is minimized if the error is made orthogonal to the data samples used in the estimate equation F.3.

$$E[e(i,j,n)g(i+k_1,j+k_2,n+k_3)] = 0 \ \forall\ k_1,k_2,k_3 \quad \text{(F.5)}$$

Substituting for $e(i,j,n)$ in equation F.5 gives

$$E[\hat{I}(i,j,n)g(i+k_1,j+k_2,n+k_3)] = E[I(i,j,n)g(i+k_1,j+k_2,n+k_3)] \ \forall\ k_1,k_2,k_3 \quad \text{(F.6)}$$

The filter coefficients are then chosen to satisfy the condition of equation F.6. Substituting for $\hat{I}(i,j,n)$ in F.6 yields,

$$\sum_{l_1,l_2,l_3} a(l_1,l_2,l_3)E[g(i+l_1,j+l_2,n+l_3)g(i+k_1,j+k_2,n+k_3)] = E[I(i,j,n)g(i+k_1,j+k_2,n+k_3)] \ \forall\ k_1,k_2,k_3,l_1,l_2,l_3 \quad \text{(F.7)}$$

The expectations can be recognized as terms from the autocorrelation and crosscorrelation sequences of the observed image sequence $g(i,j,n)$ and the actual sequence $I(i,j,n)$. The solution involves a separate equation for each coefficient, i.e. an infinite set of equations. However, using the 3D DFT gives a tractable result in terms of the power spectra concerned. From equation F.6 the following expressions result assuming stationary statistics.

$$\sum_{l_1}\sum_{l_2}\sum_{l_3} a(l_1,l_2,l_3)R_{gg}(l_1-k_1,l_2-k_2,l_3-k_3)] = R_{ig}(k_1,k_2,k_3) \quad \text{(F.8)}$$

Taking Fourier transforms yields

$$A(\omega_1,\omega_2,\omega_3)P_{gg}(\omega_1,\omega_2,\omega_3) = P_{ig}(\omega_1,\omega_2,\omega_3) \quad \text{(F.9)}$$

The only unknown quantity here is the cross power spectrum P_{ig}. However, using the original assumption of uncorrelated noise and image frames, implies that

$$P_{gg} = P_{ii} + P_{\eta\eta} \tag{F.10}$$
$$P_{ig} = P_{ii} \tag{F.11}$$

From these two equations, an expression for the 3D frequency domain Wiener filter, $A(\omega_1, \omega_2, \omega_3)$ is as follows.

$$A(\omega_1, \omega_2, \omega_3) = \frac{P_{gg}(\omega_1, \omega_2, \omega_3) - P_{\eta\eta}(\omega_1, \omega_2, \omega_3)}{P_{gg}(\omega_1, \omega_2, \omega_3)} \tag{F.12}$$

The estimated signal, $\hat{I}(i, j, n)$ is then given by the expression below.

$$\hat{I}(i, j, n) = \text{IDFT}[A(\omega_1, \omega_2, \omega_3)G(\omega_1, \omega_2, \omega_3)] \tag{F.13}$$

The IDFT is the inverse 3D DFT, and $G(\omega_1, \omega_2, \omega_3)$ is the 3D DFT of the observed signal, $g(i, j, n)$. The filter is therefore defined by the signal power spectrum and the noise variance.

Using the 3D DFT in this way makes the Wiener filter computationally attractive. However, in the temporal direction there are often not many frames involved. Therefore, the temporal Fourier component is less meaningful, having been derived from only a few samples. In the book, 3 frames are involved, hence the DFT in the temporal direction involves just 3 samples. More frames can be included but at the expense of stationarity since effects such as occlusion and uncovering are more likely to appear. The assumption of an infinite support volume is violated. This phenomenon is also applicable to the spatial components since the image is only stationary over small areas. Therefore, in a practical situation, the 3D IIR filter implemented in the frequency domain in this way, is no longer IIR since it operates on a finite volume of input data only and not on any past outputs.

The problem may be overcome by considering two further forms of the filter which allow for finite support. One form follows from the FIR filter expression in equation F.3, the other results from a matrix formulation of the filter.

F.2 The 3D FIR Wiener filter

The FIR filter result follows again by using the theorem of orthogonality. The expression F.4 for the filter is substituted into the error term in the orthogonality expression F.5 using equation F.2, except allowing for the finite support. This yields

$$\sum_{l_1, l_2, l_3} a(l_1, l_2, l_3) E[g(i + l_1, j + l_2, n + l_3) g(i + k_1, j + k_2, n + k_3)] =$$
$$E[I(i, j, n) g(i + k_1, j + k_2, n + k_3)] \tag{F.14}$$
$$\text{for} \begin{cases} -N_1 \leq k_1,\ l_1 \leq N_1 \\ -N_3 \leq k_2,\ l_2 \leq N_2 \\ -N_3 \leq k_3,\ l_3 \leq N_3 \end{cases}$$

Because the support of the FIR filter is finite, it is better to describe the filter expression in terms of offset vectors $\mathbf{q}_k$ as was done in the case of the 3DAR framework described in Chapter 2 and Appendix A. This results in a much simpler formulation[1].

$$\begin{aligned} E[\hat{I}(\mathbf{x})g(\mathbf{q}_k)] &= \sum_{l=0}^{N} a_l E[g(\mathbf{x}+\mathbf{q}_l)g(\mathbf{x}+\mathbf{q}_k)] \\ &= E[I(\mathbf{x})g(\mathbf{x}+\mathbf{q}_k)] \text{ for } 0 \le k \le N \end{aligned} \tag{F.15}$$

Here N vectors, $\mathbf{q}_i$ define the support volume of the filter, and so the expression above defines a set of N equations each for one filter coefficient. The solution for the coefficients can then follow from the matrix equation below.

$$\mathbf{R}_{gg}\mathbf{a} = \mathbf{r}_{ig} \tag{F.16}$$

The terms in each matrix are defined below.

$$\mathbf{a} = [a_0\ a_1\ a_2\ \dots\ a_N] \tag{F.17}$$

$$\mathbf{R}_{gg} = \begin{bmatrix} r_{gg}(\mathbf{q}_0,\mathbf{q}_0) & r_{gg}(\mathbf{q}_0,\mathbf{q}_1) & \dots & r_{gg}(\mathbf{q}_1,\mathbf{q}_N) \\ r_{gg}(\mathbf{q}_2,\mathbf{q}_1) & r_{gg}(\mathbf{q}_2,\mathbf{q}_2) & \dots & r_{gg}(\mathbf{q}_2,\mathbf{q}_N) \\ r_{gg}(\mathbf{q}_3,\mathbf{q}_1) & r_{gg}(\mathbf{q}_3,\mathbf{q}_2) & \dots & r_{gg}(\mathbf{q}_3,\mathbf{q}_N) \\ \vdots & \vdots & \vdots & \vdots \\ r_{gg}(\mathbf{q}_N,\mathbf{q}_1) & r_{gg}(\mathbf{q}_N,\mathbf{q}_2) & \dots & r_{gg}(\mathbf{q}_N,\mathbf{q}_N) \end{bmatrix} \tag{F.18}$$

$$\mathbf{r}_{ig}^T = [r_{ig}(\mathbf{q}_0,\mathbf{q}_0)\dots r_{ig}(\mathbf{q}_0,\mathbf{q}_1)\dots r_{ig}(\mathbf{q}_0,\mathbf{q}_N)] \tag{F.19}$$

$$\mathbf{q}_0 = [0\ 0\ 0] \tag{F.20}$$

Although this matrix solution is virtually identical to the solution for 3D AR model coefficients (See Appendix A), note the important difference that this solution includes an estimate for the coefficient at the zero lag tap, $\mathbf{q}_0$. This explains why the 3D AR formulation is called a *predictor* and this formulation a *filter.*

Again the only information not available for the solution of equation F.16 is the cross correlation vector, $\mathbf{r}_{ig}$. This vector involves correlations between the unknown original signal and the observed noisy samples. However, the assumption that the image and noise frames are uncorrelated allows a series of useful equalities[2].

$$\begin{aligned} r_{ig}(\mathbf{q}_0,\mathbf{q}_h) &= E[I(\mathbf{x})g(\mathbf{x}+\mathbf{q}_h)] \\ &= E[I(\mathbf{x})\{I(\mathbf{x}+\mathbf{q}_h)+\eta(\mathbf{x}+\mathbf{q}_h)\}] \\ &= \begin{cases} E[I(\mathbf{x})I(\mathbf{x}+\mathbf{q}_h)] & \text{for } \mathbf{q}_h \neq [0\ 0] \\ E[I(\mathbf{x})I(\mathbf{x})] & \text{for } \mathbf{q}_h = [0\ 0] \end{cases} \end{aligned} \tag{F.21}$$

(F.22)

---

[1]Note that the 3D FIR filter described here uses a mask volume of size $N_1 \times N_2 \times N_3$.

[2]In practice the data volume is finite and so a sufficiently large data volume is necessary to make measurements of correlation terms.

Multiplying equation F.1 by $g(\mathbf{x}+\mathbf{q}_h)$ and taking expectations gives

$$\begin{aligned} E[g(\mathbf{x})g(\mathbf{x}+\mathbf{q}_h)] &= E[I(\mathbf{x})I(\mathbf{x}+\mathbf{q}_h)] \\ &\quad +E[I(\mathbf{x})\eta(\mathbf{x}+\mathbf{q}_h)] + E[\eta(\mathbf{x})I(\mathbf{x}+\mathbf{q}_h)] \\ &\quad +E[\eta(\mathbf{x})\eta(\mathbf{x}+\mathbf{q}_h)] \end{aligned} \tag{F.23}$$

Substituting for $E[I(\mathbf{x})I(\mathbf{x}+\mathbf{q}_k)]$ from equation F.23 into equation F.22, gives a usable expression for the required cross correlation term using the observed signal and an estimate of the noise variance.

$$r_{\iota g}(\mathbf{q}_0, \mathbf{q}_h) = \begin{cases} E[g(\mathbf{x})g(\mathbf{x}+\mathbf{q}_h)] & \text{for } \mathbf{q}_h \neq [0,\ 0] \\ \sigma_{gg}^2 - \sigma_{\eta\eta}^2 & \text{for } \mathbf{q}_h = [0,\ 0] \end{cases} \tag{F.24}$$

σ_{gg}^2 is the observed signal variance, and $\sigma_{\eta\eta}^2$ the noise variance. For non–zero lags, it is intuitively reasonable to equate the crosscorrelation of the clean signal and the observed signal by the autocorrelation of the observed signal since the effect of the noise on the terms would be small. A solution for equation F.16 then follows using the inverse of $\mathbf{R}_{gg}$.

Having found the coefficients of the filter, the filter can then be applied following equation F.4 to achieve noise reduction.

F.3 The matrix formulation of the 3D Wiener filter

The last approach to be considered is also motivated by the need to deal with a finite volume of sequence data. In this form the signals are ordered into column vectors. The observation equation for each frame, n is then

$$\mathbf{g}_n = \mathbf{s}_n + \eta_n \tag{F.25}$$

Here $\mathbf{s}_n$ is used to represent the ordered values from the unknown clean signal $I(i,j,n)$. The vectors for each frame may then be stacked upon each other in another larger column vector. If the size of each frame is $M \times M$ pixels and there are N frames, then the new vectors are NM^2 long and the new equation is as follows.

$$\mathbf{g} = \mathbf{s} + \eta \tag{F.26}$$

The task of the Wiener filter is to operate on the observed signal, $\mathbf{g}$, to suppress noise. The 3D FIR filter above gives one framework for operating on finite data, but for all possible taps to be taken into account, the matrix solution, given below, is useful.

$$\hat{\mathbf{s}} = \mathbf{H}\mathbf{g} \tag{F.27}$$

$\mathbf{H}$ is a matrix operator of size $NM^2 \times NM^2$. It is necessary to find the coefficients of the matrix operator, $\mathbf{H}$, to give the estimated signal. Again, the idea is to choose this matrix to minimize the expectation of the squared error as below.

$$E[\mathbf{e}^T\mathbf{e}] = E[(\mathbf{s} - \hat{\mathbf{s}})^T(\mathbf{s} - \hat{\mathbf{s}})] \tag{F.28}$$

A substitution from equation F.27 can then be made into F.28 which is then minimized with respect to **H**. Again making the assumption that the image and noise frames are uncorrelated, finally allows the estimate for **H** to be derived as

$$\begin{aligned} \mathbf{H} &= E[\mathbf{s}\mathbf{s}^T](E[\mathbf{s}\mathbf{s}^T] + E[\eta\eta^T])^{-1} \\ &= \mathbf{R}_{ss}(\mathbf{R}_{ss} + \mathbf{R}_{\eta\eta})^{-1} \end{aligned} \tag{F.29}$$

The complete derivation is not shown here but can be found in [71, 107, 160, 45]. The important thing about this expression is that it involves the inverse of very large matrices. The correlation matrices are all of size $NM^2 \times NM^2$, for a typical block size of about 16×16, and 3 frames, this is 768×768. This is not a practical solution and so several authors have introduced approximations to reduce the computation necessary to calculate this inverse, [135, 160, 45, 67]. One of these approximations is discussed in Appendix G.

Appendix G

Reducing the Complexity of Wiener Filtering

Although the frequency domain Wiener filter is computationally simple, requiring only 3 separable DFT implementations, the matrix formulation of the filter is computationally quite heavy. This formulation is potentially more accurate than the frequency domain implementation since the form takes into account the finite temporal size of the data volume involved. Özkan et al [135, 42] have developed an efficient solution for this filter form which takes advantage of some approximations about the circulant nature of the correlation matrix of the signals. It is possible to derive their result by an alternative route which is described here after a brief overview of their result.

G.1 Efficient Wiener filtering via 2D DFT diagonalization

The signal model is as stated in the previous appendix on Wiener filtering.

$$\mathbf{g} = \mathbf{s} + \eta \tag{G.1}$$

The column vectors are made up of the data from N frames ordered and stacked. Therefore,

$$\mathbf{g} = [\mathbf{g}_1 \mathbf{g}_2 \cdots \mathbf{g}_N]^T \tag{G.2}$$

and so on for the other data vectors, where the frames are indexed from 1 to N.

The matrix Wiener solution was given in the previous appendix. Following Galatsanos et al [45], Özkan et al [135] noted that the correlation matrices involved were block circulant. The solution for the transformation matrix, **H** is

$$\mathbf{H} = \mathbf{R}_{ss}(\mathbf{R}_{ss} + \mathbf{R}_{\eta\eta})^{-1} \tag{G.3}$$

The Wiener estimate for s is then

$$\hat{\mathbf{s}} = \mathbf{R}_{ss}(\mathbf{R}_{ss} + \mathbf{R}_{\eta\eta})^{-1}\mathbf{g} \tag{G.4}$$

The structure of each matrix can be represented as below,

$$\mathbf{R}_{ss} = \begin{bmatrix} \mathbf{R}_{ss:11} & \dots & \mathbf{R}_{ss:1N} \\ \vdots & \ddots & \vdots \\ \mathbf{R}_{ss:N1} & \dots & \mathbf{R}_{ss:NN} \end{bmatrix} \tag{G.5}$$

$$\mathbf{R}_{\eta\eta} = \begin{bmatrix} \mathbf{R}_{\eta\eta:11} & \dots & \mathbf{R}_{\eta\eta:1N} \\ \vdots & \ddots & \vdots \\ \mathbf{R}_{\eta\eta:N1} & \dots & \mathbf{R}_{\eta\eta:NN} \end{bmatrix} \tag{G.6}$$

$$\text{Where } \mathbf{R}_{ss:hk} = E[\mathbf{s}_h\mathbf{s}_k^T] \tag{G.7}$$

$$\text{and } \mathbf{R}_{\eta\eta:hk} = E[\eta_h\eta_k^T] \tag{G.8}$$

(G.9)

Assuming that the image and noise frames are statistically homogeneous, the sub-matrices in each correlation matrix are block Toeplitz. As stated in [135, 42], the sub-matrices may be diagonalized [67] using the 2–D DFT. A matrix **W** may be defined such that $\mathbf{W}^{-1}\mathbf{g}$ stacks the 2–D DFTs of the ordered data vector **g**. Following Özkan, premultiplying both sides of equation G.4 by $\mathbf{W}^{-1}$ yields

$$\mathbf{W}^{-1}\hat{\mathbf{s}} = \mathbf{W}^{-1}\mathbf{R}_{ss}\mathbf{W}[\mathbf{W}^{-1}\mathbf{R}_{ss}\mathbf{W} + \mathbf{W}^{-1}\mathbf{R}_{\eta\eta}\mathbf{W}]^{-1}\mathbf{W}^{-1}\mathbf{g} \tag{G.10}$$

The sub-matrices in the equation have now been diagonalized via the 2–D DFT. The equation may be written as

$$\hat{\mathbf{F}}_{\hat{s}} = \mathbf{P}_{ss}[\mathbf{P}_{ss} + \mathbf{P}_{\eta\eta}]^{-1}\mathbf{F}_g \tag{G.11}$$

$$\hat{\mathbf{F}}_{\hat{s}} = \mathbf{P}_{ss}\mathbf{Q}^{-1}\mathbf{F}_g \tag{G.12}$$

(G.13)

Here, $\mathbf{F}_{\hat{s}}$, $\mathbf{F}_g$ are column vectors containing the stacked 2D DFTs of the signal estimate and observed noisy signal respectively. The matrices $\mathbf{P}_{ss}$, $\mathbf{P}_{\eta\eta}$ are the 2D cross power spectral densities of the various frames involved. Note that P_{ss} is not diagonal but its sub-matrices, $P_{ss:hk}$, are diagonal. They can be defined as below.

$$\mathbf{P}_{ss} = \begin{bmatrix} \mathbf{P}_{ss:11} & \dots & \mathbf{P}_{ss:1N} \\ \vdots & \ddots & \vdots \\ \mathbf{P}_{ss:N1} & \dots & \mathbf{P}_{ss:NN} \end{bmatrix} \quad \text{(G.14)}$$

$$\text{where } \mathbf{P}_{ss:hk} = \begin{bmatrix} P_{ss:hk,1} & \dots & 0 \\ \vdots & \ddots & \vdots \\ 0 & \dots & P_{ss:hk,M^2} \end{bmatrix} \quad \text{(G.15)}$$

$$\mathbf{Q} = \begin{bmatrix} \mathbf{Q}_{11} & \dots & \mathbf{Q}_{1N} \\ \vdots & \ddots & \vdots \\ \mathbf{Q}_{N1} & \dots & \mathbf{Q}_{NN} \end{bmatrix} \quad \text{(G.16)}$$

$$\text{where } \mathbf{Q}_{hk} = \begin{bmatrix} Q_{hk,1} & \dots & 0 \\ \vdots & \ddots & \vdots \\ 0 & \dots & Q_{hk,M^2} \end{bmatrix} \quad \text{(G.17)}$$

$$\text{and } \mathbf{Q}_{hk} = \mathbf{P}_{ss:hk} + \delta_{hk}\mathbf{P}_{\eta\eta:hk} \quad \text{(G.18)}$$

$$\forall\, h,\ k = 1 \dots N \quad \text{(G.19)}$$

Note that the noise frames are assumed uncorrelated, hence the inclusion of the Kronecker delta function[1] in the expression for $\mathbf{Q}_{hk}$ above.

The expressions are somewhat cumbersome, but the result has been to diagonalize the sub-matrices of the correlation matrices used in the Wiener estimate. The elements along the main diagonal of the sub-matrices $\mathbf{Q}_{hk}$ and $\mathbf{P}_{\mathbf{ss:hk}}$ are the ordered 2D DFT bins of the relevant correlation functions. Therefore, the element $P_{ss:hk,l}$ is the amplitude of the lth ordered frequency in the 2D Cross Power spectral density of the original, clean signal in frame h, and k.

The matrix $\mathbf{Q}$ is very large, but because of the diagonalization transformation, it is a block matrix with diagonal blocks. Therefore, Özkan et al proposed an efficient solution by invoking a rule that implies that the inverse of a block matrix with diagonal blocks is itself a block matrix with diagonal blocks. The inverse of the $NM^2 \times NM^2$ matrix $\mathbf{Q}$ was then derived from the inverse of M^2 smaller matrices of size $N \times N$. This in a sense is the crux of the solution presented by Özkan et al.

---

[1] $\delta_{hk} = 1$ for $h = k$

Define the inverse of $\mathbf{Q}$ to be $\mathbf{Z}$. Then define sub-matrices $\mathbf{Q}_k$ and $\mathbf{Z}_k$ as follows.

$$\text{Let } \mathbf{Z} = \begin{bmatrix} Z_{11} & \dots & Z_{1N} \\ \vdots & \ddots & \vdots \\ Z_{N1} & \dots & Z_{NN} \end{bmatrix}$$

$$\text{and } \mathbf{Z}_{hk} = \begin{bmatrix} Z_{hk:1} & \dots & 0 \\ \vdots & \ddots & \vdots \\ 0 & \dots & Z_{hk:M^2} \end{bmatrix}$$

$$\mathbf{Q}_k = \begin{bmatrix} Q_{11:k} & \dots & Q_{1N:k} \\ \vdots & \ddots & \vdots \\ Q_{N1:k} & \dots & Q_{NN:k} \end{bmatrix} \tag{G.20}$$

$$\mathbf{Z}_k = \begin{bmatrix} Z_{11:k} & \dots & Z_{1N:k} \\ \vdots & \ddots & \vdots \\ Z_{N1:k} & \dots & Z_{NN:k} \end{bmatrix} \tag{G.21}$$

(G.22)

Therefore the two sets of $N \times N$ sub-matrices, $\mathbf{Q}_k$ and $\mathbf{Z}_k$, are subsampled versions of the major matrices. Then the lemma invoked by Özkan at al. shows that the components of $\mathbf{Z}_k$ are equal to the inverse of the $N \times N$ submatrix, $\mathbf{Q}_k$. There are M^2 such sub matrices hence the inverse operation is much reduced in complexity. In the case of $N = 2$, or 3, an analytic solution can easily be formulated.

The Wiener estimate for the ith frame then results from substituting this inverse into equation G.12 as follows:

$$\mathbf{F}_{\hat{s}:i,k} = \sum_{p=1}^{N} P_{ss:ip,k} \sum_{q=1}^{N} Z_{pq,k} F_{g:q,k} \tag{G.23}$$

for each of the M^2 frequencies involved indexed by k.

G.2 An alternative derivation

The situation above can be viewed in a different light to give the same result. Consider again that there are $2N+1$ frames of size $(2M+1) \times (2M+1)$ in the observation equation G.2. Construct the required estimate in the spatiotemporal domain such that

$$\hat{I}(x_1, x_2, x_3) = \sum_{i,j,l} a(i,j,l) g(x_1 + i, x_2 + j, x_3 + l) \tag{G.24}$$

In this expression, (x_1, x_2, x_3) and (i, j, l) are the spatial and temporal coordinates and offsets of a pixel, where $-M \leq i,\ j \leq M$ and $-N \leq l \leq N$.

The equation G.24 may be written explicitly as the sum of separate convolutions of the observed frames g_i with a set of different coefficients for each frame. If there were 3 frames required for the FIR filter the equation would read:

$$\begin{aligned}\hat{I}(x_1,x_2,x_3) &= \sum_{i,j} a(i,j,-1)g(x_1+i,x_2+j,x_3-1)\\ &+\sum_{i,j} a(i,j,0)g(x_1+i,x_2+j,x_3)\\ &+\sum_{i,j} a(i,j,+1)g(x_1+i,x_2+j,x_3+1) \qquad \text{(G.25)}\end{aligned}$$

for a non–causal filter using a frame previous to and after the current frame x_3. This equation can then be written in the frequency domain, using the 2D DFT, as

$$\begin{aligned}F_{\hat{s}:x_3}(\omega_1,\omega_2) &= A_{-1}(\omega_1,\omega_2)F_{g:x_3-1}(\omega_1,\omega_2)\\ &+A_0(\omega_1,\omega_2)F_{g:x_3}(\omega_1,\omega_2)\\ &+A_1(\omega_1,\omega_2)F_{g:x_3+1}(\omega_1,\omega_2) \qquad \text{(G.26)}\end{aligned}$$

Here, $F_{\hat{s}:h}(\omega_1,\omega_2)$ and $F_{g:h}(\omega_1,\omega_2)$ represent the 2D DFT of the hth frame estimate and noisy observed frame respectively. Similarly $A_q(\omega_1,\omega_2)$ represents the 2D DFT of the coefficient mask in the qth offset frame, so that it is applied to frame x_3+q in this case. The transformed coefficient masks, $A_k(\omega_1,\omega_2)$ etc., are required.

The 2D DFT of the estimate of the clean signal in frame v can then be represented by

$$F_{\hat{s},v}(\omega_1,\omega_2) = \sum_{q=-N}^{N} A_q(\omega_1,\omega_2)F_{g,v+q}(\omega_1,\omega_2) \qquad \text{(G.27)}$$

Returning to the spatiotemporal domain, the MMSE (Minimum Mean Squared Error) estimate for the coefficients $a(i,j,l)$ is found by solving the following set of equations (from equation F.15).

$$\begin{aligned}\sum_{i,j,l} a(i,j,l)E[g(x_1+i,x_2+j,x_3+l)g(x_1+k_1,x_2+k_2,x_3+k_3)] &=\\ E[I(x_1,x_2,x_3)g(x_1+k_1,x_2+k_2,x_3+k_3)] & \qquad \text{(G.28)}\\ \text{for}\begin{cases} -N\leq\ k_3\ \leq N\\ -M\leq\ k_1,\ k_2\ \leq M\end{cases}\end{aligned}$$

This equation can then be simplified to make clear the correlation measurements that are required between frames.

$$\sum_{i,j,l} a(i,j,l)r_{gg:x_3+l,x_3+k_3}(i-k_1,j-k_2) = r_{sg:x_3,x_3+k_3}(k_1,k_2) \qquad \text{(G.29)}$$

$$\text{for}\begin{cases} -N\leq\ k_3\ \leq N\\ -M\leq\ k_1,\ k_2\ \leq M\end{cases}$$

Two assumptions are made, spatial homogeneity and that the noise and image frames are not correlated with each other. The solution may be simplified by taking the 2D DFT of the expression to yield

$$P_{sg:x_3,x_3+k_3}(\omega_1,\omega_2) = \sum_{q=-N}^{N} A_q(\omega_1,\omega_2) P_{gg:x_3+q,x_3+k_3}(\omega_1,\omega_2) \quad \text{(G.30)}$$

$$\text{for } -N \leq k_3 \leq N \quad \text{(G.31)}$$

Here, $P_{gg:x_3+q,x_3+k_3}(\omega_1,\omega_2)$ represents the Cross Power Spectral Density of the noisy frames x_3+q and x_3+k_3, and $P_{sg:x_3,x_3+k_3}(\omega_1,\omega_2)$ is the required cross power spectral density of the noisy and original frames. Here s is used instead of i as the index to avoid confusion with the use of i as the horizontal offset used in the filter expression.

Equation G.30 represents a set of N equations for the N coefficients $A_q(\omega_1,\omega_2)$, at each spatial frequency (ω_1,ω_2). The set of equations may be written as, (using a shortened form for the Power Spectral Densities concerned)

$$\begin{bmatrix} P_{gg:-N,-N}(\omega_1,\omega_2) & \cdots & P_{gg:-N,N}(\omega_1,\omega_2) \\ \vdots & \ddots & \vdots \\ P_{gg:N,-N}(\omega_1,\omega_2) & \cdots & P_{gg:N,N}(\omega_1,\omega_2) \end{bmatrix} \begin{bmatrix} A_{-N}(\omega_1,\omega_2) \\ \vdots \\ A_N(\omega_1,\omega_2) \end{bmatrix} = \begin{bmatrix} P_{sg:-N}(\omega_1,\omega_2) \\ \vdots \\ P_{sg:N}(\omega_1,\omega_2) \end{bmatrix} \quad \text{(G.32)}$$

$$\begin{aligned} \text{where } P_{gg:hk}(\omega_1,\omega_2) &= P_{gg:x_3+h,x_3+k}(\omega_1,\omega_2) \\ &= P_{ss:x_3+h,x_3+k}(\omega_1,\omega_2) \\ &\quad +\delta_{x_3+h,x_3+k} P_{\eta\eta}(\omega_1,\omega_2) \\ \text{and } P_{sg:h}(\omega_1,\omega_2) &= P_{sg:x_3,x_3+h}(\omega_1,\omega_2) \\ &= P_{ss:x_3,x_3+h} \end{aligned}$$

$P_{ss:hk}(\omega_1,\omega_2)$ is the Cross Power Spectrum of the frames in the clean original signal $I(i,j,n)$, similarly for $P_{\eta\eta}(\omega_1,\omega_2)$). The solution to this matrix equation at each spatial frequency would then yield the 2D DFT of the coefficients in each frame. This result can then be substituted into the frequency domain expression for the filter, equation G.27, to yield the 2D DFT of the estimate for the required frame.

The equation G.32 may be expressed in a more compact form:

$$\tilde{\mathbf{Q}}\mathbf{a} = \mathbf{p} \quad \text{(G.33)}$$

From this equation the estimate of the clean signal at spatial frequency (ω_1,ω_2) in the vth frame may be written as

$$F_{\hat{s},v}(\omega_1,\omega_2) = \mathbf{a}^T \mathbf{F}_g \quad \text{(G.34)}$$

where $\mathbf{F}_g$ is as defined by the terms required in G.27.

Letting $\tilde{\mathbf{Z}}(\omega_1, \omega_2)$ represent the required inverse matrix in equation G.32, $\tilde{\mathbf{Q}}^{-1}$, the following definition is helpful.

$$\tilde{\mathbf{Z}}(\omega_1, \omega_2) = \begin{bmatrix} Z_{-N,-N}(\omega_1, \omega_2) & \dots & Z_{-N,N}(\omega_1, \omega_2) \\ \vdots & \ddots & \vdots \\ Z_{N,-N}(\omega_1, \omega_2) & \dots & Z_{N,N}(\omega_1, \omega_2) \end{bmatrix} \tag{G.35}$$

Note that the inverses defined in equation G.35 and G.21 are identical matrices because the $\mathbf{Q}_k$ matrix introduced previously in equation G.20 is the same as the left hand matrix introduced in equation G.32. There is an offset in the indices, but this is due to the definition of the filter.

Given this matrix inverse then, it is possible using G.34 to write the 2D DFT of the estimate in the vth frame as

$$F_{\hat{s}:v}(\omega_1, \omega_2) = \sum_{p=-N}^{N} P_{ss:v,v+p}(\omega_1, \omega_2) \sum_{q=-N}^{N} Z_{pq}(\omega_1, \omega_2) F_{g:q}(\omega_1, \omega_2) \tag{G.36}$$

Here the 2D DFT of the required estimate of the original clean signal is denoted by $F_{\hat{s}:v}(\omega_1, \omega_2)$. Note that again $\hat{s}$ is used as a synonym for $\hat{I}$. This solution is identical to the one given previously in G.23 except for the range of the indices. This depends on the definition of the the filtering operation but does not affect the outcome.

This alternative approach highlights the basic difference between this efficient Wiener solution and the matrix solution discussed previously in Appendix F. It strikes a balance between the fully 3D Frequency approach, which depends on the validity of the DFT of a small number of samples in the temporal direction, and the fully spatiotemporal approach which is computationally intensive. Özkan et al presented this efficient method with respect to global interframe motion, they went on to incorporate the motion itself into the filter solution.

G.3 A final refinement

Özkan et al recognized that the inverse operation that was required for this new solution could also be computed efficiently. This was allowed if the approximation was made that the power spectra for the noise frames was identical for each frame and that

$$P_{ss:p,q}(\omega_1, \omega_2) = F_{s:p}(\omega_1, \omega_2) F^*_{s:q}(\omega_1, \omega_2) \tag{G.37}$$

Which means that the cross power spectral densities of frames p and q can be calçulated by the magnitude of the product of the 2D DFT's of the frames. They show that under these conditions, the inverse of the matrix $\mathbf{Q}_k$, defined in equation G.20, can be solved analytically. The final result for the 2D DFT of the required estimate of frame v is stated below. A complete derivation can be found in [135]. The equation stated here uses their notation for the implementation of the filter with respect to the indices used to the image frames. This was defined at the start of this appendix.

$$F_{\hat{s}:v}(\omega_1, \omega_2) = \frac{F_{s:v}(\omega_1, \omega_2) \sum_{q=1}^{N} F_{s:q}^{*}(\omega_1, \omega_2) F_{g:q}(\omega_1, \omega_2)}{\sum_{l=1}^{N} |F_{s:l}(\omega_1, \omega_2)|^2 + P_{\eta\eta}(\omega_1, \omega_2)} \tag{G.38}$$

References

[1] I. Abdelqader and S. Rajala. Energy minimisation approach to motion estimation. *Signal Processing*, 28:291–309, 1992.

[2] Bilge Alp, Petri Haavisto, Tiina Jarske, Kai Öistämö, and Yrjö Neuvo. Median-based algorithms for image sequence processing. In *SPIE Visual Communications and Image Processing*, pages 122–133, 1990.

[3] Georgios Angelopoulos and Ioannis Pitas. Least-squares multichannel filters in color image restoration. In *Proceedings European Conference on Circuit Theory and Design (ECCTD89)*, September 1989.

[4] Georgios Angelopoulos and Ioannis Pitas. Multichannel Wiener filters in color image restoration based on ar color image modelling. In *Proceedings IEEE International Conference on Acoustics Speech and Signal Processing (ICASSP)*, April 1991.

[5] A. Antoniadis and G. Oppenheim. *Wavelets and Statistics*. Springer–Verlag, 1995.

[6] G.R. Arce. Multistage order statistic filters for image sequence processing. *IEEE Transactions on Signal Processing*, 39:1146–1161, May 1991.

[7] G.R. Arce and R.E. Foster. Multilevel median filters, properties and efficacy. In *International Conference on Acoustics Speech and Signal Processing*, pages 824–826, 1988.

[8] G.R. Arce and E. Malaret. Motion preserving ranked-order filters for image sequence processing. In *IEEE Int. Conference Circuits and Systems*, pages 983–986, 1989.

[9] S. Armstrong, A. Kokaram, and P. J. W. Rayner. Non-linear interpolation of missing data using min-max functions. In *IEEE International Conference on Nonlinear Signal and Image Processing*, July 1997.

[10] J. Astola, P. Haavisto, and Y. Nuevo. Vector median filters. *Proceedings of the IEEE*, 78:678–689, April 1990.

[11] M. Adams B. Levy and A. Willsky. Solution and linear estimation of 2D nearest neighbour models. *Proceedings of the IEEE*, 78:627–641, April 1990.

[12] M. Barni. A fast algorithm for 1-norm vector median filtering. *IEEE Transactions on Image Processing*, 6(10):1452–1455, October 1997.

[13] C. Bergeron and E. Dubois. Gradient–based algorithms for block oriented map estimation of motion and application to motion–compensated temporal interpolation. *IEEE Transactions on Circuits and Systems for Video Technology*, 1:72–85, March 1991.

[14] J. M. Bernardo and A. F. M. Smith. *Bayesian Theory.* Wiley, 1995.

[15] J. Besag. On the statistical analysis of dirty pictures. *Journal of the Royal Statistical Society B*, 48:259–302, 1986.

[16] Bhavesh Bhatt, David Birks, and David Hermreck. Digital television: making it work. *IEEE Spectrum*, pages 19–28, October 1997.

[17] J. Biemond, L. Looijenga, D. E. Boekee, and R.H.J.M. Plompen. A pel–recursive Wiener based displacement estimation algorithm. *Signal Processing*, 13:399–412, 1987.

[18] M. Bierling. Displacement estimation by heirarchical block matching. In *SPIE VCIP*, pages 942–951, 1988.

[19] G. Wooi Boon, M. N. Chong, S. Kalra, and D. Krishnan. Bidirectional 3D autoregressive model approach to motion picture restoration. In *IEEE International Conference on Acoustics and Signal Processing*, pages 2275–2278, April 1996.

[20] L. Böröczky, J. N. Driessen, and J. Biemond. Adaptive algorithms for pel–recursive displacement estimation. In *Proceedings SPIE VCIP*, pages 1210–1221, 1990.

[21] L. Böröczky, K. Fazekas, and T. Szabados. Convergence analysis of a pel–recursive Wiener based motion estimation algorithm. In *Time Varying Image Processing and Moving Object Recognition 2*, pages 38–45. Elsevier, 1990.

[22] G. E. P. Box and G. C. Tiao. *Bayesian Inference in Statistical Analysis.* Addison-Wesley, 1973.

[23] J. Boyce. Noise reduction of image sequences using adaptive motion compensated frame averaging. In *IEEE ICASSP*, volume 3, pages 461–464, 1992.

[24] D. B. Bradshaw, N. G. Kingsbury, and A. C. Kokaram. A gradient based fast search algorithm for warping motion compensation schemes. In *IEEE International Conference on Image Processing*, pages 187–190. IEEE, October 1997.

[25] P. Burt and E. Adelson. The Laplacian pyramid as a compact image code. *IEEE transactions on Communications*, 31:532–540, April 1983.

[26] C. Cafforio and F. Rocca. Methods for measuring small displacements of television images. *IEEE transactions on Information Theory*, 22:573–579, 1976.

[27] John Canny. A computational approach to edge detection. *IEEE Transactions on Pattern Analysis and Machine Intelligence*, 8(6):679–698, 1986.

[28] M. N. Chong, P. Liu, W. B. Goh, and D. Krishnan. A new spatio-temporal MRF model for the detection of missing data in image sequences. In *IEEE International Conference on Acoustics and Signal Processing*, volume 4, pages 2977–2980, April 1997.

[29] E. Coyle, J. Lin, and M. Gabboui. Optimal stack filtering and the estimation and structural approaches to image processing. *IEEE Transactions on Acoustics, Speech and Signal Processing*, 37:2037–2065, December 1989.

[30] S. C. Dabner. Real time motion vector measurement hardware. In *3rd International Workshop on HDTV*, Aug. 1989.

[31] T. Dennis. Nonlinear temporal filter for television picture noise reduction. *IEE Proceedings*, 127:52–56, April 1980.

[32] H. Derin and P. Kelly. Discrete–index Markov–type random processes. *Proceedings of the IEEE*, 77:1485–1509, October 1989.

[33] J. Driessen, R. Belfor, and J. Biemond. Backward predictive motion compensated image sequence coding. In *Signal Processing V: Theories and Applications*, pages 757–760, 1990.

[34] J. Driessen, J. Biemond, and D. Boekee. A pel–recursive segmentation and estimation algorithm for motion compensated image sequence coding. In *IEEE ICASSP*, pages 1901–1904, 1989.

[35] J. Driessen, L. Boroczky, and J. Biemond. Pel–recursive motion field estimation from image sequences. *Visual Communication and Image Representation*, 2:259–280, 1991.

[36] E. Dubois and S. Sabri. Noise reduction in image sequences using motion compensated temporal filtering. *IEEE Transactions on Communications*, 32:826–831, July 1984.

[37] S. Efstratiadis and A. Katsagellos. A model based, pel–recursive motion estimation algorithm. In *Proceedings IEEE ICASSP*, pages 1973–1976, 1990.

[38] S. Efstratiadis and A. Katsaggelos. A multiple–frame pel–recursive Wiener–based displacement estimation algorithm. In *SPIE VCIP IV*, pages 51–60, 1989.

[39] David P. Elias and Nick G. Kingsbury. The recovery of a near optimal layer representation for an entire image sequence. In *Proceedings IEEE International Conference on Image Processing*, 1997.

[40] David P. Elias and K. K. Pang. Obtaining a coherent motion field for motion based segmentation. In *Picture Coding Symposium*, pages 541–546, 1996.

[41] W. Enkelmann. Investigations of multigrid algorithms for the estimation of optical flow fields in image sequences. *Computer Vision Graphics and Image Processing*, 43:150–177, 1988.

[42] A. Erdem, M. Sezan, and M. Özkan. Motion–compensated multiframe Wiener restoration of blurred and noisy image sequences. In *IEEE ICASSP*, volume 3, pages 293–296, March 1992.

[43] J. Patrick Fitch, Edward J. Coyle, and Neal C. Gallagher. Median filtering by threshold decomposition. *IEEE Trans. Acoustics and Signal Processing*, 32:1183–1188, December 1984.

[44] S. Fogel. The estimation of velocity vector fields from time–varying image sequences. *Computer Vision Graphics and Image Processing: Image Understanding*, 53:253–287, May 1991.

[45] N. Galatsanos and R. Chin. Digital restoration of multichannel images. *IEEE Transactions on ASSP*, 37:415–421, March 1989.

[46] Neal C. Gallagher and Gary L. Wise. A theoretical analysis of the properties of median filters. *IEEE Trans. Acoustics and Signal Processing*, 29:1136–1141, December 1981.

[47] S. Geman and D. Geman. Stochastic relaxation, gibbs distributions and the Bayesian restoration of images. *IEEE Transactions on Pattern Analysis and Machine Intelligence*, 6:721–741, 1984.

[48] S. Geman and D. McClure. A nonlinear filter for film restoration and other problems in image processing. *CVGIP, Graphical Models and Image Processing*, 54:281–289, July 1992.

[49] M. Ghanbari. The cross–search algorithm for motion estimation. *IEEE Transactions on Communications*, 38:950–953, July 1990.

[50] B. Girod. Motion–compensating prediction with fractional–pel accuracy. *IEEE Transactions on Communications*, Accepted March 1991.

[51] S. Godsill. *Restoration of Degraded Audio Signals.* PhD thesis, Cambridge University Engineering Dept., 1994.

[52] S. Godsill and A. Kokaram. Joint interpolation, motion and parameter estimation for degraded image sequences with missing data. In *Signal Processing VIII*, volume I, pages 1–4, September 1996.

[53] S. Godsill and A. Kokaram. Restoration of image sequences using a causal spatio-temporal model. In *The Art and Science of Bayesian Image Analysis*, pages 189–194, July 1997.

[54] S. Godsill and P. Rayner. A Bayesian approach to the detection and correction of error bursts in audio signals. In *IEEE ICASSP*, volume 2, pages 261–264, 1992.

[55] S. J. Godsill. Bayesian enhancement of speech and audio signals which can be modelled as ARMA processes. *International Statistical Review*, 65(1):1–21, 1997.

[56] P.J. Green. Reversible Jump MCMC computation and Bayesian model determination, September 1994.

[57] Tom R. Halfhill. CDs for the gigabyte era. *Byte magazine*, pages 139–144, October 1996.

[58] Barry G. Haskell, Atul Puri, and Arun N. Netravali. *Digital Video: An Introduction to MPEG-2.* Chapman Hall, 1997.

[59] David J. Heeger and James R. Bergen. Pyramid-based texture analysis/synthesis. In *IEEE International Conference on Image Processing*, pages 648–651, October 1995.

[60] F. Heitz and Bouthemy. Motion estimation and segmentation using a global bayesian approach. In *Proceedings ICASSP*, 1990.

[61] F. Heitz and Bouthemy. Multimodal motion estimation and segmentation using markov random fields. In *Proceedings International Conference on Pattern Recognition*, pages 378–383, 1990.

[62] H. Higgins. The interpretation of a moving retinal image. *Proceedings of the Royal Society, London*, B 208:385–397, 1980.

[63] H. Higgins. The visual ambiguity of a moving plane. *Proceedings of the Royal Society, London*, B 223:165–175, 1984.

[64] Anil N. Hirani and Takashi Totsuka. Combining frequency and spatial domain information for fast interactive image noise removal. In *Proceedings SIGGRAPH*, pages 269–276, 1996.

[65] B. Horn and B. Schunck. Determining optical flow. *Artificial Intelligence*, 17:185–203, 1981.

[66] T. Huang. *Image Sequence Analysis.* Springer–Verlag, 1981.

[67] B. Hunt. The application of constrained least squares estimation to image resoration by digital computer. *IEEE Transactions on Computers*, 22:805–812, September 1973.

[68] Keith Jack. *Video Demystified.* Hightext, 1993.

[69] A. Jain. Partial differential equations and finite-difference methods in image representing, Part 1: Image representation. *Journal of Optimization Theory and Application*, pages 65–91, September 1977.

[70] A. Jain and J. Jain. Partial differential equations and finite difference in image processing Part 2: Image restoration. *IEEE Transactions on Automatic Control*, pages 817–834, October 1978.

[71] A.K. Jain. *Fundamentals of Digital Image Processing.* Prentice Hall, 1989.

[72] Ronald J. Jurgen. Digital video. *IEEE Spectrum*, pages 24–30, March 1992.

[73] D. Kalivas and A. Sawchuk. Motion compensated enhancement of noisy image sequences. In *IEEE ICASSP*, volume 1, pages 2121–2124, 1990.

[74] R. H. Kallenberger and G. D. Cvjetnicanin. *Film into Video.* Focal Press, 1994.

[75] S. Kalra, M. N. Chong, and D. Krishnan. A new autoregressive (AR) model based algorithm for motion picture restoration. In *IEEE International Conference on Acoustics and Signal Processing*, pages 2557–2560, April 1997.

[76] S. Karunasekera and N. Kingsbury. A distortion measure for blocking artifacts in images based on human visual sensitivity. *IEEE Transactions on Image Processing*, 4(6):713–724, June 1995.

[77] A. Katsagellos, J. Driessen, S. Efstratiadis, and R. Lagendijk. Spatio–temporal motion compensated noise filtering of image sequences. In *SPIE VCIP*, pages 61–70, 1989.

[78] J. Kearney, W.B. Thompson, and D. L. Boley. Optical flow estimation: An error analysis of gradient based methods with local optimisation. *IEEE Transactions on Pattern Analysis and Machine Intelligence*, pages 229–243, March 1987.

[79] R. Kleihorst. *Noise Filtering of Image Sequences*. PhD thesis, Information Theory Group, Delft University of Technology, The Netherlands, 1994.

[80] R. Kleihorst, G. de Haan, R. Lagendijk, and J. Biemond. Motion compensated noise filtering of image sequences. In *Signal Processing VI*, pages 1385–1388. Elsevier Science, 1992.

[81] A. Kokaram. 3D Wiener filtering for noise suppression in motion picture sequences using overlapped processing. In *Signal Processing V, Theories and Applications*, pages 1780–1783, September 1994.

[82] A. Kokaram and W. J. Fitzgerald. Image processing applied to ancient manuscripts. In *20th International Congress of Papyrologists.*, August 1992.

[83] A. Kokaram and S. Godsill. A system for reconstruction of missing data in image sequences using sampled 3D AR models and MRF motion priors. In *European Conference on Computer Vision 1996*, pages 613–624. Springer–Verlag, April 1996.

[84] A. Kokaram and S. Godsill. Joint detection, interpolation, motion and parameter estimation for image sequences with missing data. In *IEEE International Conference on Image Processing*, pages 191–194. IEEE, October 1997.

[85] A. Kokaram and S. Godsill. Joint detection, interpolation, motion and parameter estimation for image sequences with missing data. In *Image Analysis and Processing*, volume 2, pages 719–725. Springer–Verlag, September 1997.

[86] A. Kokaram, R. Morris, W. Fitzgerald, and P. Rayner. Detection of missing data in image sequences. *IEEE Image Processing*, pages 1496–1508, November 1995.

[87] A. Kokaram, R. Morris, W. Fitzgerald, and P. Rayner. Interpolation of missing data in image sequences. *IEEE Image Processing*, pages 1509–1519, November 1995.

[88] A. Kokaram, N. Persad, J. Lasenby, W. Fitzerald, A. McKinnon, and M. Welland. Restoration of images from the scanning tunnelling microscope. *Applied Optics*, 34(23):5121–5131, August 1995.

[89] A. Kokaram and P. Rayner. An algorithm for line registration of TV images based on a 2–D AR model. In *Signal Processing VI, Theories and Applications*, pages 1283–1286, August 1992.

[90] A. Kokaram and P. Rayner. Removal of impulsive noise in image sequences. In *Singapore International Conference on Image Processing.*, pages 629–633, September 1992.

[91] A. Kokaram and P. Rayner. A system for the removal of impulsive noise in image sequences. In *SPIE Visual Communications and Image Processing*, pages 322–331, November 1992.

[92] A. Kokaram and P. Rayner. Detection and interpolation of replacement noise in motion picture sequences using 3D autoregressive modelling. In *IEEE International Symposium on Circuits and Systems*, pages 21–24, June 1994.

[93] A. Kokaram and P. Rayner. Removal of replacement noise in motion picture sequences using 3D autoregressive modelling. In *Signal Processing V, Theories and Applications*, pages 1760–1763, September 1994.

[94] A. Kokaram, P. M. B. Roosmalen, P. J. W. Rayner, and J. Biemond. Line registration of jittered video. In *IEEE International Conference on Acoustics Speech and Signal Processing*, pages 2553–2556, April 1997.

[95] A. Kokaram, A. Stark, and W. J. Fitzgerald. Enhancement and restoration of ancient manuscripts. In *SPIE Conference on Applications of Digital Image Processing XV.*, pages 322–331, July 1992.

[96] A. C. Kokaram. *Motion Picture Restoration.* PhD thesis, Cambridge University, England, May 1993.

[97] A. C. Kokaram. Reconstruction of severely degraded image sequence. In *Image Analysis and Processing*, volume 2, pages 773–780. Springer–Verlag, September 1997.

[98] J. Konrad and E. Dubois. Bayesian estimation of motion vector fields. *IEEE Transactions on Pattern Analysis and Machine Intelligence*, 14(9), September 1992.

[99] E. E. Kuruoglu. Signal estimation in non-gaussian noise using polynomial filters. Master's thesis, Cambridge University, England, 1995.

[100] E. E. Kuruoglu, P. J. W. Rayner, and W. J. Fitzgerald. Impulsive noise elimination using polynomial iteratively reweighted least squares. In *IEEE Workshop on Digital Signal Processing*, September 1996.

[101] E. E. Kuruoglu, P. J. W. Rayner, and W. J. Fitzgerald. Least l_p-norm estimation of autoregressive model coefficients of symmetric α-stable processes. *IEEE Signal Processing Letters*, pages 201–204, july 1997.

[102] R. L. Lagendijk and J. Biemond. *Iterative Identification and Restoration of Images.* Kluwer Academic Publishers, 1991.

[103] S. Z. Li. Discontinuity-adaptive MRF prior and robust statistics: A comparitive study. *Image and Vision Computing*, pages 227–223, April 1995.

[104] S. Z. Li. *Markov Random Field Modeling in Computer Vision.* Springer–Verlag, 1995.

[105] S. Z. Li. On disontinuity-adaptive smoothness priors in computer vision. *IEEE Transactions on Pattern Analysis and Machine Intelligence*, pages 576–586, june 1995.

[106] J. S. Lim. Image restoration by short space spectral subtraction. *IEEE ASSP*, 28:191–197, April 1980.

[107] Jae S. Lim. *Two-Dimensional Signal and Image Processing.* Prentice-Hall, 1990.

[108] J. Lin, T. Sellke, and E. Coyle. Adaptive stack filtering under the mean absolute error criterion. *IEEE Transactions on Acoustics, Speech and Signal Processing*, 38:938–954, June 1991.

[109] J. Liu, W. H. Wong, and A. Kong. Covariance structure of the Gibbs sampler with applications to the comparison of estimators and augmentation schemes. *Biometrika*, 81:27–40, 1994.

[110] J. F. A. Magarey, A. C. Kokaram, and N. G. Kingsbury. Optimal schemes for motion estimation on colour image sequences. In *IEEE International Conference on Image Processing*, pages 187–190. IEEE, October 1997.

[111] J. F. A. Magarey, A. C. Kokaram, and N. G. Kingsbury. Robust motion estimation using chrominance information in colour image sequences. In *Image Analysis and Processing*, volume 2, pages 486–493. Springer–Verlag, September 1997.

[112] Julian F. A. Magarey. *Motion estimation using complex wavelets.* PhD thesis, Cambridge University, England, 1997.

[113] S.G. Mallat. Multifrequency channel decompositions of images and wavelet models. *IEEE Transactions on Acoustics, Speech and Signal Processing*, pages 2091–2110, December 1989.

[114] D. Marr. *Vision.* W. H. Freeman and Company, 1982.

[115] D. Martinez and J. Lim. Implicit motion compensated noise reduction of motion video scenes. In *IEEE ICASSP*, pages 375–378, 1985.

[116] D. Martinez and J. Lim. Spatial interpolation of interlaced television pictures. In *IEEE ICASSP*, volume M9.21, pages 1886–1889, 1989.

[117] D. M. Martinez. *Model-based motion estimation and its application to restoration and interpolation of motion pictures.* PhD thesis, Massachusetts Institute of Technology, 1986.

[118] R. D. Morris. *Image Sequence Restoration using Gibbs Distributions.* PhD thesis, Cambridge University, England, 1995.

[119] R. D. Morris and W. J. Fitzgerald. Detection and correction of speckle degradation in image sequences using a 3D markov random field. In *Proceedings International Conference on Image Processing : Theory and Applications (IPTA '93).* Elsevier, June 1993.

[120] R. D. Morris and W. J. Fitzgerald. Replacement noise in image sequences, detection and interpolation by motion field segmentation. In *IEEE International Conference on Acoustics and Signal Processing (ICASSP)*, 1994.

[121] R. D. Morris and W. J. Fitzgerald. Stochastic and deterministic methods in motion picture restoration. In *Proceedings International Workshop on Image Processing*, June 1994.

[122] R. D. Morris and W. J. Fitzgerald. Discontinuous motion and occlusion estimation: Theory and application. In *Proceedings International Conference for Young Computer Scientists (ICYCS)*, September 1996.

[123] M. J. Nadenau and S. K. Mitra. Blotch and scratch detection in image sequences based on rank ordered differences. In *5th International Workshop on Time-Varying Image Processing and Moving Object Recognition*, September 1996.

[124] H. Nagel. Constraints for the estimation of displacement vector fields from image sequences. In *International Joint Conference on Artificial Intelligence.*, pages 945–951, 1983.

[125] H. Nagel. Displacement vectors derived from second order intensity variations in image sequences. *Computer Vision, Graphics and Image Processing*, 21:85–117, 1983.

[126] H. Nagel. Recent advances in image sequence analysis. In *Premiere Coloque Image Traitment, Synthese, Technologie et Applications.*, pages 545–558, May 1984.

[127] H. Nagel. Spatio–temporal modeling based on image sequences. In *International Symposium on Image Processing and its Applications.*, pages 222–252, 1984.

[128] H. Nagel. Towards the estimation of displacement vector fields by 'oriented smoothness' constraints. In *International Joint Conference on Pattern Recognition*, pages 6–8, 1984.

[129] H. Nagel. Image sequences – ten (octal) years – from phenomenology towards a theoretical foundation. In *IEEE ICASSP*, pages 1174–1185, 1986.

[130] H. Nagel. On a constraint equation for the estimation of displacement rates in image sequences. *IEEE Transactions on Pattern Analysis and Machine Intelligence*, 11:13–30, January 1989.

[131] H. Nagel and W. Enkelmann. An investigation of smoothness constraints for the estimation of displacement vector field from image sequences. *IEEE Transactions on Pattern Analysis and Machine Intelligence*, 8:565–592, September 1986.

[132] A. Netravali and J. Robbins. Motion–compensated television coding: Part 1. *The Bell System Technical Journal*, 58:631–670, March 1978.

[133] A. Netravali and J. Robbins. Motion–compensated coding: Some new results. *The Bell System Technical Journal*, 59:1735–1745, November 1980.

[134] A. Nieminen, P. Heinonen, and Y. Nuevo. A new class of detail–preserving filters for image processing. *IEEE Transactions on Pattern Analysis and Machine Intelligence*, 9:74–90, January 1987.

[135] M. Özkan, A. Erdem, M. Sezan, and A. Tekalp. Efficient multiframe Wiener restoration of blurred and noisy image sequences. *IEEE Transactions on Image Processing*, 1:453–476, October 1992.

[136] M. Özkan, M. Sezan, and A. Erdem. LMMSE restoration of blurred and noisy image sequences. In *SPIE VCIP*, pages 743–754, 1991.

[137] M. Özkan, M. Sezan, and A. Tekalp. Motion–adaptive weighted averaging for temporal filtering of noisy image sequences. In *SPIE Image Processing Algorithms and Techniques III*, pages 201–212, February 1992.

[138] F. Paisan and A. Crise. Restoration of signals degraded by impulsive noise by means of a low distortion, non–linear filter. *Signal Processing*, 6:67–76, 1984.

[139] I. Pitas and P. Tsakalides. Multivariate ordering in colour image filtering. *IEEE Transactions on Circuits and Systems for video technology*, September:247–259, October 1991.

[140] I. Pitas and A. N. Venetsanopoulos. *Nonlinear Digital Filters Principles and Applications.* Kluwer Academic Publishers, 1990.

[141] W. K. Pratt. *Digital Image Processing.* Wiley, 1978.

[142] William H. Press, Saul A. Teukolsky, William T. Vetterling, and Brian P. Flannery. *Numerical Recipes in C, The Art of Scientific Computing.* Cambridge University Press, 2nd edition, 1992.

[143] Gregory C. Reinsel. *Elements of Multivariate Time Series Analysis.* Springer–Verlag, 1993.

[144] J. Riveros and K. Jabbour. Review of motion analysis techniques. *IEE Proceedings*, 136:397–404, December 1989.

[145] J. Robbins and A. Netravali. *Image Sequence Processing and Dynamic Scene Analysis.*, chapter Recursive Motion Compensation: A Review, pages 76–103. Springer–Verlag, 1983.

[146] Peter M. B. Van Roosmalen, R. L. Lagendijk, and J. Biemond. Correction of intensity flicker in old film sequences. *Submitted to: IEEE Transactions on Circuits and Systems for Video Technology*, December 1996.

[147] Peter M. B. Van Roosmalen, R. L. Lagendijk, and J. Biemond. Flicker reduction in old film sequences. In *Time-varying Image Processing and Moving Object Recognition 4*, pages 9–17. Elsevier Science, 1997.

[148] Peter M. B. Van Roosmalen, S. J. P. Westen, R. L. Lagendijk, and J. Biemond. Noise-reduction for image sequences using an oriented pyramid thresholding technique. In *IEEE International Conference on Image Processing*, volume 1, pages 375–378. IEEE, 1996.

[149] A. Rosenfeld, editor. *Univariate and Multivariate random fields for images*, pages 245–258. Academic Press, 1981.

[150] A. Rosenfeld and A. Kak. *Digital Picture Processing.* Academic Press, 1982.

[151] J. J. O Ruanaidh and W. J. Fitzgerald. *Numerical Bayesian Methods Applied to Signal Processing.* Springer–Verlag, Springer Series in Statistics and Computing, 1996.

[152] Theodore S. Rzeszewski, editor. *Digtal Video (Concepts and Applications across Industries).* IEEE, 1995.

[153] R. Schalkoff. *Digital Image Processing and Computer Vision.* Wiley, 1989.

[154] B. Schunck. Image flow: fundamentals and future research. In *IEEE ICASSP*, pages 560–571, 1985.

[155] B. Schunck. The image flow constraint equation. *Computer Vision, Graphics and Image Processing*, 35:20–46, 1986.

[156] M. Sezan, M. Özkan, and S. Fogel. Temporally adaptive filtering of noisy image sequences using a robust motion estimation algorithm. In *IEEE ICASSP*, volume 3, pages 2429–2431, May 1991.

[157] M. Ibrahim Sezan and R. L. Lagendijk. *Motion Analysis and Image sequence Processing.* Kluwer Academic, 1993.

[158] Thomas Sikora. Mpeg digital video-coding standards. *IEEE Signal processing magazine*, pages 82–100, September 1997.

[159] P. S. Spencer. *System identification with application to the restoration of archived Gramophone recordings.* PhD thesis, Cambridge University, England, 1990.

[160] C. Srinivas. A stochastic model–based approach for simultaneous restoration of multiple misregistered images. In *SPIE VCIP*, pages 1416–1427, 1990.

[161] R. Srinivasan. Image restoration by spatial filter design. In *SPIE VCIP*, pages 193–197, 1986.

[162] C. Stiller. Motion–estimation for coding of moving video at 8kbit/sec with gibbs modelled vectorfield smoothing. In *SPIE VCIP.*, volume 1360, pages 468–476, 1990.

[163] R. Storey. Electronic detection and concealment of film dirt. *UK Patent Specification No. 2139039*, 1984.

[164] R. Storey. Electronic detection and concealment of film dirt. *SMPTE Journal*, pages 642–647, June 1985.

[165] P. Strobach. Quadtree–structured linear prediction models for image sequence processing. *IEEE Transactions on Pattern Analysis and Machine Intelligence*, 11:742–747, July 1989.

[166] T. Strohmer. Computationally attractive reconstruction of bandlimited images from irregular samples. *IEEE Image Processing*, pages 540–548, April 1997.

[167] Kaijun Tang, Jaakko Astola, and Yrjö Neuvo. Adaptive nonlinear multivariate image filtering for mixed noise removal. In *Proceedings International Conference of Circuits and Systems*, pages 427–430, 1993.

[168] Martin. A. Tanner. *Tools for Statistical Inference.* Springer–Verlag, 1996.

[169] A. Murat Tekalp. *Digital Video Processing.* Prentice Hall, 1995.

[170] Jim Teter. Precise film recorders for motion: They came from hollywood. *Advanced Imaging Magazine*, pages 40–41, January 1992.

[171] C. Therrien. Statistical model–based algorithms for image analysis. *IEEE Proceedings*, 74:532–551, April 1986.

[172] C. Therrien. *Decision Estimation and Classification.* Wiley, 1989.

[173] G. Thomas and H. Lau. Generation of high quality slow motion replay using motion compensation. In *IEE Conference (IBC 1990)*, Sept. 1988.

[174] G. A. Thomas. TV Picture motion measurement. *UK Patent Specification No. GB2 188510A*, Sept. 1987.

[175] G. A. Thomas. Distorting the time axis; motion compensated image processing in the studio. In *IEE Conference (IBC 1988)*, pages 256–259, Sept. 1988.

[176] W. B. Thompson, K. M. Mutch, and V. A. Berzins. Dynamic occlusion analysis in optical flow fields. *IEEE Transactions on Pattern Analysis and Machine Intelligence*, 7:374–383, July 1985.

[177] P. E. Trahanias and A. N. Venetsanopoulos. Vector directional filters. *IEEE Transactions on Image Processing*, 2:528–534, October 1993.

[178] E. Trundle. *TV and Video Technology*. Newnes, 1996.

[179] J. Tukey. *Exploratory Data Analysis*. Addison–Wesley, 1977.

[180] Adrian Turner, Helen MacKintosh, and David Francis OBE. *Guide to the Museum of the Moving Image*. British Film Institute, 1988.

[181] K. Uomori, A. Morimura, and H. Ishii. Electronic image stabilisation system for video cameras and VCR's. *SMPTE Journal*, pages 66–75, February 1992.

[182] S. V. Vaseghi. *Algorithms for the Restoration of Archived Gramophone Recordings*. PhD thesis, Cambridge University, England, 1988.

[183] S. V. Vaseghi and P. J. W. Rayner. Detection and suppression of impulsive noise in speech communication systems. *Proceedings IEE*, 137:38–46, 1990.

[184] R. Veldhuis. *Restoration of lost samples in Digital Signals*. Prentice Hall, 1980.

[185] Martin Vetterli and Jelena Kovacevic. *Wavelets and Subband Coding*. Prentice Hall, 1995.

[186] T. Vlachos. Improving the efficiency of MPEG-2 coding by means of film unsteadiness correction. In *Proceedings SPIE Conference on Digital Compression Technology and Systems*, volume 2952, pages 534–538, October 1996.

[187] T. Vlachos. Simple method for the estimation of global motion parameters using sparse translational motion vector fields. *To appear in Electronic Letters*, 1998.

[188] T. Vlachos and G. A. Thomas. Motion estimation for the correction of twin-lens telecine flicker. In *Proceedings IEEE International Conference on Image Processing*, volume 1, pages 109–112, September 1996.

[189] D. Walker and K. Rao. Improved pel–recursive motion compensation. *IEEE transactions on communications*, 32:1128–1134, October 1984.

[190] John Y. A. Wang and Edward H. Adelson. Representing moving images with layers. *IEEE Transactions on Image Processing*, 5(3):625–638, September 1994.

[191] G. Winkler. *Image Analysis, Random Fields and Dynamic Monte Carlo Methods.* Springer–Verlag, 1995.

[192] J. Woods. Two-dimensional discrete Markovian fields. *IEEE Transactions on Information Theory*, pages 232–240, March 1972.

[193] J. Woods. Markov image modelling. *IEEE Transactions on Automatic Control*, pages 846–850, October 1978.

[194] E. M. Yeh, A. C. Kokaram, and N. G. Kingsbury. Perceptual distortion measure for edge-like artifacts in image sequences. In *SPIE Human Vision and Electronic Imaging III*, volume 3299. SPIE, January 1998.

[195] Olli Yli-Harja. *Median Filters: Extensions, Analysis and Design.* PhD thesis, Lappeenranta University of Technology, 1989.

[196] R. Young and N. Kingsbury. Video compression using lapped transforms for motion estimation/compensation and coding. In *SPIE VCIP*, pages 276–288, 1992.

[197] A. Zaccarin and B. Liu. Fast algorithms for block motion estimation. In *IEEE ICASSP*, volume 3, pages 449–452, 1992.

Index

MIX
Papier aus verantwortungsvollen Quellen
Paper from responsible sources
FSC® C105338

If you have any concerns about our products,
you can contact us on
ProductSafety@springernature.com

In case Publisher is established outside the EU,
the EU authorized representative is:
Springer Nature Customer Service Center GmbH
Europaplatz 3, 69115 Heidelberg, Germany

Printed by Libri Plureos GmbH
in Hamburg, Germany